SOIL SCIENCES

SOIL SCIENCES

Editor

Lauro Nardo

scitus
academics

Soil Sciences

Edited by **Lauro Nardo**

This Book is Printed on Acid-Free Paper. ♾

ISBN: 9781681179186
Library of Congress Control Number: 2017948947

SCITUS Academics LLC,
616, Corporate Way, Suite 2, 4766,
Valley Cottage, NY 10989

www.scitusacademics.com

Notice

Printed in United States of America.

Preface

We all know that soil is a key element of agriculture. Without it we wouldn´t be able to grow plants, which are used as food for both humans and animals. Soils are the most complex and ecologically significant biogeochemical systems on Earth. In many parts of the world, soils are "threatened by urbanization, deforestation, poor agricultural practices, pollution and overgrazing, leaving them bare, degraded and unproductive," says the article. It is estimated that the world loses 24 billion tons of fertile soil each year. Agriculture is a prime cause of land degradation and soil depletion, and this poses serious problems for smallholder farmers across the world. Soil science will be fundamental in overcoming the growing threat of global food and fuel crop shortages as the world's population continues to rise. Soils play multiple roles in the quality of life throughout the world, not only as the resource for food production, but also as the support for our structures, the environment, the medium for waste disposal, water, and the storage of nutrients. A healthy soil can sustain biological productivity, maintain environmental quality, and promote plant and animal health. Understanding the impact of land management practices on soil properties and processes can provide useful indicators of economic and environmental sustainability.

This book 'Soil Sciences' is intended to provide an up-to-date account of the current state of knowledge in recent practices and assessments in soil science. Moreover, it presents a comprehensive evaluation of the effect of residue/waste application on soil properties and the mechanism of plant adaptation and plant growth. It introduces several innovative approaches for soil remediation and risk assessment, including advances in phytoremediation and implementation of metabolomics in soil sciences. We hope that this book will continue to meet the expectations and needs of all interested in varied fields with proficiency in soil science, health, toxicology, and other disciplines who contribute and share their findings to take this area forward for future investigations.

Table of Contents

CHAPTER 1

The Role of Soil Properties in Plant Endemism – A Revision of Conservation Strategies

María Luisa Bárcenas-Argüello[1], Ma. del Carmen Gutiérrez-Castorena[2] and Teresa Terrazas[3]

[1] Botany Program, Postgraduate College, Texcoco, , Mexico

[2] Soils Program, Postgraduate College, Texcoco, Mexico

[3] Botany Department-Biology Institute, National Autonomous University of Mexico, México City, Mexico

1. INTRODUCTION

The soil as the support of all terrestrial ecosystems is distributed as a continuous landscape and varies according to drainage, geomorphology and litho-climatic conditions [1]. There are five major factors that control the formation of soil: parent materials, climate, biota, topography and time [2]. These factors present interdependence; for example contrasting climatic regimes are likely to be associated with contrasting types of vegetation. Nonetheless in certain situation one of the factors had the dominant influence in determining differences among of set of soil [3]. In addition processes of soil genesis are operating under the influence of environmental factors; therefore we can understand the relationship between particular soils and the landscape and ecosystem in which they function [4]. Climate is perhaps the most influential of the four factor acting on large geographical areas (large scale), contributing to the development of specific types of soils and vegetation patterns [5], such as Gelisols (tundra vegetation: lichens, grasses and low shrub), Histosols (water loving plants: pond weeds, cattails, sedges, reed, mosses), Spodosols (mainly coniferous species), Alfisols (deciduous forest), Mollisols (grasses), Aridisols (xerophytic plants), and Oxisols (tropical rain forest). However, at regional level (medium scale) soil variability is often related to small changes in topography and thickness of parent materials or to the effect of organism. In Trans-Mexican Volcanic Belt, is common finding soil sequences or catenas related with specific vegetation species such as Andosols (Pines-volcanic ashes), Cambisols (deciduous trees-colluvial material), fluvisols (crop lands-alluvial material) y solonchack (halophyte grasses-lacutrine material) [6]. Finally at local scales, variation in edaphic characters often provides the best statistical explanation for variation in floristic composition. Frequently, systematic variations in the parent material are closely related to endemism [7]. Endemic species have relatively narrow tolerance to changes in their environment, and can be dependant on certain geologic and edaphic features. There are studies that provide detailed information on species and plant communities, some include relief features, and others report physical and chemical analyses which allow us to infer about soil fertility [8,9] while some other studies have performed statistical analysis to correlate such variables [10-12]. This chapter presents a revision of the current knowledge on the role of soil properties as for pH, H_2O_2 reaction, and reactions to HCl in the distribution of endemic plant species and synthetize in table 1 studies that give details examples of this relationship (plant-soil endemicity).

2. ENDEMISM

Endemic, in botany, means that a plant species is considered native to the country –region-where it can be found [13] and the term is applied to the distribution of organisms [14]. Although climatic factors are the most studied [15,16]; endemism is a non-ecological [17], geological event. Climate limits the flora [18,19], while geological characteristics largely define habitat diversity [4]. Moreover, edaphically severe habitats commonly support edaphic endemics, which are plant species that do not occur elsewhere [20]. Although that might be not enough to recognize the endemic species, edaphological characters (macro and micro) are essential to establish phylogenetic hypothesis of endemic taxa and areas of endemism, looking for consistency with geological models [21-23].

TABLE 1. Studies in which species distribution and soil features are reported.

Species/References	Soi/Substrate type	Technique	Results/Conclusions	Localities
Calochotusobispoensis, C. tiburonensis, C. pulchellus (Liliaceae) [40]	Serpentine	Concentrations of Ni and Cu in plant tissue	Three species are endemics	Coastal of California USA
Ariocarpus kotschoubeyanus (Cactaceae) [64]	Silty, dry lake beds	Cartographic method by conglomerates	This plant is an edaphic specialist	Chihuahuan Desert, Mexico
Satureja arkansana (Lamiaceae) and *Coreopsis lanceolata* (Asteraceae) [25]	Sandstone glade with alkaline soil	Soil pH	*Satureja arkansana* was absent on soil with a pH less than 6.1 In areas with soil pH greater than 6.1 *C. lanceolata* was absent	Arkansas USA
Hemizonia pungens ssp. *pungens* (Asteraceae) [20]	Alkali pools	Reciprocal transplant greenhouse experiment	This plant grows better on non-alkali soil when grown without competition	Yolo, California, USA
Guaiacum unijugum (Zygophyllaceae) [67]	Coastal dune and arroyo environments	Genetic analysis using microsatellite	The current extraction of gravel its habitat which could pose a direct threat	Baja California Sur, Mexico
Agave bracteosa, A. victoria-reginae, A. albopilosa *Brahea berlandieri* *Dasylirion berlandieri* *Hesperaloe funifera* var. *funifera* *Yucca filifera*	Litosols stoniness soils, summit, slopes, Stoniness soils depth soils	Sampling of 39 plots (1 Ha each one) all species were recorded and particle size, pH, depth and organic matter	It was observed that soil characteristics drastically alter the conformation of the vegetation and therefore species are not present	Northern Gulf Coastal Plain in northeastern Mexico

(All monocotyledons) [71]				
Lupinus subcarnosus (Leguminosae) [72]	Sandy soil	Allozymic variation in enzymes and other proteins	Edaphically restricted species is less genetically variable	East-central Texas USA
Lasthenia californica (Asteraceae) [73,74]	Serpentine soils	Genetic, physiological and phylogenetic studies	Races A and C of *L. californica* coexist on serpentine soil, but inhabit soil of differing physical and chemical properties	Palo Alto, California in the Santa Cruz Mountains USA
Coccoloba cereifera (Polygonaceae), [75]	Sandy soils, gravelly soils and quartzitic outcrops	Each 25 m^2 quadrant was classified according to soil types and were sampled for chemical and granulometric analyses	The spatial distribution of *Coccoloba* was largely related to the arrangement of sandfields	Serra Do Cipó, southeastern Brazil
Erigonum nervulosum (Polygonaceae) *Streptanthus brachatus* and *S. morrisonii* (Brassicaceae) [76]	Serpentine barren soils	Low concentrations of Ca, Mg, P, N	Differences in Ca and Mg between serpentine soils allow distinct species distribution	Lake County, California, USA

3. SOIL

Soil, in soil taxonomy [24] is a natural body comprised of solids (minerals and organic matter), liquid, and gases that occurs on the land surface. Soil occupies space, and is characterized By one of the following: horizons or layers, that are distinguishable from the initial materials a result of additions, losses, transfers, and transformations of energy and matter or the ability to support rooted plants in a natural environment.

We intend to clarify that for some plant species, soil has played an essential role in their evolution and current distribution. Therefore, it is necessary to indicate that according to soil taxonomy is not possible to classify the earthy materials used in pots in greenhouses. In the same sense, plants even grow on trees, but trees are regarded as non soil. Soil covers the earth's surface as a continuum, except on bare rock. Some endemics are restricted to a particular geological formation or to one type of rock or rock outcrops. Endemicity found on rocky outcrops, either calcareous or otherwise, has been reported by several researchers [25-31].

In order to understand the relevance of soil for plant endemism, it is necessary to highlight that the geological processes as genesis of unique soil types may provide the necessary isolation for the genesis of unique biota, and the edaphic factors are used to draw the link between environments and taxa [32]. If the environmental scenario is potentially multi-dimensional [33], some soil properties should be also considered as predictors. Previous research [2] considers that the distinction between soil and environment is arbitrary; and might constitute a theoretical artifact which does not represent natural processes.

3.1. Soil Genesis

The transformation of rock into soil is designated as soil formation. Climate, organisms, relief, rocks, and the time are soil forming factors. Therefore, soil can be considered as a particular combination of its forming factors. For a given combination of factors there is only one soil type [2]. Soil properties such as pH, clay content, porosity, etc, are determined by the combination of these factors. The smallest change in any one of the properties, gives rise to a new soil.

Climate is usually considered the dominant soil forming factor, and cannot be described by a single index [4]; for example, the high proportion of smectite in soil, indicate a highly seasonal semiarid subtropical climate [34].

Living organism as bacterial species are able to fix N2, dissolved P, weathered extrusive igneous rock, marble, and limestone, and significantly mobilized useful minerals, such as P, K, Mg, Mn, Fe, Cu, and Zn in rock minerals [35]. Additionally, plant root systems alter the structure of the surrounding soil [36], and the roots of some plants have the ability to exude low-molecular-weight organic acids that produce changes in the availability of nutrients [37].

Relief modifies the water relationships in soils, affected by slope processes such as erosion, landslides and other mass movement [38-39]. However, little is known about the dynamics of soil mosaics formed by slope processes; whereby mountainous regions are characterized by high soil diversity.

The soil formation is regulated by the origin of parent material and by the age of the exposed surface. The nature of the parent material profoundly influences soil characteristics such as the chemical weathering and the quantity and type of clay minerals. While the total composition of the parent rocks is only one of the factors involved in soil formation, it is of considerable interest to analyze some of these rocks, and of soils of similar origin. The time of soils formation refers to the age of the exposed surface. The soils change with time and undergo a process of evolution [2].

3.2. Soil Properties

Physical, chemical and mineralogical analyses are used in soil taxonomic criteria. For standard laboratory methods descriptions see Appendix of Keys to Taxonomy of Soils [24]. The diversity of soil properties resulted in the diversity of soil use and soil ecological functions [38,39]. In edaphic islands such as serpentine and limestone outcrops upon which plant grow are necessary specific analyses; per example, in ultramafic soil trace elements as Mn, Cu, Zn, Cr and the heavy metals Ni and Co are extracted, as well as in calcareous soils are measured soluble and exchangeable P [40,41]. Overall, previous authors agree that vegetation differences are strongly associated with differences in the bedrock [29,42].

4. SAMPLING APPROACH

The existing literature comprises a wide range of sampling techniques to obtain vegetation samples. The plant populations may vary in size and in number of individuals per species. According to our field observations, it is necessary to select populations that adequately represent the endemic area (Figure 1). Soil sampling must be performed in the same locations where vegetation has been previously sampled. The overlap of both sampling activities allows correlating changes in vegetation and soil [43]. Thus, the resulting plots are clearly representative of the surrounding area. Accordingly, geology and topography are essential characteristics that need to be thoroughly examined. For instance, spatial variability studies have indicated that even when the relief of the site is gently rolling, erosion processes can affect soil properties [44, 45], and therefore alter the results of sample analysis. We do not recommend to combined soil samples from which the different edaphic variables are measured [11], this method can give erroneous conclusions because information at the micro-scale level is lost.

The geology and the topography are factors that influence the formation of particular soil type and the establishment of specific biological forms, [38]. The geological origin of rocks can be identified visually and mineralogical composition by X-ray diffraction or other methods [31]. Undisturbed rock samples must be collected, without showing any chemical or physical weathering. The slope and rock outcrops are some landscape features easy to distinguish. The slope is important by the sediments mobilized by slope processes as for landslides, colluviation, and accumulation of the material that eroded from upper landscape positions [39], this can be recorder in degrees or percent. The geomorphic position can register as summit, shoulder, back slope, toe slope, and floodplain if any (Figure 2).

Figure 1. Delimiting study area in slope or plane environments. (A, B) Nuevo León, México. Habitats for several succulent species endemic to the Chihuahuan Desert.

Figure 2. Geomorphic position of *Cephalocereus* species. (A) *C. nizandensis* on summit; (B) *C. senilis* on summit and shoulder.

The slope aspect would influence insolation, temperature and moisture. Isolation, temperature and moisture also must be into accounted because they have effect on the bedrock or with the vegetation [29, 77]. In the surface it is necessary to test soil and bedrock reactions to HCl as a measure of their calcareousness. Moreover, the soil properties of depth, stoniness, root distribution, H_2O_2 reaction, structure, and Munsell color should be recorded [31,54]. If the characteristics of the site permit, three or more bulk soil or rhizosphere soil and rock samples along the geomorphic position per site should be collected (Figure 3). In the cases where there was no soil, organic debris must be collected [54].

Finally, the abundance of certain minerals in the soil can influence the physiological response and metabolism in plant species, such as heavy metals accumulation and the synthesis and accumulation of biominerals [40, 53]. The most common biominerals in plant tissues are calcium oxalates (Figure 4), and their abundance is associated with calcareous soils

[78]. We recommend isolate crystals from the plant tissue for better analyses and carried out X-ray diffraction (XDR), chemical composition and morphology with scanning electron microscopy (SEM). We also suggest use the petrographic microscope to know the optical properties of the crystals for be able to identify in the sand fraction of soil [57]. Additionally to relate the soil elements that plants take up soil and incorporate in their tissues, use energy dispersive X-ray (EDX) on crystals analyses. In order to determine the importance of the biominerals in the soil properties, is necessary that at least three hundred grains from the sandy fraction will count on a grain mount by line counting method using a petrographic microscope [54, 57].

Figure 3. Samples. (A) rock and organic debris for *Cephalocereus apicicephalium*; (B) soil for *C. totolapensis*.

5. EDAPHIC ENDEMISM

The interest for environmental conservation has been growing over the last decades. However, there has been little or no consensus on priority species or conservation strategies Table 2. The conservation of soil diversity greatly overlaps with such for plants, and both become endangered as a result of land use [46]. The fact that soil taxa are geographically restricted is important for planning and conservation efforts. Soil characteristics often play an essential role in determining plant community distributions [47]. The endemism of native plant species in edaphically specialized habitats suggest that these native endemic species are uniquely specialized to survive and grow better under the conditions prevalent in these harsh areas. For instance, several authors [48] reported there are almost five edaphically-restricted or -endemic butterflies, mostly associated with serpentine soils. Some species are absolutely limited by the edaphic restriction of their host plants. These are a better argument for biodiversity preservation [49].

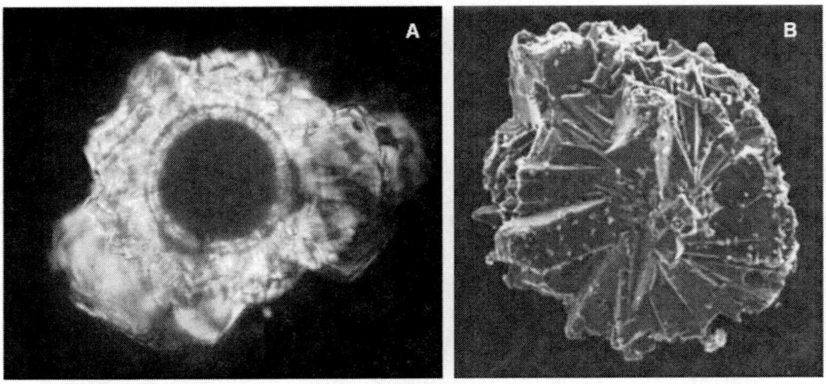

Figure 4. Calcium oxalate crystals isolated from *Cephalocereus* species. (A) petrographic microscope; (B) scanning electron microscopy.

Table 2. Studies in which species endemism are related to soil features.

Species	Evidence	Reference
Calochotus (3 spp.)	Soil studies	40
Cephalocereus (5 spp)	Soil-rock studies	54, 57
Calulanthus amplexicaulis var *barbarae*	Soil studies	59
Mammillaria fraileana	Soil-rock studies	31

Some explanations for the close relationship soil-plant have been looking at morpho-anatomical changes and physiological response to variations in soil parameters. Soil nutrient status determines leaves with glands or without glands: sclerophyllous (leaves without glands) plants exist almost exclusively on oligotrophic soil; whereas orthophyllous (leaves with glands) growth on more or less equally on both oligotrophic and eutrophic soil [50]. In this sense, exceptionally high levels of species turnover were found along all three soil fertility gradients which reflect the high degree of edaphic specialization of the flora [51]. Soil fertility is difficult to quantify, because it dependent not only of the nitrogen (N) and phosphorus (P) status of the soils, but also on their availability. Differences between species in ability to solubilize mineral nutrients could affect the ability or inability of plants to grow in particular soils. In calcareous soils, species suffer lime-chlorosis by Fe deficiency, and their growth is affected by inability to solubilize the native phosphate [52]. In ultramaphic soils, the vegetation accumulates large quantities of heavy metals in their tissues. So, endemic plants have developed strategies to grow successfully in these unusual conditions. For example, to minimize water requirements and excessive water loss, serpentine plants are able to reduce water potentials to levels lower than found on nonserpentine soils, as well as keep stomata closed or nearly closed [53].

Soil constitutes the main source of nutrients for plants. However, for plants growing on bare rock mycorriza have been described as an important factor in promoting edaphic specialization [35, 37]. This high degree of host specificity of symbiotic microbes could enhance nutrient uptake in the infertile soils or rocks. In rhizoplane of cacti, several bacterial species were isolated. This bacterium fixed N_2, dissolved P, weathered extrusive igneous rock, marble and limestone, and significantly mobilized useful minerals, such as P, K, Mg, Mn, Fe, Cu and Zn in rock minerals [35]. Other rock-colonizing cacti usually grow in cracks or fissures that are deeply penetrated by the root system: *Mammillaria fraileana* [31] and *Cephalocereus apicicephalium* and *Cephalocereus nizandensis* (Figure 5) [54]. Little is known about weathering mechanisms, except that the roots of these species can exude low-molecular organic acids (LOAs). The LOAs in root exudates may play an important role in the solubilization and plant availability of mineral nutrients in the rock [55]. Additionally, [20] mentioned that rock outcrop represent refuges from competition with other (often exotic) species.

Figure 5. Rocky habitats. (A) *Cephalocereus apicicephalium* grows in fissures; (B) *C. nizandensis* grows in cracks.

The edaphic endemics are now restricted to unusual and sometimes contaminated soils, but may have been able to withstand large concentrations of metals in their tissues or large quantities of calcium oxalate crystals [40]. The ability to tolerate excessively high levels of nickel and other heavy metals may be a physiological adaptation of the genus *Calochortus* and not necessarily an evolutionary response by several species to life on an ultramafic substrate. The amount of crystalline Ca oxalate in the oldest leaves of *Eucalyptus diversicolor* may be related in part to the high levels of exchange-able soil calcium [56]. *Cephalocereus* species could accumulate great quantities of calcium oxalate crystals even if there is low calcium soluble in the soil [54]. *Cephalocereus nizandensis* and *C. apicicephalium* grow on limestone outcrops, where the Ca is precipitated and *C. totolapensis* preferred acid soils from andesites, siltstones or mica schist (soluble Ca is 19-72 parts per million). The amount of soluble Ca is also very low in where grow *C. columna-trajani* (63-229 parts per million) and *C. senilis* (82-100 parts per million) [57]. The last two species have the larger epidermal crystals of the genus (Figure 6) [58].

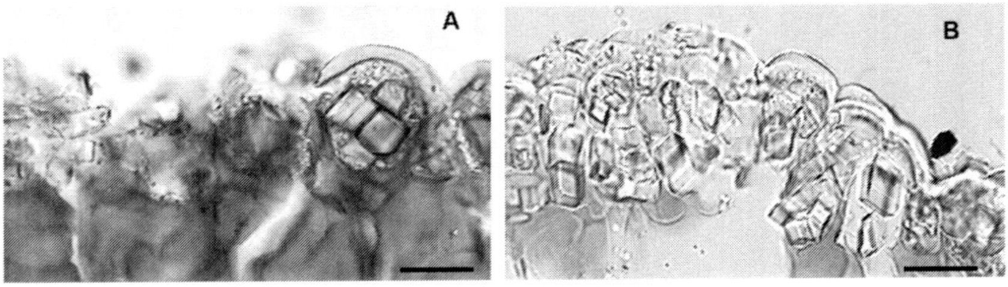

Figure 6. Prismatic calcium oxalate crystals in epidermal cells. (A) *Cephalocereus columna-trajani*; (B) *C. senilis*. Scale A = 40 μm, B = 50 μm.

For studies of evolution of a soil-type adapted endemic species is necessary to perform phylogenetic analysis. The existence of genetically compatible taxa with such distinct edaphic requirements presents a unique opportunity for intensive study of the genetic basis of tolerance to soil-type. A group plant may have evolved in a very dynamic selective context with strong edaphic selective pressures. In [59] they examined phylogenetic relationships of the rare serpentine endemic taxon *Caulanthus amplexicaulus* var. *barbarae*. They found that the serpentine taxa were nonmonophyletic evolving independently at least three times, suggesting that tolerance to serpentine may be gained or lost through relatively few genetic changes. In other case, [50] construct phylogeny for the *Pentaschistis* clade with 82 species in three genera. They investigated the association between leaf anatomy type and soil nutrient type on which species grow. Despite there is little phylogenetic constrain in soil nutrient type. However, only orthophyllous-leaved species diversify on eutrophic soils. Nevertheless, modern phylogenies on a number of phanerogam genera occurring on New Caledonia (*Acianthus*, *Cupaniopsis*, *Guioa*, *Morinda*, and *Oxera*) have shown a shift in soil preference (from non serpentine to serpentine soils and vice versa). Thus [60] concluded that the ability to grown on serpentine soil is either a plesiomorphic or a very homoplasious character and therefore the hypothesis that serpentine soils preserve the indigenous flora in New Caledonia against competition with immigrant species cannot be supported for these groups. Rarely are made specific soil studies, the data are taken often of general charts. We suggest making detailed studies as in *Cephalocereus*, according to plant species and soil type [57]. Therefore, it might be possible to infer the role of soil in the evolution of endemic plant species using the phylogenetic analysis.

6. CONSERVATION STRATEGIES

In the landscape, the vegetational differences often serve to delineate the geologic discontinuities of an area even to the casual observer. The remarkable differences often observed in plant cover for different soil types in adjacent areas, have naturally led to attempts to explain these phenomena in terms of the physical or chemical properties of the soil, or of the physiological characteristics of the plants [61]. These areas should be priority sites for conservation to preserve the unique interaction between soil and plant species as well as the microbiota and fauna. For example, the halophytic and gypsophytic vegetation of the Ebro-Basin at Los Monegros [62] or flora of the Coastal Calcareous Hills of the Biosphere Reserve Baconao in Cuba [63] are excellent to demonstrate the varied adaptations of plant types and life-forms as strategies to survive on edapho-climatic harsh conditions of various kinds. In the Chihuahuan Desert region it was found that several Cactaceae species, particularly many members of the Cacteae tribe often inhabit extremely specialized habitats, such as gypsum and other unusual soil formations (Figure 7) [64]. The patches of edaphic endemism also frequently exist as refuges for native species in highly invaded ecosystems [20], ultramafic substrates act as sites in which *Pinus balfouriana* escapes of the competition [65]. Moreover, the work with *Helianthus exilis* showed the need to protect specialized microhabitat found only within the large serpentine outcrops, the species cannot survived outside the narrow conditions proper of its habitat [66]. However, the scarcity of conclusive studies on role of soil to determine the prevalence of endemic plants hampers the efforts of public and private organizations to preserve such areas.

Figure 7. A) *Aztekium ritterii* grows in outcrops of steep slopes of crystalline grypsum; (B) *Turbinicarpus valdezianus* grows in calcarious rocks.

The population size can greatly vary among populations of the same specie generating micro-endemic nature. The small populations of some species consist of adult individuals that may be 3 as in *Cephalocereus totolapensis* [54] or 4 as in *Guaiacum unijugum* [67]. These population sizes are not reported sometimes considerer no significant, but many regions show a unique assemblage of species or a higher level of species richness or other associated species which could serve to protect this ecosystem. In Australian alpine vegetation the analyses of the relationships between physiognomic variation and environment indicate that edaphic factors are more important than climatic factors in differentiating formations [18]. Thus, edaphic discontinuities should be determining the size and population distribution and should be considered when proposing conservation areas.

Frequently, endemic species are less widely distributed and are less well represented in protected areas than other threatened species [68]. Baconao Biosphere Reserve presents high floristic composition and which endemic species represent 21%; however they are restricted at limestone hills, parent material that cover only 6.6% of total area [63]; and the same applies when registering Asteraceae endemic species to the Mexican state of Oaxaca with 53.4% and many of them not found in any Biosphere Reserve [69]; in the same sense, 12 endemic plant species are restricted to serpentine soil in Puerto Rico, all rare, uncommon and very localized within their limited distribution, and only two have been placed on the United States Federal list as threatened or endangered [70], and most species endemic to the state of Nuevo Leon (Northeast of Mexico) present in the submontane scrub have restricted distribution and specially cacti are not located within any protected area [71]. Therefore it is important to reconsider the extension of the sites identified as irreplaceable for various members of the flora endemic worldwide.

The effects of growing human populations on natural communities, on ecosystems, and on some endemic plant populations results in degraded state of sites due to human activity as roadway, tourist development, extraction of mineral as gravel, sand, and others. However, an understanding of the interrelations between soil or bedrock and occurrence of endemics, becomes even more important in the context of restoration ecology and the reversal of land degradation. In addition, the role of soil in the determination of endemic plants has not been sufficiently studied; thus, public and private organizations have not intensified their efforts to preserve such areas. Table 1 show some examples of the different soils supporting endemic species and how through different techniques of study has been able to establish a close relationship between soil and plant. As can be seen, the soil-plant endemism is not exclusive to one type of soil or a plant family.

7. CONCLUSIONS

Diverse studies have demonstrated that soil characteristics are correlated with differences in bedrock. Although incipient and timid studies intend to respond to which extent do soil characteristics correlate with vegetation patterns, this is a question that should not be forgotten in plant endemic studies.

We cannot do random sampling hoping to find a relationship between habitat and population size or presence / absence of some species. Efforts should be directed to characterize the habitat or habitats where the species grows in order to determine whether the type of soil, rock, and bedrock are the most important factors for endemic species development. When the limiting factor is the substrate, and abrupt limit in abundance is expected, as well as in population parameters. Narrow endemism in plants is frequently related to soil specificity, and many endemic plants are found in patches of certain soil within a different soil matrix. Additionally as with the soil, vegetation usually changed abruptly at the contact zone.

Edaphic endemic plants are highly vulnerable to extinction due to stochastic events, habitat degradation, climatic change, and invasion by weedy species. Finally, the protected areas may be unable to maintain regional species diversity and representativeness, especially if additional fragments are lost and fragmented landscapes are left unmanaged and they are fragile to soil erosion or degradation by chemical contamination.

REFERENCES

1. Cotler AH. El Uso de la Información Edáfica en los Estudios Ambientales. Gaceta Ecológica 2003;(68) 33-42.
2. Jenny H. Factors of Soil Formation. A of System Quantitative Pedology, New York: Dover Press;1994.
3. Boul SW, Hole FD, McCraken RJ: Génesis y Clasificación de Suelos. México: Trillas; 1998.
4. Brady NC, Weil RR. The Nature and Properties of Soils. New Jersey: Prentice-Hall; 1999.
5. USDA-NRCS.Global Soil Regions 1:1 million scale. Soil Survey Division, World Soil Resources. 1998.
6. Cajuste BL, Gutiérrez-Castorena MdelC. El Factor Relieve en la Distribución de Suelos en México. In: Krasilnikov P, Jiménez NFJ, Reyna TT, García CNE. (eds.) Geografía de Suelos de México. México: Las Prensas de Ciencias; 2011. p73-86.
7. Billings WD. The Environmental Complex in Relation to Plant Growth and Distribution. The Quarterly Review of Biology 1952;27(3) 251-265.
8. Misra A, Tyler G. Influence of Soil Moisture on Soil Solution Chemistry and Concentrations of Minerals in the Calcicoles *Phleum phleoides* and *Veronica spicata* Grown on a Limestone Soil. Annals of Botany 1999;84(3) 401-410.

9. Patterson TB, Givnish TJ. Geographic Cohesion, Chromosomal Evolution, Parallel Adaptative Radiations, and Consequent Floral Adaptations in *Calochortus* (Calochortaceae): Evidence from a cDNA Phylogeny. New Phytologist 2004;161(1) 253–264.
10. Méndez-Larios I, Ortiz E, Villaseñor JL. Las Magnoliophyta Endémicas de la Porción Xerofítica de la Provincia Florística del Valle de Tehuacán-Cuicatlán, México. Anales del Instituto de Biología, Universidad Nacional Autónoma de México. Serie Botánica 2004;75(1) 87-104.
11. Ruedas M, Valverde T, Zavala-Hurtado JA. Analysis of the Factors that Affect the Distribution and Abundance of Three *Neobuxbaumia* Species (Cactaceae) that Differ in their Degree of Rarity. Acta Oecologica 2006;29(2) 155-164.
12. Svenning J-C. Microhabitat Specialization in a Species-Rich Palm Community in Amazonian Ecuador. Journal of Ecology 1999;87(1) 55-65.
13. Font Quer P. Diccionario de Botánica. Barcelona: Labor; 1977.
14. Esparza-Olguín LG. ¿Qué Sabemos de la Rareza en Especies Vegetales? Un Enfoque Genético-Demográfico. Boletín de la Sociedad Botánica de México 2004;(75) 17-32.
15. Davies TJ, Barraclough TG, Savolainen V, Chase MW. Environmental Causes for Plant Biodiversity Gradients. Philosophical Transactions of the Royal Society B, 2004;359(1450) 1645-1656.
16. Giordani P, Incerti G. The Influence of Climate on the Distribution of Lichens: a Case Study in a Borderline Area (Liguria, NW Italy). Plant Ecology 2008;195(2) 257-272.
17. Luna-Vega I. Aplicaciones de la Biogeografía Histórica a la Distribución de las Plantas Mexicanas. Revista Mexicana de Biodiversidad 2008;79(1) 217-241.
18. Kirkpatrick JB, Bridle KL. Environmental Relationships of Floristic Variation in the Alpine Vegetation of Southeast Australia. Journal of Vegetation Science 1998;9(2) 251–260.
19. Yesson C, Culham A. Phyloclimatic Modeling: Combining Phylogenetics and Bioclimatic Modeling. Systematic Biology 2006;55(5) 785-802.
20. Veblen KE, Young TP. A California Grasslands Alkali Specialist, *Hemizonia pungens* spp. *pungens*, Prefer Non-Alkali Soils. Journal of Vegetation Science 2009;20(1) 170-176.
21. Andrés HAR, Morrone JJ, Terrazas T, López ML. Análisis de Trazos de las Especies Mexicanas de *Rhus* Subgénero *Lobadium* (Angiospermae: Anacardiaceae). Interciencia 2006;31(12) 900-904.
22. Morrone JJ, Márquez J. Halffter's Mexican Transition Zone, Beetle Generalized Tracks, and Geographical Homology. Journal of Biogeography 2001;28(5) 635-650.
23. Webb CO, Ackerly DD, McPeek MA, Donoghue MJ. Phylogenies and Community Ecology. Annual Review of Ecology and Systematics 2002;33 475-505.
24. Soil Survey Staff. Keys to Soil Taxonomy, 11th ed., Washington: USDA-Natural Resources Conservation Service; 2010.
25. Jeffries DL. Analysis of the Vegetation and Soils of Glades on Calico Rock Sandstone in Northern Arkansas. Bulletin of the Torrey Botanical Club 1985;112(1) 70-73.
26. Collins SL, Mitchell GS, Klahr SC. Vegetation-Environment Relationships in a Rock Outcrop Community in Southern Oklahoma. The American Midland Naturalist 1989;122(2) 339-348.
27. Pérez-García E A, Meave J, Gallardo C. Vegetación y Flora de la Región de Nizanda, Istmo de Tehuantepec, Oaxaca, México. Acta Botanica Mexicana 2001;(56) 19-88.
28. Pérez-García EA, Sevilla AC, Meave J, Scariot A. Floristic Differentiation in Limestone Outcrops of Southern Mexico and Central Brazil: a Beta Diversity Approach. Boletín de la Sociedad Botánica de México 2009;84(2) 45-58.
29. Searcy KB, Wilson BF, Fownes JH. Influence of Bedrock and Aspect on Soils and Plant Distribution in the Holyoke Range, Massachusetts. Journal of the Torrey Botanical Society 2003;130(3) 158-169.
30. Müller JV. Herbaceous Vegetation of Seasonally Wet Habitats on Inselbergs and Lateritic Crusts in West and Central Africa. Folia Geobotanica 2007;42(1) 29-61.

31. Lopez BR, Bashan Y, Bacilio M, De la Cruz-Agüero G. Rock-Colonizing Plants: Abundance of the Endemic Cactus *Mammillaria fraileana* Related to Rock Type in Southern Sonoran Desert. Plant Ecology 2009;201(2) 575-588.
32. Kruckeberg AR, Rabinowitz D. Biological Aspects of Endemism in Higher Plants. Annual Review of Ecology and Systematics 1985;(16) 447-479.
33. Elith J, Leathwick JR. Species Distribution Models: Ecological Explanation and Prediction Across Space and Time. Annual Review of Ecology, Evolution, and Systematics 2009;40 677-697.
34. Perales R, Serrano H, García BA, Hernández FM. Inferencias Paleoambientales de Mioceno Medio de Somosaguas (Pozuelo de Alarcón, Madrid) Basadas en la Estructura de Tamaños Corporales de su Fauna de Mamíferos. Paleolusitana 2009;(1) 317-325.
35. Puente ME, Li CY, Bashan Y. Microbial Populations and Activities in the Rhizoplane of Rock-Weathering Desert Plants. II. Growth Promotion of Cactus Seedlings. Plant Biology 2004;6(5) 643-650.
36. Kearney M, Porter W. Mechanistic Niche Modelling: Combining Physiological and Spatial Data to Predict Species' Ranges. Ecology Letters 2009;12(4) 334-350.
37. Bar-Yosef B. Root Excretions and their Environmental Effects: Influence on Availability of Phosphorus. In: Waisel Y, Eshel A, Kafkafi U. (eds.), Plant Roots: The Hidden Half. New York: Marcel Dekker; 1996. p.581-605.
38. Krasilnikov P, García-Calderón NE, Fuentes-Romero E. Pedogenesis and Slope Processes in Subtropical Mountain Areas, Sierra Sur de Oaxaca, México. Revista Mexicana de Ciencias Geológicas 2007;24(3) 469-486.
39. Krasilnikov P, García CNE, García PMdelS. Soils Developed on Different Parent Materials. Terra Latinoamericana 2007;25(4) 335-344.
40. Fiedler PL. Heavy Metal Accumulation and the Nature of Edaphic Endemism in the Genus *Calochortus* (Liliaceae). American Journal of Botany 1985;72(11) 1712-1718.
41. Zohlen A, Tyler G. Soluble Inorganic Tissue Phosphorus and Calciole-Calcifuge Behaviour of Plants. Annals of Botany 2004;94(3) 427-432.
42. Méndez-Mendoza C, Reyes-Agüero JA, Aguirre-Rivera JR, Peña-Valdivia CB. Distribución Geográfica y Ecológica de *Ephedra* L. en el Altiplano Potosino. Revista Chapingo Serie Horticultura 2000;6(1) 131-138.
43. Montaño-Arias NM, García-Sánchez R, Ochoa-de la Rosa G, Monroy-Ata A. Relación entre la Vegetación Arbustiva, el Mezquite y el Suelo de un Ecosistema Semiárido en México. Terra Latinoamericana 2005;24(2) 193-205.
44. Lozano PZ, Bravo C, Ovalles F, Hernández RM, Moreno B, Piñango L, Villanueva JG. Selección de un Diseño de Muestreo en Parcelas Experimentales a Partir del Estudio de la Variabilidad Espacial de los Suelos. Bioagro 2004;16(1) 61-72.
45. Cristobal AD, Alvarez SME, Hernández AE, Maldonado TR, Pérez GM, Castro BR. Variabilidad Espacial de Propiedades Químicas del Suelo y su Uso en el Diseño de Experimentos. Terra Latinoamericana 2008;26(4) 317-324.
46. Amundson R, Guo Y, Gong P. Soil Diversity and Land Use in the United States. Ecosystems 2003;6(5) 470-482.
47. Guo Y, Gong P, Amundson R. Pedodiversity in the United States of America. Geoderma, 2003;117(1-2) 99-115.
48. Gervais BR, Shapiro AM. Distribution of Edaphic-Endemic Butterflies in the Sierra Nevada of California. Global Ecology and Biogeography 1999;8(2) 151-162.
49. Raven PH. Catastrophic Selection and Edaphic Endemism. Evolution 1964;18(2) 336-338.
50. Galley C, Linder HP. The Phylogeny of the *Pentaschistis* Clade (Danthonioideae, Poaceae) Based on Chloroplast DNA, and the Evolution and Loss of Complex Characters. Evolution 2007;61(4) 864-884.
51. Cowling RM. Diversity Components in a Species-rich Area of the Cape Floristic Region. Journal of Vegetation Science 1990;1(5) 699-710.

52. Ström L, Olsson T, Tyler G. Differences Between Calcifugue and Acidifugue Plants in Root Exudation of Low-Molecular Organic Acids. Plant and Soil 1994;167(2) 239-245.
53. Brady KU, Kruckeberg AR, Bradshaw HD.. Evolutionary Ecology of Plant Adaptation to Serpentine Soils. Annual Review of Ecology, Evolution and Systematics 2005;36 243-266.
54. Bárcenas-Argüello ML, Gutiérrez-Castorena MdelC, Terrazas T, López-Mata L. Rock-Soil Preferences of Three *Cephalocereus* (Cactaceae) of Tropical Dry Forest. Soil Science Society of American Journal 2010;74(4) 1374-1384.
55. Stöm L. Root Exudation of Organic Acids: Importance to Nutrient Availability and the Calcifuge and Calcicole Behaviour of Plants. Oikos 1997;80(3) 459-466.
56. O'Connell AM, Malajczuk N, Gailitis V. Occurrence of Calcium Oxalate in Karri (*Eucalyptus diversicolor* F. Muell.) Forest Ecosystems of South Western Australia. Oecologia 1983;56(2/3) 239-244.
57. Bárcenas-Argüello ML. Distribución Ecológica del Subgénero *Neodawsonia* Backeb. del Género *Cephalocereus* Pfeiff. (Cactaceae), en el Istmo de Tehuantepec, México. PhD thesis. Colegio de Postgraduados México; 2011.
58. Bárcenas-Argüello ML. Filogenia del género *Cephalocereus* Pfeiff. (Cactaceae) *sensu* Anderson con Base en Caracteres Estructurales. MSc thesis. Colegio de Postgraduados México; 2006.
59. Pepper AE, Norwood LE. Evolution of *Caulanthus amplexicaulis* var. *barbarae* (Brassicaceae), a Rare Serpentine Endemic Plant: A Molecular Phylogenetic Perspective. American Journal of Botany 2001;88(8) 1479-1489.
60. De Kok R. Are Plant Adaptations to Growing on Serpentine Soil Rare or Common? A Few Case Studies from New Caledonia. Adansonia 2002;24(2) 229-238.
61. Whittaker RH. The Ecology of Serpentine Soils. Ecology 1954;35(2) 258-288.
62. Breckle SW. Halophytic and Gypsophytic Vegetation of the Ebro-Basin at Los Monegros. In Melic A, Blasco-Zumeta J. (eds) Manifiesto Científico por Los Monegros Boletín de la Sociedad Entomológica Aragonesa 1999;(24) 101-104.
63. Figueredo CLM, Reyes DOJ, Acosta CF, Fagilde EMC. Estudio Florístico de los Cerros Calizos Costeros de la Reserva de la Biósfera Baconao, Cuba. Polibotánica 2009;(28) 69-117.
64. Hernández HM, Gómez-Hinostrosa C, Hoffman G. Is Geographical Rarity Frequent among the Cacti of the Chihuahuan Desert. Revista Mexicana de Biodiversidad 2010;81(1) 163-175.
65. Eckert AJ. Influence of Substrate Type and Microsite Availability on the Persistence of Foxtail Pine (*Pinus balfouriana*, Pinaceae) in the Klamath Mountains, California. American Journal of Botany 2006;93(11) 1615-1624.
66. Wolf A. Conservation of Endemic Plants in Serpentine Landscapes. Biological Conservation. 2001;100(1) 35-44
67. McCauley RA, Cortés-Palomec AC, Oyama K. Distribution, Genetic Structure, and Conservation Status of the Rare Microendemic Species, *Guaiacum unijugum* (Zygophyllaceae) in the Cape Region of Baja California, Mexico. Revista Mexicana de Biodiversidad 2010;81(3) 745-758.
68. Aguirre GJ, Duivenvoorden JF. Can we Expect to Protect Threatened Species in Protected Areas? A Case Study of the Genus *Pinus* in Mexico. Revista Mexicana de Biodiversidad. 2010;81(3) 875-882.
69. Suárez-Mota ME, Villaseñor JL. Las Compuestas Endémicas de Oaxaca, México: Diversidad y Distribución. Boletín de la Sociedad Botánica de México. 2011;88(1) 55-66.
70. Cedeño-Maldonado JA, Breckon GJ. Serpentine Endemism in the Flora of Puerto Rico. Caribbean Journal of Science 1996;32(4) 348-356.
71. Estrada-Castillón E, Villareal-Quintanilla JA, Jurado-Ybarra E, Cantú-Ayala C, García-Aranda MA, Sánchez-Salas J, Jiménez-Pérez J, Pando-Moreno M. Clasificación,

Estructura y Diversidad del Matorral Sub-Montano Adyacente a la Planicie Costera del Golfo Norte en el Noreste de México. Botanical Sciences 2012;90(1) 37-52.

72. Babbel GR, Selander RK. Genetic Variability in Edaphically Restricted and Widespread Plant Species. Evolution 1974;28(4) 619-630.

73. Rajakaruna N, Bohm BA. The Edaphic Factor and Patterns of Variation in *Lasthenia californica* (Asteraceae). American Journal of Botany. 1999;86(11) 1576-1596.

74. Rajakaruna N, Baldwin BG, Chan R, Desrochers AM, Bohm BA, Whitton J. Edaphic Races and Phylogenetic Taxa in the *Lasthenia californica* Complex (Asteraceae: Heliantheae): an Hypothesis of Parallel Evolution. Molecular Ecology 2003;12(6) 1675-1679

75. Ribeiro KT, Fernandes GW. Patterns of Abundance of a Narrow Endemic Species in a Tropical and Infertile Montane Habitat. Plant Ecology 2000;147(2) 205-218.

76. McCarten NF. Rare and Endemic Plants of Lake County Serpentine Soils Habitats. Sacramento:Report for the Endangered Plant Project; 1988.

77. Nyssen J, Vermeersch D. Slope Aspects Affects Geomorphic Dynamics of Coal Mining Soil Heaps in Belgium. Geomorphology 2010;123(1-2) 109-121.

78. Garvie LAJ. Decay of Cacti and Carbon Cycling. Naturwissenschaften 2006;93(3) 114-118.

CHAPTER 2

Potassium in Soils of Glacial Origin

Mariusz Fotyma[1], Piotr Ochal[1] and Jan Łabętowicz[2]

[1] *Institute of Soil Science and Plant Cultivation, Puławy, Poland*

[2] *Life Science University, Warsaw, Poland*

1. INTRODUCTION

1.1. Soils in Poland

Poland occupies a territory of 312,7 km2 with 99,7% of its area lying in the Baltic Sea catchment. The average altitude is 173 m and about 90% of the territory is situated below 300m above sea level, hence there is the predominance of lowlands. The majority of the country's territory is drained by two big rivers Vistula (55,7%) and Oder (33,9%) and few small Coastal rivers discharging directly to the Sea (9,3%). Vistula is exclusively the Polish river with springs in Carpathian Mountains in the South, flowing across the middle of the country and discharging the water to Baltic Proper, while river Oder borders the territories of Poland and Germany (Figure 1).

Considering the area and the population, Poland constitute an average sized (38 million inhabitants) country according to European standards. The rural area comprise around 190 km2 i.e. about 60% of the country's territory. From those area around 160 km2 is dedicated to agricultural activities and the rest to rural infrastructure. Poland is a country with the highest ratio of agricultural land compared to other European countries.

According to European standards the country is quite densely afforested, with about 30% of forests land. A substantial territory is submerged under the lakes and rivers, including world well known Mazurian and Pomeranian regions. The majority of soils in the country are of glacial origin. The first so-called Narwian glaciation, covering the small area only occurred already in early Pleistocene. The main glaciation, South Poland and Middle Poland occurred in proper Pleistocene era and were split into five sub-periods (named after rivers: Nidian, Sanian 1, Sanian 2, Odranian and Wartanian). The youngest so-called Vistulian glaciation followed in late Pleistocene, dated back to 100 thousands of years (Figures 2, 3).

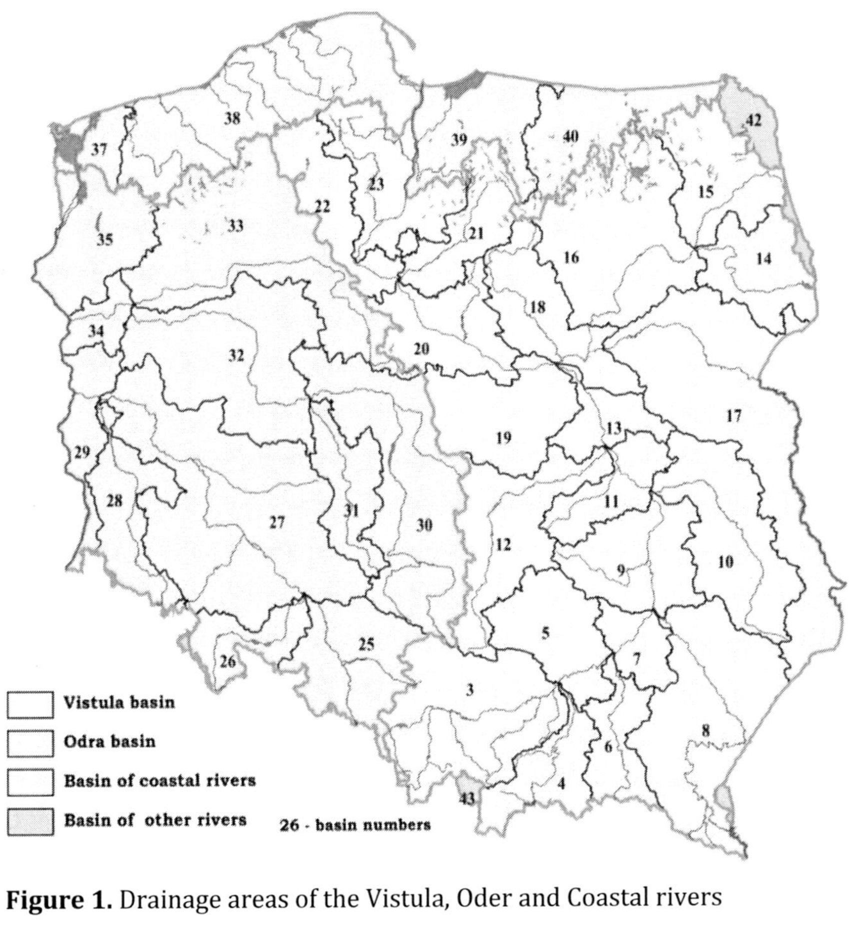

Figure 1. Drainage areas of the Vistula, Oder and Coastal rivers

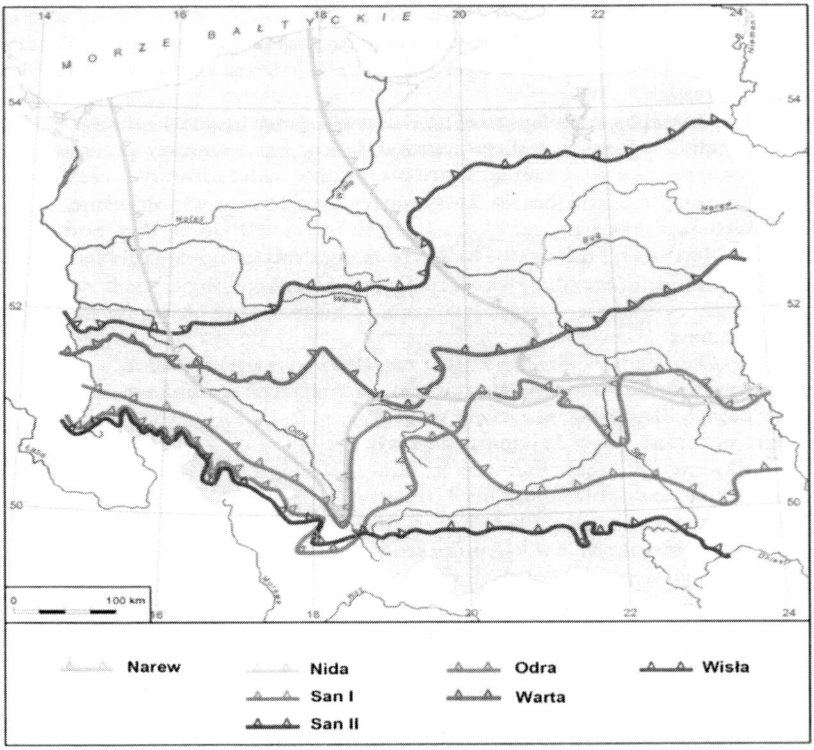

Figure 2. The borders of sequential glaciations in Poland [1]

Crossing the country from the north (Baltic Sea) to the south (Carpathian and Sudety Mountains) the following zones of the soils can be distinguished [2]:

- poorly sorted clays and sands of moraines (Cambisols, Stagno-gleyic Luvisols, Cambic Arenosols)
- fluvi-glacial sands (Podzols)
- broad, flat periglacial zone (Luvisols, Podzols, Arenosols, Fluvisols, Gleysols,Histosols)
- old eroded mountains and hills covered by loess and glacial deposits (Luvisols, Phaezems,Rendzinas)
- mountaineous zone

Table 1. Sequence of glaciations in Poland [1]

Pleistocene era												
		South Poland glaciation						Middle Poland glaciation				
Narew glaciation	Early interglaciation	Nida glaciation	Interglaciation 1	San 1 glaciation	Interglaciation 2	San 2 glaciation	Great intergalaciation	Oder glaciation	Interglaciation 3	Warta glaciation	Late interglaciation	Vistula glaciation
Thousands of years B.C.												
1000	850	760	660	630	560	530	430	330	230	210	130	110

According to geological history above 70% of Polish mineral soils have been formed from Pleistocene boulder clay and sand, strongly bathed and sorted by glacial waters. About 28% of soils are formed from loose and slightly loamy sand as well as from gravel. Soils of loess and alluvial origin cover a very small area of agricultural land (Table 1). It is not surprising that the ratio of soils of low fertility and productivity is estimated at 40% of the area.

For agricultural purposes, the soils in Poland are classified according to their productivity into soil suitability complexes and according to their buffer capacity (with respect to water and nutrients) into soil categories. Both classifications partly overlap. Productivity complexes, altogether thirteen on arable soils and three on grassland, are distinguished on the base of the soil parent's rocks, climate, soil water properties and the position in relief. This classification focuses on the possibility of growing different crops and on potential crop yield. On the very high quality complexes, numbered 1, 2 and 10, all crops can be grown satisfactorily, on the high ones, numbered 3, 4, 8 and 11 practically all crops can be grown as well but yields are significantly lower. On medium quality soil complex numbered 5 triticale, rye, and maize may be grown. Low quality soil complexes, numbered 6, 9 and 12 are suitable for rye, oats and potato and the very low quality ones, numbered 7 and 13 for rye and lupine only. The very high quality and high quality soil suitability complexes cover roughly 50% of arable soils, medium quality complex about 16% and the low and very low quality ones about 34% of arable land. Soil classification into categories is based on the content of fine fraction, i.e. soil particles below 0,02 in diameter. In this study four soil categories have been distinguished: very light up to 10% fine fraction, light 11-20% fine fraction, medium 21-35% fine fraction and heavy above 35% fine fraction. The very light soils cover 23,4% of arable area, light soils 36,1% of area, medium 29,4% and the heavy ones 11,1% of area only. For the purposes of this paper soil classification into such categories will be used extensively. Other factors limiting the fertility and productivity of Polish soils are the low content of soil organic matter SOM and strong acidification of most of the soils. According to the newest survey 7,6% of arable soils show low content of SOM (below 1%), 47,1% soils, medium content (1,1-2,0% SOM), 29,3% high content (2,1-3% SOM) and only 3% of soils, very high content of organic matter (above 3,0%) [3]. The summary of the last four-years period of agrochemical soils monitoring program reveals that 20,2% of soils are very acid (pH_{KCl} below 4,5), 29,4% acid (pH 4,5-5,5), 28% slightly acid (pH5,5-6,5), 14,7% neutral (pH 6,5-7,2) and 7,7% alkaline (pH above 7,2) [4]. There is a strong relationship between soil pH and soil's category. Very light and light soils are simultaneously very acid and acid while medium and heavy soils are much less acidified.

Table 2. Parent rocks of soils in Poland [3]

Parent rocks of soils	Total area thousands ha	% share in relation to	
		Total area	Agricultural land
Light loam	2562	15,8	18,8
Medium and heavy loam	2062	10,4	14,2
Loess	1396	3,3	4,8
Alluvial	788	4,7	5,8
Medium sand	2476	10,2	12,4
Sand and slightly loamy sand	4262	34,6	24,8
Very fine sandy soil	739	4,2	4,6
Rendzina	235	1,1	1,6
Massive rocks	599	6,1	3,9
Peat and muck	1414	8,5	9,6
Gravel	88,4	0,9	0,5

2. SOIL POTASSIUM–CONCEPTUAL–FUNCTIONAL APPROACH

Potassium is the seventh most abundant element in the Earth's crust representing on average 2,8% of total elemental composition. The average concentration of potassium in mineral soil is 1,4%, ranging between 0,01-3,7% [5] and, hence it is the fourth or fifth the most abundant element. The enrichment ratio, i.e. the relation between K in soil and K in Earth's crust, for potassium is 0,67 and is similar to the one for another important plant nutrient, phosphorus. In plant nutrition potassium is the second, after nitrogen element appearing in the highest concentration. The most important, potassium bearing minerals are primary aluminosilicates (feldspars, biotite and micas-muscovite) and secondary aluminosilicates called phyllosilicates. The last group of minerals is formed either through weathering the primary minerals or through the precipitation of secondary silicates [5]. Phyllosilicates and associated clay minerals play the most important role in the soil processes governing the availability of potassium for plants. Secondary clay minerals make up the greatest part of clay fraction of the soil characterized by particle less then 0,002 mm and, hence soils rich in clay fraction are generally abundant in potassium.

Researches on potassium in agriculture, concerning soil are as a rule based on conceptual-functional approach [6, 7, 8]. In this approach, four forms or pools of potassium have been distinguished (Figure 3). The potassium in soil solution plays the pivotal role in plant nutrition. This pool represents only up to 5% of the average plant demand for this element and not more than 0,1–0,2% of the total content of potassium in the upper soil horizon. The exchangeable or readily available potassium refers to the form of this element reversibly adsorbed on the edges of soil secondary minerals making up about 1-2% of the total potassium. Non-exchangeable or fixed potassium pool is slowly available for plants but represents a considerable 1–10% share of the total amount of this element. This form of potassium is adsorbed on wedge sites of secondary clay minerals, showing selectivity for potassium ions. The remaining pool called structural or lattice potassium is held in the structure of the primary soil minerals, feldspars and micas and becomes available for plants throughout very slowly running weathering processes. A conceptual – functional approach is based on the recognition of soil mineralogy and potassium transformation processes occurring in the soil. The availability of potassium for plants is, in this approach recognized as well.

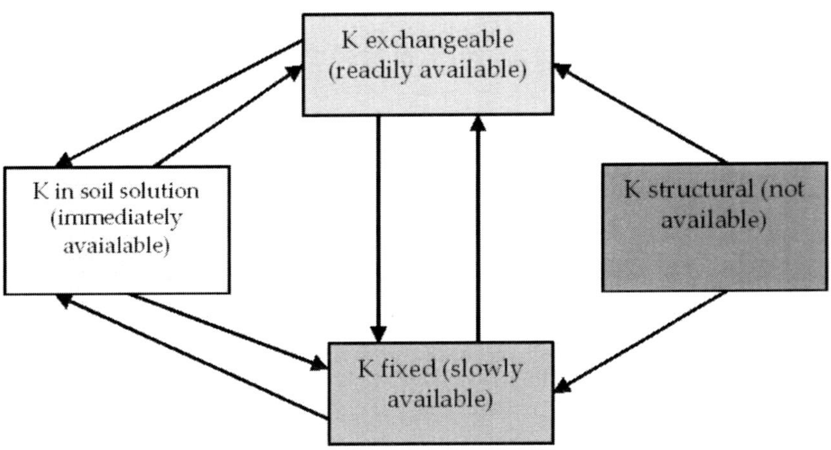

Figure 3. Conceptual–functional approach, forms (pools) of potassium in soil [8]

In the soil's solution, potassium occurs as a monovalent cation K+ in the hydrated form. The diameter of this cation, 0,66 nm is smaller than the diameter of sodium Na+, 0,72 nm. The hydration number of potassium ion, i.e. the number of water particles surrounding K+, being 10,5 is smaller as this number for sodium ion, 16,6. The K+ charge density, as the relation of its charge to the radius, is 0,0075 C·pm-1. Potassium does not form any chelates, ion pairs and/or ion complexes in soil solution. The concentration of potassium in the soil solution is in the range 0,1–1 mmol·L-1 K+ i.e. up to 40 mg K·dm-3. In Poland, in the previous research not referred in this paper, the average concentration of potassium in 136 soil samples of glacial origin was 30 mg K·L-1 and in 80% of samples it did not exceed 50 mg K·L-1 [10]. Due to very high solubility of potassium compounds applied in fertilizers, potassium ions are almost instantly removed from the soils solution by adsorption and plant uptake processes and the solution is very rarely saturated with potassium ions. In the upper 30 cm soil layer, saturated to field water capacity the content of potassium in soil solution is in the range 5–25 kg K·ha-1. This content is far below the potassium requirements of high-yielding crops. However, this pool is rather quickly replenished from the other pools, mainly from the exchangeable potassium pool.

The exchangeable potassium is held at the non-specific adsorption sites of clay minerals and at the phenolic and carboxylic groups of soil, organic matter. Potassium ion is adsorbed non-specifically in the hydrated form. These non-specific adsorption sites occur in planar "p" and edge "e" position of clay minerals and amorphous iron and aluminum hydroxides (Figure 4) [11,13,14].

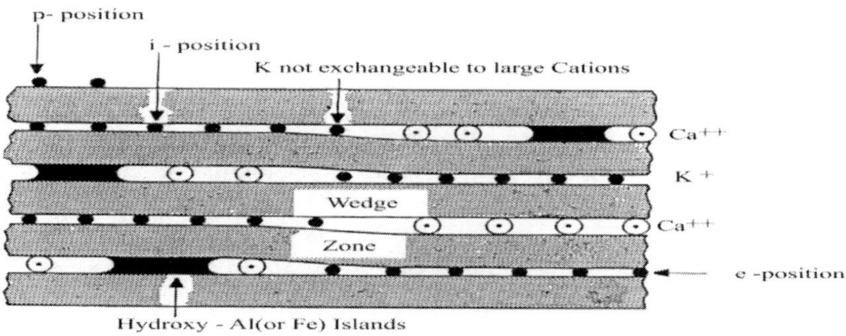

Figure 4. Model of 1:2 type clay mineral (illite) showing the positions of K^+ adsorption [11].

Clay minerals have a sheet-like structure and are built of Si tetrahedral layer and Al octahedral layer(s) in either 1:1 or 1:2 arrangement. Clay minerals are charged negatively due to the isomorphous substitution of ions pairs having the same size and coordination number, but differing in valences, i.e. Al3+ - Si4+ in tetrahedral layer and/or Mg2+-Al3+ in tetrahedral layer. The arising negative charges of clay minerals are balanced by exchangeable cations, e.g. potassium held at planar or edge positions. The 1:1 clays have very little isomorphous substitution and hence low negative charge and cation exchange capacity. Negative charges occur also at broken edges of mineral's crystals at the surfaces of amorphous and oxide minerals and soil organic matter [5,16]. The structure and properties of clay minerals and soil organic matter SOM are presented in Table 3. In Poland the prevailing soil – layered minerals are hydrated micas and illite [Długosz et al. 2005], non-expanding minerals of low cation exchange capacity.

Table 3. Structure and properties of clay minerals and SOM [12, 13].

Clay mineral	Isomorphous substitution	Cation exchange capacity cmol(+) kg^{-1}
Kaolinite 1: 1	None	1 – 15
Montmorillonite 1: 2 (expanding)	Mg for Al, Al for Si	80 – 150
Vermiculite (limited expanding)	Al for Si	140 – 200
Hydrous mica (illite) (non expanding)	Al for Si, Mg for Al	40 – 70
Amorphous oxides		2 – 4
Soil organic matter		200

The soil solution as well as the soil adsorption complex contains not only potassium but other cations as well. The total amount of cations for a certain element is partitioned amongst the soil solution and the fraction of the element adsorbed as a complex. The prevailing cation in well managed agricultural soils is calcium. To describe the relations between potassium and calcium several kinetic equations of cation exchange have been developed [5, 14]. The best-known and the most often used is Gapon equation [13]:

$$K_G = [\text{K-soil}]n(Ca^{2+})^{1/2}/[Ca_{1/2}\text{soil}]n(K^{+})$$

Where: K_G is equilibrium constant often called selectivity coefficient, [K-soil] and [$Ca_{1/2}$ soil] are expressed in mol·kg^{-1} and $(Ca^{2+})^{1/2}$ and (K^+) in mol·L^{-1}

The value of K_G varies between the pair of exchanging cations and on the binding site at the clay mineral. As lower the value of K_G as easier is potassium exchanged for calcium cation. According to [11] the K_G for p-position is 2,21 $(mM·L^{-1})^{-1/2}$, for e-position is 102 $(mM·L^{-1})^{-1/2}$ and for i-position is infinite (Figure 4). The binding selectivity of kaolinite type 1:1 minerals and for the organic matter are close to this at p-position. Potassium adsorbed in p-position and e-position at clay minerals and on the soil organic matter is, therefore, easily exchangeable and available for plants and adsorbed in i-position is practically fixed and hence slowly available only.

Non-exchangeable or fixed potassium is held between tetrahedral layers of clay minerals like hydrous micas and vermiculite (Figure 4). Potassium ion in non-hydrated form has almost the same size as oxygen ion and fits perfectly into the spaces between sheets of 2:1 types of clay minerals [15].The binding forces between K+ and mineral's surfaces are greater than between these cations themselves. It results in the partial collapse of clay structure and "entrapping" potassium ions at wedge position between tetrahedral layers. The ions of the similar size as NH_4^+, but not much larger Ca^{2+} and Mg^{2+}, fit into these positions and can replace (exchange) entrapped K^+.

Structural or lattice potassium is covalently bonded with crystal structure of potassium rich primary silicates minerals like micas (biotite, muscovite) showing layered structure and feldspars (orthoclase, microcline) showing framework structure. Micas contain 8,7–9,8% of potassium. These minerals are found in coarser fractions of soils, silt, sand and here included potassium is freed in the long-term weathering processes (Figure 5). This pool of potassium is practically unavailable for plants, at least in the perspective of crop rotation. The typical weathering process runs from muscovite through hydro muscovite to illite and mixed clay minerals kaolinite-illit or illit-montmorollinite. In these processes potassium ion is substituted by hydroksyanion H_3O^+ and, hence each next product contains less potassium [16]. Clay minerals can be formed as well as a secondary ones as a result of synthesis from the end-products of weathering the feldspars.

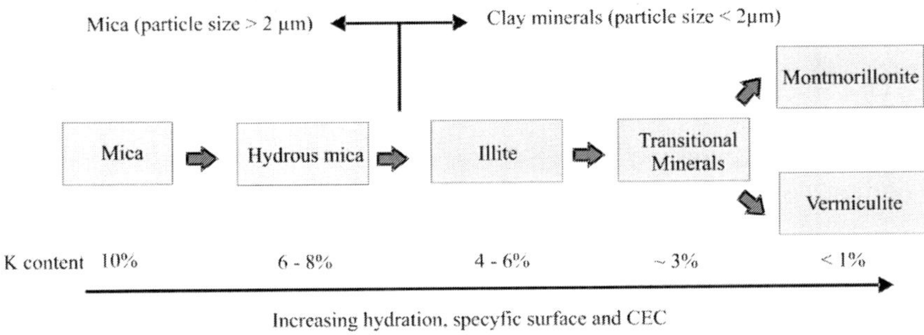

Figure 5. Model of freeing potassium in the sequence of weathering processes the primary soil minerals [17].

Table 4. Analytical form of potassium in the soils of glacial origin, mg K·kg⁻¹ soil [18, 19]

Potassium form	Brief description of method	Remarks
Water soluble K_{H2O}	Extraction with water at soil/water ratio 1: 5	Close to soil solution, form immediately available for plants
Available (exchangeable) K_{ex}	Extraction with ammonium acetate buffered to pH 7,0	Classical method of estimation , form recognized as available
Available K_{DL}	Extraction with calcium lactate buffered with HCl to pH 3,5	Official method in Poland and Latvia
Available K_{CAL}	Extraction with calcium acetate and calcium lactate, buffered with acetate acid to pH 3,7	Official method in Austria and Germany
Available K_{AL}	Ammonium lactate, buffered with acetate acid to pH 3,7	Official method in Lithuania, Slovenia and Hungary
Available K_{Meh}	CH_3COOH,0,25NH_4NO_3, NH_4F, HNO_3, EDTA	Mehlih-3 method, official in Estonia, Czech Rebublic, Slovakia
Reserve K_{res}	Extraction with boiling mol·dm⁻³ HNO_3	Close to fixed, form slowly available for plants
Nominal total K_{sem}	Extraction with hot HCl and HNO_3 (Aqua Regia)	Very slowly available for crops
Total K_{tot}	Fluorescence atomic spectrometry	Total content of potassium, unavailable for crops
Soil texture	Except the soils from KALIFERT project soil texture was evaluated by the "finger" method directly in the field. In Kalifert, the full particle composition of soil was analyzed quantitatively by laser method.	

3. METHODS AND MATERIALS

3.1. Operational–Analytical Approach

Conceptual – functional approach is theoretically oriented, and its focus is on the chemistry of soil potassium and principles of potassium availability for crops. From agricultural point of view more interesting is, however analytical approach focused on distinguishing by laboratory methods the potassium forms and linking these forms with crop's potassium requirements. In this approach, five analytical forms of potassium have been distinguished : water soluble, available, reserve, nominal total and total (Table 4).

Analytical forms of potassium, however, based on the conceptual pools differ from them quite substantially. The amount of potassium extracted with water is higher than the amount in soil solution. For this reason, it is expressed in mg $K \cdot kg^{-1}$ soil and not in mg $K \cdot L^{-1}$. Available potassium includes the water soluble one and a part of exchangeable form of this element. Otherwise the relations between exchangeable and available forms are statistically very close. Reserve potassium has an analytical meaning only, although conceptually reflects the potassium pool entrapped in layers of non-expanding clay minerals. In the analytical approach two forms of "total" potassium is distinguished. Nominal total K_{sem} is extracted with boiling Aqua Regia and, apart from previously mentioned three forms includes a part of structural potassium of soil's primary and secondary minerals. Total potassium can be measured either by fusion with alkali or by roentgen spectrometry, and includes all, already distinguished forms of this element and the whole potassium in soil minerals and organic substance. The analysis for nominal potassium is much easier than for the total one and, therefore it is often applied in research on soil potassium, which sometimes is a source of misunderstanding. Operational - analytical approach was applied in the own investigation on potassium in soils of Poland, executed in three research projects funded by Polish Ministry of Science and Higher Education. These projects have been running in the years 2002-2005 (1st project), 2006 – 2008 (2ed project) and 2009-2012 (3rdproject). The 1st project was focused on so called available forms of potassium and the relation between water soluble potassium K_{H2O} and available one K_{DL} and the 2ed on different forms of potassium in Polish soils. The 3rd project is still running in scope of international collaboration with 10 countries belonging to MOEL group (Mttelosteuropaische Lander). This collaboration was established already in 1998 for exchanging information and standardizing methods of soil fertility investigation and fertilizer recommendations [23]. The main aim of this project including not only the soils of glacial origin from Estonia, partly Germany, Latvia, Lithuania and

Poland but also soils from Austria, Czech Republic, Hungary, Slovakia and Slovenia was to follow long-term changes in the soil's potassium forms depending on fertilization. Another aim was to compare the methods of estimation the content of so called available potassium in the soils (see table 4) in different countries.

The results of all three projects have been partly published [8,18,19,21] but rather in a reports-like way. In this paper the cross-synthesis of all projects, subordinated to the specific problems is presented. In operational-analytical approach, the content of potassium forms is linked with the soil texture and not with the mineralogical composition of soil. Mineralogical soil analysis is very cumbersome, and depends considerably on the method of extraction the clay minerals. Otherwise it is well known that the content of potassium increases from the coarse sand to the clay soil's fractions. According to Brogowski et al. [20] in soil composed of 38% of sand, 52% silt and 10% of clay the share of total potassium in the fraction of sand was merely 13% in the fraction of silt 64% and in the fraction of clay 23%. The conclusion is that total potassium is under-proportionally located in sand fraction, almost proportionally in silt and over-proportionally in clay fraction.

4. RESULTS

4.1. Water Soluble K_{h2o}–Versus Available K_{dl} Potassium

In the years 2006 and 2007 each of the 17 Agrochemical Laboratories operating in Poland collected about 1.600 representative soil samples and analyzed them for soil texture, soil pH and the content of available potassium K_{DL}. In the same soil samples, the content of water soluble K_{H2O} potassium was estimated in the laboratory of the Institute of Soil Science and Plant Cultivation at Puławy using method described in Table 4. In the population of almost 24.000 soil samples the content of both potassium forms was significantly differentiated depending on soil texture (soil category) and soil pH. The mean values of the K_{H2O} and K_{DL} contents in the classes of soil texture and soil reaction are presented in tables 5 and 6. Due to the far from normal distribution of the data, medians were presented along with average values.

Table 5. The content of potassium forms depending on the soil category [19]

Soil category	No of samples	K_{DL} mg K·kg^{-1} soil		K_{H2O} mg K·kg^{-1} soil		% K_{H2O} in K_{DL}	
		average	median	average	median	average	median
very light	2044	86,3	78,0	27,5	24,2	33,3	32,1
light	10290	115,2	108,7	32,1	28,6	28,7	27,4
medium	9492	130,7	124,5	28,2	24,5	21,9	20,8
heavy	1832	135,2	120,8	25,2	21,1	19,2	17,8
all soils	23658	120,4	112,0	29,6	26,1	25,6	24,2

Table 6. The content of potassium forms depending on soil pH [19]

Soil reaction,pH	No of samples	K_{DL} mg K·kg^{-1} soil		K_{H2O} mg K·kg^{-1} soil		% K_{H2O} in K_{DL}	
		average	median	average	median	average	median
very acid	4012	87,9	75,5	21,2	18,4	26,9	25,4
acid	6934	117,8	111,2	27,9	25,2	25,0	23,5
slightly acid	6488	132,7	126,2	32,2	29,8	25,0	23,9
neutral	3668	137,3	128,2	34,7	31,1	25,4	24,4
alkaline	2554	123,6	115,4	33,0	29,1	26,9	25,5

The content of available potassium increases in the direction from the very light and light soils to the medium and heavy soils and from the very acid soils to the neutral ones. Alkaline soils contain less available potassium than the slightly acid and neutral ones. The content of water soluble potassium is admittedly the lowest in the very light soils, but soils of remaining categories show rather similar content of this potassium form. The content of water soluble potassium depends more on the soil's acidity and increases significantly in the direction from the very acid to alkaline soils. In the whole population of soil samples, the mean share of water soluble potassium makes about a quarter of the available form of this element. This share decreases significantly in the direction from the very light to heavy soils. Due to the opposing tendency in changing the content of available (increase) and water soluble (decrease) content of potassium in conformity with soil pH, the share of K_{H2O} in K_{DL} is practically independent of this soil characteristic. Generally, the heavier soils are more abundant in available forms of potassium, but this element is less accessible for the crops in comparison with the coarse-textured soils. Increasing soil pH influences, however, positively both the content of available and water-soluble forms of potassium. The latter is easily accessible for the crops. These statements are very important in Poland due to the prevalence the very light and light soils and the high soil acidity. Between the content of available potassium and potassium soluble in water exists a significant correlation and this relation was quantified by linear regression models. Two approaches were applied here: multiplicative regression analysis for each soil category separately and model of comparison the linear regressions including four soils categories jointly. The calculations were performed using two adequate procedures, a multiplicative regression model and comparison of regression lines model, offered by statistical package Statgraphic 5+. The regression equations for multiplicative model are presented below in frame.

$$\text{Very light soils: } K_{DL} = 4{,}7689 \cdot K_{H2O}\text{^}0{,}8698 \quad R^2 = 0{,}74$$

$$\text{Light soils : } K_{DL} = 6{,}9677 \cdot K_{H2O}\text{^}0{,}8068 \quad R^2 = 0{,}71$$

$$\text{Medium soils: } K_{DL} = 12.5141 \cdot K_{H2O}\text{^}0{,}7023 \quad R^2 = 0{,}65$$

$$\text{Heavy soils: } K_{DL} = 10{,}8869 \cdot K_{H2O}\text{^}0{,}7788 \quad R^2 = 0{,}70$$

On the base of these equations the content of available potassium K_{DL} can be calculated for a given range of potassium soluble in water K_{H2O} contents, separately for soil categories. Such calculation has been performed in a new approach to calibrate the content of available potassium K_{DL} in soils of glacial origin in Poland (Table 7). For a better visualization of the dependence on the relation $K_{DL}K_{H2O}$ the comparison of regression lines model offered by Statgraphic 5+ package has been applied (Figure 6) as well.

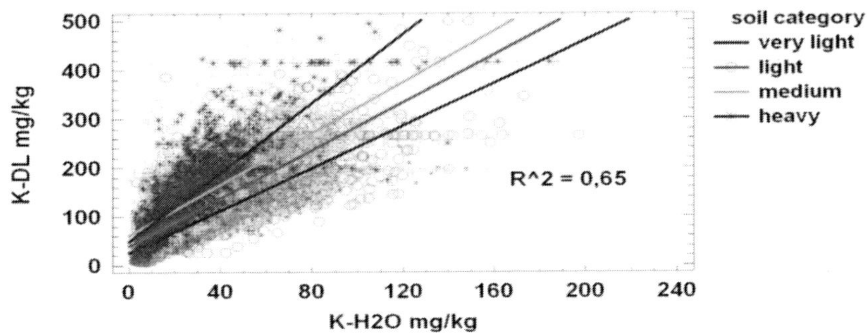

Figure 6. Relation between available K_{DL} and water soluble K_{H2O} potassium depending on soil category (comparison of regression lines)

From this model the conclusion is easily drown, that as heavier is a soil as steeper is the slope of the regression line corresponding to this relation. In practice it means that by the same content of available potassium K_{DL} the concentration of potassium ions, hence its accessibility for plants is much higher in the very light against the heavy soils. This model, however gives a transparent picture, is limited to straight regression lines only and therefore its predictability (one correlation coefficient only) is lower than the already presented multiplicative regression model (in frame). By interpretation the already presented data, one must remember that the number of soil's samples and therefore its representation was the highest for light and medium soils as well as for soils characterized by acid and slightly acid reaction. Data for the very light and very acid soils (Tables 5, 6) were less representative.

Since 1985, the five classes of available potassium content K_{DL} (along with phosphorus) are implemented in Poland. The ranges of potassium content in such classes are originating from the previous partition into three classes and have not been changed ever since. Grounding on the large number of soil samples (24.000) in which the content of water soluble and available soil potassium has been measured simultaneously the modification of official classes has been suggested. The idea underlying this suggestion was to link the content of available potassium K_{DL} with the content of water soluble K_{H2O} one, which is directly accessible to plant roots, according to the following procedure. The whole range of data concerning K_{H2O},for each soil category separately, was split into pentiles, around the median value. For the lower and upper limits of each pentile the corresponding K_{DL} have been calculated using the regression equations included above in the frame. Such calculated ranges of K_{DL} values are proposed as the new classes of available potassium content in soils of glacial origin in Poland. The details of this idea have been presented in separate publication [19]. For the sake of further consideration the official and proposed critical values of available potassium content are, however, presented in Table 7.

From the comparison of the official and proposed classes of available potassium content appears that official system undervalues potassium contents in very light and light soils as well as in the very low and low classes of K_{DL}. Overvalued are, indeed values for high and very high values of potassium content, independently of the soil texture. There is resemblance of both classifications in the range of medium potassium content and, with the exception of heavy soils quite similar are medians for proposed and the official systems. The author's proposition is focused on improving the officially accepted and being in use for over 25 years system of soil classification for available potassium content. In this proposition, the special position has the medium class for K_{DL}. Medium class is proposed as the threshold range for potassium content, which should be accomplished and kept on in the sustainable system of fertilization. However this proposition needs the further development, particularly in the very light and heavy soils categories for which the number of samples was not fully representative.

Table 7. The official and proposed [19] classes of available potassium in soils of Poland

Soil category	K_{DL} mgK·kg^{-1} soil, official classes					K_{DL} mgK·kg^{-1} soil, proposed classes				
	Very low	Low	Medium*	High	Very high	Very low	Low	Medium	High	Very high
very light	< 20	21-62	63-103	104-145	>146	< 50	51-70	71-90	91-120	> 121
light	< 41	41-83	84-124	125-166	> 167	< 65	66-90	91-115	92-140	>141
medium	< 62	63-104	105-166	167-207	>208	<85	86-110	87-140	141-170	>171
heavy	<83	84-124	125-207	208-249	>250	< 90	91-120	92-155	156-200	>201

*threshold range focused on in the fertilizer recommendation system in Poland

4.2. The Content of Available Potassium K_{dl} in Soils of Poland

In Poland, 17 Agrochemical Laboratories subjected to Ministry of Agriculture and Rural Development and covering with their activity the whole country's territory, are operating. The main task of these Laboratories is soil and plant testing and launching fertilizer recommendations. Every year several thousand soil samples are being analyzed for the soil pH and the content of available forms of potassium, phosphorus and magnesium. The results are making available for farmers, but they are collected in the data bank as well. Every fourth year the data are subjected to statistical analysis and published with the aim to monitoring the soil fertility status in Poland. The synthesis of the data concerning the content of available potassium, including the results of over 950.000 soil samples analyzed over the years 2004–2008 is presented in the following tables [4]. The synthesis reveals that the content of available potassium K_{DL} depends on the soil category and soil pH (Table 8) which is in accordance with the data already presented for 24.000 soils samples.

Table 8. The content of available potassium in soils of Poland [4]

Soil category	mg K·kg^{-1} soil		Soil acidity	mg K·kg^{-1} soil	
	average	median		average	median
very light (37.170)*	91,2	76,0	very acid	88,9	74,7
light (376.602)	126	112	acid	118	105
medium (413.098)	159	143	slightly acid	138	126
heavy (130.681)	195	183	neutral	144	129
total (957.551)	148	130	alkaline	134	120

*number of soil samples

The numeric data have been categorized and presented in the tables of contingency (Tables 9 and 10), which better visualize the link between potassium content and soil categories, and/or soil pH. Categorization of samples is based on the official classes of potassium content (Table 7)

Table 9. The ratio of soils samples in different classes of potassium content depending on soil category [4]

Soil category	% samples in the classes of available potassium content				
	very low	low	medium	high	very high
very light	7,06	42,97	28,84	12,29	8,84
light	12,20	31,19	27,66	15,78	13,18
medium	17,25	24,45	31,53	12,86	13,91
heavy	15,88	20,02	41,46	9,67	12,97
total	14,68	27,21	31,26	13,55	13,30

As follows from Table 9 one-half of the area of the very light soils is very poor in available potassium, while 20% of these soils show high and very high content of this element. Most of the medium and heavy soils contains medium to very high content of potassium.

Table 10. The ratio of soils samples in different classes of potassium content depending on soil acidity [4]

Soil acidity	% samples in the classes of available potassium content					total
	very low	low	medium	high	very high	
very acid	25,85	37,08	24,37	7,49	5,21	20,21
acid	13,97	28,76	32,47	13,47	11,33	29,39
slightly acid	9,96	22,35	34,30	16,56	16,83	28,00
neutral	10,48	21,69	32,52	15,83	19,48	14,74
alkaline	13,30	23,66	31,22	14,44	17,38	7,66

The data in Table 10 prove that very acid and acid soils are simultaneously poor in available potassium, while over 60 % of neutral and alkaline soils contain medium to very high K content. In the whole Poland around 42 % of soils shows the very low and low content of available potassium. On this area the recommended rates of potassium fertilizers exceed the of-take of this element with the crop yields. The surplus of potassium would contribute to increase the content of available nutrient and should bring the soil in the future to the medium class. The content of available potassium is substantially differentiated among the country's regions (Figure 7). In eight regions (administrative units), the ratio of soils poor in potassium is over 40% and only in three regions it does not exceed 30%. These differences are partly grounded on pedological soil origin but also on different management practices in agriculture. It concerns, mainly, different among regions consumption of potassium fertilizers and limestone, which influence the soil acidity and indirectly the content of available potassium forms. This problem lays, however outside the scope of the paper.

4.3. Comparison Different Method For Estimation The Available Form Of Potassium

In the Central-Eastern European countries (see MOEL group in Materials) different methods are applied for estimation the available potassium in soil. In scope of the MOEL co-operation, the research project has been launched on comparison the methods used in ten countries (see last chapter). Altogether, 132 soil samples have been collected from long-term field experiments in the treatments without and with potassium fertilization and two soil layers, 0-25 and 25-50 cm. The samples were analyzed for the content of exchangeable potassium K_{ex}, which was included as a reference method, and available potassium by methods presented in Table 4.

Considering very different texture of individual soils the variability of data was big enough for making the reliable comparison of the applied methods (Table 11).

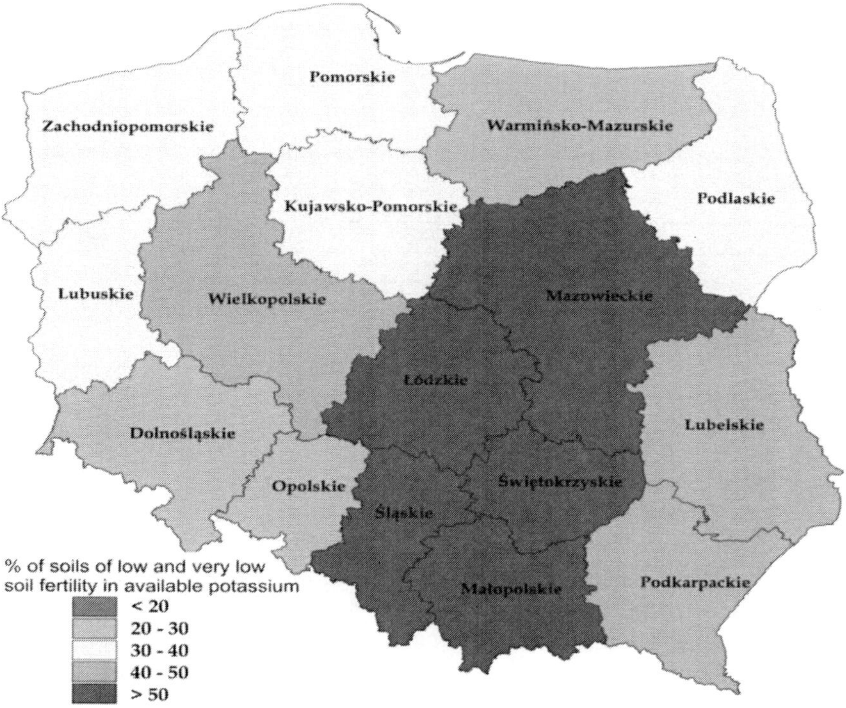

Figure 7. The percent of soils poor in potassium in regions of Poland [4]

Table 11. The content of available potassium by different methods [21]

Potassium form mg K•kg^{-1}soil	light soils(46)*		medium soil(26)		heavy soils(24)		very heavy soils(36)		total (132)	
	average	std**	average	std**	average	std**	average	std**	average	Std^2
K_{ex}	72,2	37,1	122	75,0	135	72,5	168	91,1	120	78,6
K_{DL}	75,5	35,4	119	83,9	112	69,1	116	78,2	102	66,7
K_{CAL}	54,6	32,9	90,0	61,8	83,3	47,9	93,3	60,0	77,3	52,5
K_{AL}	128	55,6	186	111	197	95,6	238	127	182	106
K_{Meh}	88,6	40,1	137	74,2	144	64,9	176	88,6	132	75,3

*number of samples, **standard deviation

The average amounts of potassium extracted by different methods were in the following decreasing order: K_{AL} > K_{ex} >K_{Meh}>K_{DL}>K_{CAL}. Exchangeable potassium K_{ex} is a form (pool) defined conceptually. Therefore, it may be presumed that the method of available potassium determination is as better as its results correlate closer with this theoretically grounded pool (Table 12).

Table 12. Relationships between pairs of examined soil tests of available potassium

Test	K_{ex}	K_{DL} (X)	K_{CAL} (X)	K_{AL} (X)	K_{Meh} (X)
K_{Meh}	0,98	0,92	0,93	0,97	-
K_{AL}	0,97	0,94	0,95	-	K_{AL}=-0,69+1,38X
K_{CAL}	0,91	0,97	-	K_{CAL}=-8,14+0,47X	K_{CAL}=-8,64+0,65X
K_{DL}	0,90	-	K_{DL}=5,96+1,24X	K_{DL}=-6,32+0,59X	K_{DL}=-6,36+0,82X
K_{ex}	-	K_{ex}=11,27+1,07X	K_{exm}=14,17+1,36X	K_{ex}=-10,2+0,71X	K_{ex}=-15,9+1,03X

The closest correlation has been found between exchangeable potassium K_{ex} and the Mehlich 3 method K_{Meh}. Besides, the slope of the regression line between these two methods is close to one, which means that both lines run parallel. However, using Mehlich 3 method somewhat higher amounts of potassium are extracted. Similar proportionality has been also found between K_{ex} and K_{DL}, although the correlation coefficient between these two potassium forms is weaker than between K_{ex} and K_{Meh}. The highest amounts of potassium are extracted using K_{AL} method, and it seems that with this method, not only exchangeable potassium but a part of reserve form K_{res} is estimated.

4.4. The Content and Proportion of Different Potassium Forms in Polish Soils

In the years 2002–2004, in the areas of the oldest glaciation periods (Sanian I and Sanian II, see Figure 2) seven hundreds soils samples representing medium and heavy soils have been collected. The samples, from the soil layers 0-25 cm and 25-50 cm were analyzed for soil pH, and the content of different forms of potassium by methods described in Table 3. Factor which the most influence the content of all potassium forms was soil categories, i.e. the content of soil's particles less then 0,02 mm (Table 13).

Table 13. The content of potassium forms in two soil layers (medians) mg K·kg^{-1} soil [18,19,22].

Soil category	No of samples	K_{H2O}		K_{ex}		K_{res}		K_{sem}	
		0-25 cm	25-50 cm	0-25 cm	25-50 cm	0-25 cm	25-50 cm	0-25 cm	25-50 cm
v.light	32	17	8,87	59,1	31,1	156	97	234	226
light	177	24,3	14,9	91,3	56.0	254	195	503	440
medium	253	22,0	13,8	112	72.6	382	204	872	789
heavy	238	21,6	10,1	166	104	655	470	2372	2256

The content of all forms of potassium, except the water soluble one increased significantly from the very light to the heavy soils. The water-soluble potassium K_{H2O} makes 0,7-5%, exchangeable potassium K_{ex} 7-18% and reserve potassium K_{res} 23-45% of the nominal total potassium. In these ranges, the ratio of available or slowly available potassium forms in the nominal total potassium was the highest in the very light and the lowest in heavy soils.Therefore the coarse-textured soils although contain considerably less potassium, the greater part of it appears in the form immediately available and readily available for crops.The relative content of potassium in the subsoil against the plow soil layer was 56%, 60%, 66% and 87% for water soluble, exchangeable, reserve and nominal total potassium forms respectively. It means that the plow soil's layer (0-25 cm) is enriched, particularly in the potassium forms available for crops.

4.5. Long–Term Changes in Potassium Forms

In 2010 in 10 countries belonging to MOEL group (see Materials) at least one or more long-term fertilizers experiments, including treatments with NPK and without NP potassium fertilization has been selected. In 2010 soil samples from these treatments and from two soil layers 0-25(30) cm and 25(30)-50(60cm) were collected and prepared for analysis. All soils from Estonia,Latvia, Lithuania and Poland were of glacial origin and have been characterized by coarse texture as light ones. These soils fall in the topic of the paper and other soils, of much heavier structure make a frame of reference. Soil analysis has been performed by methods presented in Table 4. It is worth to emphasize that in this particular project, the soil texture was analyzed by laser method and the total content (not only seemingly total) of potassium has been estimated by the fluorescence atomic spectrometry. The content of all potassium forms are included in Table 14.

Table 14. The content of potassium forms depending on soil texture (category) [8]

Potassium pool mgK•kg^{-1} soil	Soil category*				Soil level cm		Fertilization	
	light**	medium	heavy	v.heavy	0-25	25-50	NP	NPK
samples	46	26	24	36	66	66	61	71
K_{H2O}	14,6	17,0	14,8	12,3	16,8	12,0	11,8	17,0
K_{ex}	72,2	122	135	168	135	104	100	137
K_{res}	339	525	762	964	651	595	618	628
K_{sem}	1624	2903	4719	5892	3513	3692	3552	3640
K_{total}	14051	14675	15360	15877	14852	14968	14798	15009

* light < 20% of silt, medium 20-35% silt, heavy 35-50% silt, very heavy > 50% silt, silt particles less then 0,02 mm. ** soils of glacial origin friom Estonia, Latvia, Lithuania and Poland

The factor which most strongly determinates the content of all pools of potassium, except the immediately available K_{H2O}, is a soil category, i.e. the percent of silt. The differences in the content of K_{ex}, K_{res} and K_{sem} between the light and very heavy soils are two and three-folds and in the content of total potassium K_{tot} reaches almost 15%. The content of immediately available potassium K_{H2O} is practically independent of soil category and is generally higher in light and medium, than in heavy and very heavy soils. The content of two pools of potassium (K_{H2O} and K_{ex}), accessible for plants was higher in the upper soil layer in comparison to the subsoil while slowly available K_{res} and structural pools (K_{sem} and K_{tot}) seem to be more uniformly distributed in the soil profile 0-50 cm. Long term application of potassium fertilizers showed a positive effect on the content of K_{H2O} and K_{ex} pools of this element. However, the reserve K_{res} and total potassium K_{tot} contents in the soil have hardly changed in spite of long-term soil mining from this element in the control treatment. There is a strong correlation between immediately available K_{H2O} and readily available K_{ex} potassium and the weaker between K_{H2O} and K_{res}. No correlation exists between immediately available and total (K_{sem}, K_{tot}) potassium. Readily available potassium K_{ex} correlated the strongest with slowly available one K_{res} and significantly with the pools of total potassium. The strongest estimated correlation was between the slowly available and total potassium and, between both operationally distinguished pools of total potassium. The ratio of potassium forms has been calculated against the pool of total potassium K_{tot}(Figure 8) and/or the pool of nominal total potassium K_{sem} (Figure 9). Both approaches provide inconsistent results, when related to the soil texture.

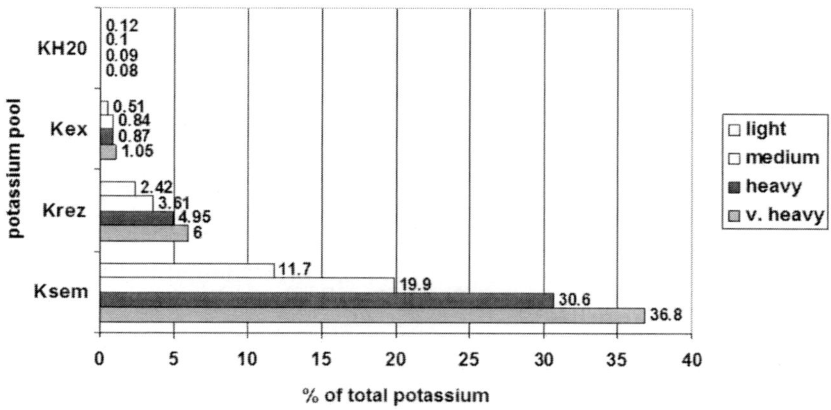

Figure 8. The ratio of potassium pools in the content of total potassium K_{tot}.

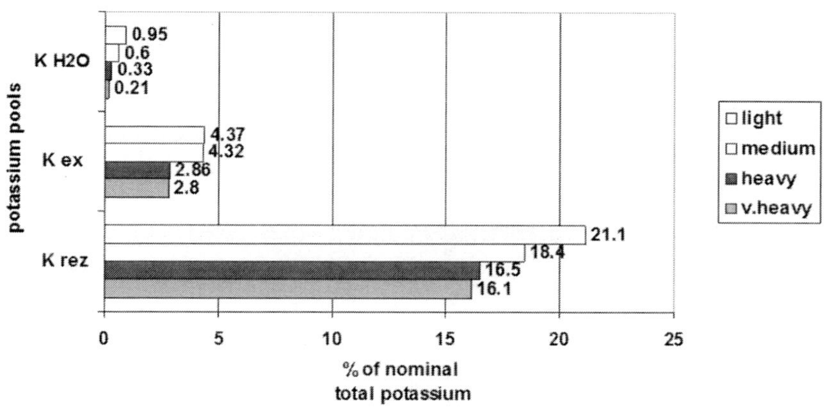

Figure 9. The ratio of potassium pools in the content of nominal total potassium K_{sem}

The relative content of all potassium forms (except the water soluble one) calculated against the total potassium K_{tot} increased systematically from the very light to very heavy soils. However, the ratio of all potassium forms calculated against the nominal total potassium, decreased in the same direction. This discrepancy has not yet deserve special attention in the available literature and needs further research. The relative content of potassium forms in light soils (mainly of glacial origin), against the nominal total potassium is very close to those found in the research conducted in scope of the 1stproject in the years 2002-2005 [18]. From this finding, it can be concluded and recommended that by relating the content of different potassium forms to the total potassium a clear distinction should be made between the nominal total K_{sem} and total K_{tot} forms. Such differentiation is particularly relevant, due to analytical problems in many research conducted on soil potassium, for which only nominal total form of this element is estimated.

It has been already mentioned that in this research, the content of clay (soil particles less than 0,002 mm) and silt (soil particles less than 0,02 mm) was quantitatively estimated by a diffraction laser method. It was therefore, possible to calculate the degree of clay and silt saturation by different forms of potassium (Table 15).

Table 15. Saturation of clay and fine fraction with different potassium forms, depending on the long-term fertilization with this element [8]

Potassium pool	mg K per 1% of clay		mg K per 1% of fine fraction**	
	without potassium NP	with potassium NPK	without potassium NP	with potassium NPK
Exchangeable K_{ex}	1,15 (0,69)*	2,03 (1,53)	0,30 (0,15)	0,50 (0,30)
Reserve K_{rez}	7,25 (3,45)	9,01 (4,88)	1,88 (0,74)	2,06 (0,76)
nominal total K_{sem}	37,7 (12,3)	47,1 (23)	10,1 (2,98)	11,6 (3,79)
Total K_{tot}	208 (123)	253 (163)	56 (30)	60,0 (32,6)

*In parenthesis the standard deviation, ** particles <0,02 mm

Results from Table 15 indicate, that the arable soils in Central-Eastern Europe, properly fertilized contain about 2 mg of readily available K_{ex} and about 9 mg of slowly available K_{res} potassium per 1% of clay and respectively 0,5 mg of readily available and about 2 mg slowly available potassium per 1% of fine fraction. Long–term soil mining from potassium resulted in diminishing the saturation of clay and silt with readily available potassium K_{ex} by about 70% and the saturation with slowly available potassium K_{rez} by about 20%.

According to the results achieved in this project, comparison of potassium content in different forms at the beginning of experiments, with the amount taken up by crops in this period of time was also preliminary considered (Table 16). This problem is beyond the scope of this paper and, besides might deserve the further research.

Table 16. The amount of potassium at the beginning of experiments, in the soil layers 0-25 cm and 25-50 cm, against its off take with crop yields in the NP treatment, kg K·ha^{-1}.[8]

K pool	Light soils			Medium soils			Heavy soils			Very heavy soils		
	0-25	25-50	off*take	0-25	25-50	off take	0-25	25-50	off take	0-25	25-50	off take
K_{H2O}	75	57		86	66		79	44		69	37	
Kex	390	259		653	483		694	464		797	614	
Krez	1368	1236	2476	2152	1938	2871	3150	2580	3193	3999	3445	3374
Ksem	6253	7115		10992	11126		17973	18404		22742	22512	
Ktot	53028	53470		55064	55560		58120	58184		60220	59910	

*sum of the potassium removed from the soil by all crops grown in experiments

Independently, of soil texture the amount of exchangeable potassium K_{ex} in the first 0-50 cm of the soils profile 0-50 cm does not meet the long-term plant's requirements for this element. Therefore, this pool needs to be constantly replenished from the pool of slowly available potassium K_{rez}. In the light soil even this pool in the soil profile 0-50 cm hardly matches the long-term plant requirements and had to be supplemented from the structural pool of potassium K_{sem}. In the medium soil, the amount of slowly available potassium K_{rez} in the upper soil level is too small against the plant requirements and only the amount in the soil profile 0-50 cm suffices to fulfill these requirements. In the heavy and very heavy soils, the upper soil level, 0-25 cm contains enough slowly available potassium to cover the long-term plant's requirements for this element. The amount of available and slowly available potassium was higher in the upper soil level in comparison to the subsoil. This regularity does not concern the nominal total K_{sem} and total K_{tot} potassium pools, which practically do not differ between the soil levels. It leads to the conclusion that upper level of arable soils is enriched with plant accessible potassium

5. SUMMARY AND CONCLUSIONS

Most of the soils in Poland are of glacial origin and are formed from poorly sorted clays and sands of moraines and fluvial-glacial sands. These soils are deeply leached down, poor in bases and, therefore, acid. Almost 60% of the soils belong to very light and light categories, i.e. contain less than 20% of fine fraction and over 50% of soils are very acid and acid. The content of organic matter is generally low and very low even at the limit established in

Poland, which is below 2,0% of SOM. In the paper, two approaches to potassium in the soil are presented. The first approach, conceptual – functional, is based upon the literature data, and it focuses on the defined potassium pools and physicochemical processes governing the potassium dislocation between these pools. The second approach, operational-analytical, concerns the potassium forms isolated from the soil samples using chemical methods. The second approach has been applied in the own research carried on in scope of three scientific projects in the years 2002 – 2012. The soils of glacial origin in Poland have been characterized as regards the content of water soluble K_{H2O}, available (in different extracts, mainly calcium lactate K_{DL}), reserve (in boiling nitrate acid K_{res}), nominal total (in Aqua Regia) and total (Roentgen spectrometer) K_{tot}potassium forms. The content of all forms of potassium, except water soluble one depended significantly on soil granulometric composition and soils pH and increases from the very light to heavy, and from very acid to neutral soils. Properly managed soils contain 2 mg K_{ex} and 9 mg of K_{res}per 1 % of clay fraction (<0,002 mm). The new calibration figures for available potassium K_{DL} by linking the K_{DL} content to water soluble potassium K_{H2O} is proposed. Applying these figures would promote more sustainable potassium fertilizer's management. The soils of glacial origin in Poland are generally poor in available potassium and the ratio of soils showing very low and low potassium (K_{DL}) content exceeds 40 %. The water-soluble potassium (K_{H2O}) makes 0,7-5%, exchangeable potassium (K_{ex}) 7-18% and reserve potassium (K_{res}) 23-45% of the nominal total potassium. In the light soils, the ratio of nominal total to total potassium is slightly over 10 %. In long-term experiments with potassium fertilization, the amounts of potassium in different forms were compared with the uptake of this element by crops grown in control (without K fertilization)treatments. Independently, of soil texture the amount of exchangeable potassium in the soil profile 0-50 cm does not suffice to cover the crop's requirements for this element. Therefore, this pool needs to be constantly replenished from the pool of slowly available potassium K_{rez} and even from the pool of nominal total potassium. The problem of potassium fertilization and its influence on the forms of potassium in the soil is beyond the scope of this paper and, besides might deserve the further research.

6. ACKNOWLEDGEMENTS

The experimental part of this paper is a summary of three research projects funded by the Ministry of Science and Higher Education in Poland : Nr 3 PO6R 087 23 "Nawkal 2002 – 2005, No 2 PO6R 48 30 "Kalpol" 2006-2008 and NN 310 204437 "Kalifert" 2009 – 2012. Author expresses sincere thanks to all colleagues from the MOEL group for providing the soil samples in scope of the last project.

REFERENCES

1. K Harasimiuk, Rodzoś, Geografia Polski. UMCS, Lublin ; 2004
2. Soil Atlas of EuropeEuropean Soil Bureau Network ;2005
3. H Terelak, S Krasowicz, T Stuczynski, srodowisko rolnicze Polski i racjonalne użytkowanie rolniczej przestrzeni produkcyjnej. In: Igras J., Pastuszak M. (ed.) Contribution of Polish agriculture to emission of nitrogen and phosphorus compounds to the Baltic Sea : Puławy; 201270
4. P Ochal, Wykorzystanie syntetycznego wskaźnika do oceny agrochemicznego stanu gleb w Polsce. PhD, manuscripts: Puławy ;2012
5. M. E Essington, Soil and water chemistry, an integrated approach. London :RC Press; 2004
6. V Romheld, E. A Kirkby, 2010Research on potassium in agriculture; needs and prospects. Plant Soil.2010; 335155180
7. Syers J,K.Soil and plant potassium in agriculture. IFS Proceedings 1998; No. 411.
8. M Fotyma, The consequences of soil mining from potassium stock. Nawozy i Nawozenie- Fertilizers and Fertilization. 201143520
9. E Epstein, A. J Bloom, 2005Mineral nutrition of plants. Principles and perspectives. Second edition; 2005. Sinauer Associates, Inc, Publishers, Sunderland, Massachusetts.
10. J Labetowicz, Skład chemiczny roztworu glebowego w zróżnicowanych warunkach glebowych i nawozowych; 1995Fundacja "Rozwój SGGW" Warszawa.
11. K Mengel, ., E.A. Kirkby, . Principles of plant nutrition third edition ;1982. International Potash Institute, Bern.
12. S. A Barber, Soil nutrient bioavailability, a mechanistic approach, second edition; 1995John Wiley & Sons,Inc.
13. H Kim, Tan. Principles of soil chemistry second edition ;1993Marcel Dekker Inco.
14. D. L Sparks, Environmental soil chemistry, second editions ;2003Academic Press.
15. Troeh,F.R., Thompson,L. M, Soils and soil fertility, sixth edition;.Blackwell Publishing
16. J Pracz, Podstawy mineralogii ;2003SGGW, Warszawa.
17. D Schroeder, Soils- facts and concepts ;1984International Potash Institute,Bern.
18. M Fotyma, 2007Content of potassium in different forms in the soils of southeast Poland. Polish J. Soil Sci; 2007. XL.11932
19. M Fotyma, 2009Forms and tests of available potassium in soils. Nawozy i Nawozenie- Fertilizers and Fertilization; 2009. 34924
20. Z Brogowski, S Uziak, T Komornicki, . Distribution of potassium in granulometric soils fractions. Polish Journal of Soil Science; 2009. XLII/1: 30-40.
21. M Fotyma, 2011Testy glebowe potasu łatwo dostępnego dla roślin. Nawozy i Nawożenie- Fertilizers and Fertilization ;2011. 44516
22. W Grzebisz, M Fotyma, Recommendations and use of potassium fertilizers in Central-Eastern Europe. The International Fertilizer Society. Proceedings;2007621
23. J Loch, years of MOEL consultative meetings in retrospection. Nawozy i Nawozenie-Fertilizers and Fertilization;200937716

CHAPTER 3

Improving Available Phosphorus in Acidic Soil Using Biochar

Bayu Dume[1] *, Dejene Ayele[2] , Alemayehu Regassa[1] and Gezahegn Berecha[3]

[1] *Department of Natural Resource Management, College of Agriculture and Veterinary Medicine Jimma University, P. O. Box: 307, Jimma, Ethiopia.*

[2] *Department of Chemistry, College of Natural Sciences, Jimma University, P. O. Box 387, Jimma, Ethiopia.*

[3] *Department of Horticulture and Plant Science, College of Agriculture and Veterinary Medicine Jimma University, P. O. Box: 307, Jimma, Ethiopia.*

ABSTRACT

Understanding of phosphorus (P) retention and release mechanisms provides crucial information for the effective management of phosphorus to enhance crop production and sustain soil. In acidic soil, available phosphorus is fixed by aluminum and iron. To overcome this problem, soils are limed to fix aluminum and iron. But this practice is not economical for small scale farmers and also it is not environmentally friendly. This study was conducted to improve phosphorus availability using biochar produced from coffee husk and corn cob to fix aluminum and iron instead of phosphorus. Acidic soil samples were mixed with biochar applied at the rates of 0, 5, 10 and 15 t ha^{-1} and incubated in laboratory for 2 months at ambient temperature. The results showed significant effects ($p<0.01$) on selected soil chemical properties by increasing soil pH and reduced exchangeable acidity, exchangeable aluminum, and exchangeable iron in a way that enhanced the availability of phosphorus. Due to the incorporation of biochar the available P level increased to a level ranging 3.64±0.34 - 23.21±0.07 mg/kg after an incubation period of 2 months and it increased by 84.3% available phosphorus when coffee husk biochar produced at 500°C temperature was applied at a rate of 15t/ha. Moreover, further field researches are needed to evaluate the effect of biochar on availability, the fate and uptake of available P in soil.

Keywords: Fixation, biochar, feedstock, acidic soil, pyrolysis temperature.

INTRODUCTION

Phosphorus is an essential macronutrient for plants next to nitrogen. Understanding of phosphorus (P) retention and release mechanisms provides crucial information for the effective management of phosphorus to enhance crop production and sustain soil. Phosphorus undergoes several geo-chemical processes in soil such as solubilization, complexation, adsorption, and precipitation that determine its mobility and fate. These chemical processes are a complex function of several soil properties including, Al and Fe oxide form and content, the amount and form of silicate clays, and $CaCO_3$ content. The impact of these properties on P retention and release depends on soil acidity or alkalinity. The recovery of phosphorus by plants from applied fertilizer is limited in acidic soils due to P fixation. In acidic soils, the P is fixed by high-energy sorption surfaces such as oxides and hydroxides of Fe and Al by formation of insoluble Fe and Al phosphates by ligand exchange and precipitation reactions (Ohno and Amirbahman, 2010). The availability of P is influenced by soil organic matter, pH, and exchangeable and soluble Al and Fe. Phosphorus is generally available to crops at soil pH of 6 to 7. When the soil pH is less than 6, P deficiency increases in most crops. Conventionally, large amounts of lime and inorganic P fertilizers such as phosphate rocks and Triple Superphosphate (TSP) are used to saturate Al and Fe ions. This approach has not been successful because it is not economical for small scale farmers. The practice is also not environmentally friendly. Because, over liming precipitates P ions with Ca as calcium phosphate, whereas excessive use of P fertilizers causes eutrophication. Although there exist some information on P sorption and fixation using organic matter, there is a lack of information on the use of biochar to reduce P fixation. This is because biochar has high affinity for Al and Fe. Their affinity enables long term chelation of Al and Fe by biochar instead of P. Hence P will become readily and timely available for efficient crop use (Ohno et al., 2007).Therefore; the objective of this study was to improve available phosphorus by using biochar produced from the pyrolysis of coffee husk and corn cob at different charring temperature on acidic soil.

MATERIALS AND METHODS

Description of The Soil Sampling Area

Soil sample was collected from Dedesa, Southwest Ethiopia. The site is located between 7°50'-8°10' N latitude and 36°30'E to 36°45' E longitude. The altitude of the area is about 2260 m above sea level. The mean annual minimum and maximum temperatures are 13 and 28°C, respectively and the mean minimum and maximum annual rainfalls are 1800 and 2200 mm respectively. The soil of the study area is dominated by Nitosol (World Reference Base, 2006) (Figure 1).

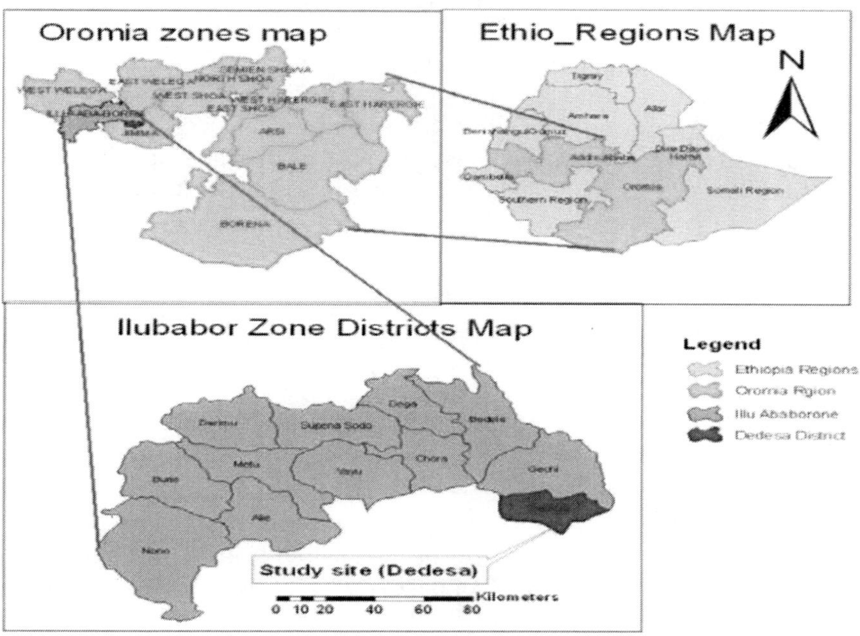

Figure 1. Map of the study area.

Preparation Of Biochar

Biochar of coffee husk and corn cob were prepared in Jimma University College of Agriculture and Veterinary Medicine (JUCAVM) by using a pyrolysis unit at two different temperatures (350 and 500°C) and 3 h of residence time. The resulting biochar materials were grounded and sieved through a 0.25 mm -mesh sieve.

Soil Sampling, Preparation And Incubated With Biochar

The top 0 to 30 cm soil samples were collected by using auger. The collected soil samples were air-dried, crushed by using mortar and pestle and then passed through a 2 mm -mesh sieve. The effect of different levels of the biochar produced from different feedstock and different temperatures on available phosphorous was examined through a laboratory incubation experiment. Two hundred fifty grams of air-dried soil (<2 mm) were weighed in different beakers and biochar was added at rates of 0, 5, 10 and 15 t/ha which is equivalent to 0, 1.366, 2.732 and 4.098, g/kg respectively and thoroughly homogenized. The moisture content of the soil-biochar mixture was maintained at field capacity throughout the incubation period, by adding distilled water whenever necessary. Three replicates of each treatment were prepared, randomly placed and incubated in the laboratory at ambient temperature for 2 months. At the end of two months, samples (≈50 g) were removed from all the treatments and analyzed for pH, OC, OM, TN and other parameters were also analyzed as per the standard methods. But available P was analyzed by following successive weeks of incubation.

Physicochemical Properties Of Biochar Materials

The surface area of biochar was estimated according to Sears's method for silica-based materials (Sears, 1956). This can be obtained by agitating 1.5 g of each of the produced biochar in 100 ml of diluted hydrochloric acid (pH 3). Then a 30 g of sodium chloride was added with stirring and the volume was made up to 150 ml with deionized water. The solution was titrated with 0.10M NaOH and the volume, V, needed to raise the pH from 4 to 9 was then recorded. S (m2/g) = 32V-25, where V is the volume of sodium hydroxide require raising the pH of the sample from 4 to 9 and S is the surface area. pH and electrical conductivity (EC) were measured in distilled water at 1:10 biochar to water mass ratio after shaking for 30 min (ASTM, 2009). Biochar organic carbon content was determined by the Walkley-Black method and total nitrogen (TN) by the Kjeldahl method as cited in Chintala et al., 2013. Available phosphorous (P) was determined by using the Olsen extraction method (Shaheen et al., 2009). Total exchangeable bases were determined after leaching the biochar with ammonium acetate.

Concentrations of Ca and Mg in the leachate were determined by atomic absorption spectrometer. K and Na were determined by flame photometer. Cation exchange capacity was determined at soil pH 7 after displacement by using 1N ammonium acetate method, and then estimated titrimetrically by distillation of ammonium that was displaced by sodium (Gaskin et al., 2008).

Physicochemical Properties Of Soil Sample And The Soil-Biochar Mixture

The particle size distribution (texture), of the soil sample was determined by the Boycouos hydrometric method after destroying OM using hydrogen peroxide (H_2O_2) and dispersing the soils with sodium hexametaphosphate $(NaPO3)_6$. Soil bulk density was determined by the undisturbed core sampling method after drying the soil samples in an oven at 105°C to constant weights.The pH of the soil and soil-biochar mixture was determined in water suspension at 1:2.5 soil/soil-biochar: liquid ratio (w/v) potentiometrically using a glass-calomel combination electrode (Van Reeuwijk, 1992). Electrical conductivity (EC) was measured from a 1:5 (w/v) soil to water ratio after a one hour equilibration time as described by (ASTM, 2009). The Walkley and Black (1934) wet digestion method was used to determine carbon content and, percent OM was obtained by multiplying percent soil OC by a factor of 1.724 following the assumptions that OM is composed of 58% carbon. Total N was analyzed using the Kjeldahl method by oxidizing the OM in (0.1N H2SO4) as described in Black (1965). Cation exchange capacity and exchangeable bases (Ca, Mg, K and Na) were determined after extracting the soil samples by (1N NH4OAc) at pH 7. Ex. Fe was determining by AAS after extracting by DTPA solution.

Exchangeable Ca and Mg in the extracts were analyzed using atomic absorption spectrometer (AAS), while Na and K were analyzed by flame photometer (Rowell, 1994). Cation exchange capacity was there after estimated titrimetrically by distillation of ammonium that was displaced by sodium from NaCl solution (Chapman, 1965). Exchangeable acidity was determined by titration method after 1N KCl solution at pH 7 used to leach exchangeable hydrogen and aluminum ion from soil sample. Available P was determined by using 1M HCl and 1M NH_4F solutions as an extractant by Bray II method (Van Reewijk, 1992). Total P was determining by spectrophotometer after digested by concentrated sulfuric acid. The sample-extractant mixtures were shaken for 30 min on a horizontal shaker (Shaheen et al., 2009), then centrifuged for 10 min at 1500 rpm and filtered by using Whatman no. 42 filter paper. The clear supernatant solutions were collected and analyzed using spectrophotometer at 882 nm.BDL=below

Table 1. Selected physicochemical properties of the studied soil and biochar produced from coffee husk and corn cob at 350°C and 500°C (Mean ±SD)

Parameters	Soil	CHB350	CHB500	CCB350	CCB500
Bulk density (gm/cm^3)	1.22 ± 0.03				
Specific surface area(m2/g)	-	14.07±0.02	26.2±0.01	4.46±0.05	18.14±0.04
pH-H_2O (1:2.5)	5.08 ± 0.06	9.62 ± .06	11.04 ±0.02	8.154 ± .01	9.44 ± 0.03
Exch. Acidity	4.5± 0.1	BDL	BDL	BDL	BDL
EC (mS/cm) (1:5)	0.03 ± 0.00	4.29 ± 0.03	6.44 ± 0.13	0.891 ± 0.23	1.81 ± 0.24
Exch. Ca (me/100g)	8.08 ± 1.32	50.48 ± 0.68	61.48 ± 0.81	37.38 ± 0.56	48.36 ± 0.06
Exch. Mg (me/100g)	1.20 ± 0.2	6.71 ± 0.11	8.21 ± 0.06	4.93 ± 0.04	6.43 ± 0.06
Exch. K (me/100g)	0.8 ± 0.02	1.96 ± 0.27	2.77 ± 0.43	1.711 ± 0.26	2.16 ± 0.14
Exch. Na (me/100g)	0.02 ± 0.00	3.43 ± 0.02	5.15 ± 0.11	0.71 ± 0.18	1.45 ± 0.19
Exch. Fe	35.54± 1.12	BDL	BDL	BDL	BDL
Exch. Al	795.00± 0.23	BDL	BDL	BDL	BDL
CEC (me/100g)	24.36 ± 1.7	64.75 ± 0.76	79.23 ± 0.33	47.52 ± 0.66	62.03 ± 0.80
Organic Carbon (%)	3.97 ± 0.23	16.45 ± 1.96	26.91 ± 7.22	13.98 ± 2.45	20.57 ± 1.40
Organic Matter (%)	6.85 ± 0.39	28.35 ± 3.38	46.39 ± 12.45	24.09 ± 4.23	35.46 ± 2.41
Nitrogen (%)	0.34 ± 0.02	1.42 ± 0.17	2.32 ± 0.62	1.2 ± 0.21	1.77 ± 0.12
Total P	16.2± 0.14	105.25±2.12	149.12±3.45	91.92±1.32	116.22±2.56
Available P (mg /kg)	4.52 ± 0.09	9.79 ± 1.34	13.87 ± 2.16	8.55 ± 1.31	10.81 ± 2.41
Texture	Clay loam				
%Sand	29.33 ± 4.16				
%Clay	30.67 ± 4.16				
%Silt	40.00 ± 0.00				

Table 2. Effect of biochar application on soil pH, exchangeable acidity, exchangeable aluminum and exchangeable iron.

Biochar materials	Rate of Biochar(t/ha)	pH	Exchangeable acidity	Exchangeable Al	Exchangeable Fe
			cmol (+)/kg		
Control	0	5.2±0.03^d	1.9± 0.1^d	795± 0.4^d	35.54± 0.1^d
	5	6.10±0.07^{abc}	1.22± 0.1^{bc}	530± 0.3^b	29.62± 0.02^b
CHB350	10	6.10±0.33^{abc}	1.19± 0.1^{bc}	373.5± 0.3^c	14.8± 0.02^c
	15	6.20±0.01^{ab}	1.15± 0.2^{bc}	186.75± 0.3^e	9.87± 0.02^e
	5	6.1±0.15^{abc}	0.16± 0.04^e	124.5± 0.01^f	17.64± 0.03^f
CHB500	10	6.3±0.10^{abc}	0.14± 0.04^e	62.25± 0.01^g	5.88± 0.03^g
	15	6.6±0.03^a	0.11± 0.04^a	20.75± 0.01^a	2.96± 0.03^a
	5	6.1±0.14^{bc}	1.34± 0.2^{bc}	560± 0.01^b	30.12± 0.12^b
CCB350	10	6.1±0.07^{abc}	1.25± 0.2^{bc}	395.5± 0.1^{bc}	20.08± 0.12^{ef}
	15	6.01±0.09^c	1.19± 0.2^{bc}	197.75± 0.1^e	15.02± 0.12^e
CCB500	5	6.01±0.19^c	0.37± 0.5^c	353.3± 0.2^c	21.09± 0.11^{ef}
	10	6.10±0.01^{abc}	0.29± 0.5^{ef}	176.6± 0.2^{ef}	10.54± 0.11^e
	15	6.14±0.02^{abc}	0.19± 0.5^e	58.88± 0.2^g	5.27± 0.11^g
Pv<0.05		<.0001	<.0001	<.0001	<.0001
LSD		0.3821	0.271	0.1911	0.1932

LSD: least significant difference, CHB350: Coffee husk biochar at 350°C, CHB500: Coffee husk biochar at 500°C, CCB350: Corn cob biochar at 350°C, CCB500: Corn cob biochar at at 500°C.

detection level, CEC: Cation exchange capacity, CHB350: Coffee husk biochar produced at 350°C, CHB500: Coffee husk biochar at 500°C, CCB350: Corn cob biochar at 350°C, CCB500: Corn cob biochar at 500°C. Selected physicochemical properties of the studied soil and biochar are shown in Table 1. In the Table it indicates that the soil is strongly acidic. This shows that, the soil might possibly be affected by Al toxicity, excessive levels of micronutrients such as Co, Cu, Fe, Mn, Zn and, deficiency of macronutrients such as Ca, K, Mg, S, N, P. The low EC value shows that the soil is non-saline which indicates that the total concentration of the major dissolved inorganic solutes (Na^+, Mg^{2+}, Ca^{2+}, K^+, Cl^-, SO_4^{2-}, HCO_3^-, NO_3^- and CO_3^{2-}) in the soil solution is low and this soil acidity can cause limited availability of macronutrients like P and micronutrients Mo which binds to iron and aluminum oxides in acidic soil (Brady and Weil, 2002).

As can be seen in the table, the coffee husk biochar was more alkaline and has higher base cation concentration relative to that of the corn cob biochar. The pH, EC, CEC, P, and base cation concentration were higher in the coffee husk biochar produced at 500 and 350°C followed by corn cob biochar produced at 500 and 350°C.The high pH values of coffee husk biochar may be due to hydrolysis undergone by carbonates and bicarbonates of base cations such as Ca, Mg, Na, and K, which were present in the feedstock's materials (Gaskin et al., 2008). The EC value of coffee husk biochar was found to be higher than that of corn cob biochar, indicating the existence of more water soluble salts in coffee husk biochar than in corn cob biochar. The CEC of coffee husk biochar was also found to be higher than that of corn cob biochar. This may be due to high negative charge potential of surface functional groups in coffee husk than in corn cob. In general, results of the characterization studies of the biochars are clear demonstrations of the significant difference in the composition of biochar produced from different feed-stocks even when they are pyrolyzed under the same temperature. This fact was also reported in a study carried out by Novak et al., 2009. Available P, organic carbon and total nitrogen were also found to be higher in coffee husk biochar than in that of corn cob biochar. The high carbon, organic matter, total nitrogen and total P content in coffee husk biochar might have enriched the soil with high organic matter.

Statistical Data Analysis

Data analysis was done using SAS version 9.2. Three way analysis of variance (ANOVA) namely two feedstock bi4ochar materials, two different pyrolysis temperatures and four application rates were performed to see the significance of differences in the effects of the various soil parameters and among each treatment, using the General Linear Model (GLM) procedure of SAS 9.2. Means separation was done using Least Significant Difference (LSD) after the treatments were found significant at $p<0.05$.

RESULTS AND DISCUSSION

Effect of Biochar Application on Soil Ph, Exchangeable Acidity, Exchangeable Aluminum And Exchangeable Iron

The biochar application significantly affects soil pH compared with the control pH (Table 2). The increase in soil pH was due to the rapid proton (H^+) exchange between the soil and the biochar (Tang et al., 1999). The reduction in exchangeable acidity, exchangeable Al, and exchangeable Fe relates to the increase in soil pH. Increase in pH resulted in the precipitation of exchangeable and soluble Al and Fe as insoluble Al and Fe hydroxides, thus reducing the concentrations of Al and Fe in the soil solution and also exchangeable acidity (Ritchie, 1994). The increase in the pH of the soil due to the application of biochar was generally attributed to an increase in ash content, because ash residues are generally dominated by carbonates of alkali and alkaline earth metals, phosphates and small amounts of organic and inorganic N (Arocena and Opio, 2003). A study by Khanna et al., 1994, has also revealed that the increase in soil pH due to the application of biochar could be because of the high surface area and porous nature of biochar that subsequently increase the CEC of the soil. But, with an increase in incubation period the pH of the soil-biochar mixture nearly constant and this could be due to buffering capacity of the soil.

Effect Of Biochar Application On Oc, Om And Tn

The application of different rates of biochar on the acidic soil significantly ($P < 0.01$) increased the mean soil organic carbon (OC), organic matter (OM), total nitrogen (TN) and total phosphorous content (Table 3). The untreated acidic soil had 3.97 ± 0.23% OC and 6.85 ± 0.39% OM level before treatment, however, due to the incorporation of biochar the OC level increased to a level ranging 3.8 ± 0.29 - 5.79% and 6.56 ± 0.51%- 9.98 ± 0.07% OM in the first month of incubation period and it increased by 34.37% OC and 34.3% OM. In the control however, the level of OC and OM were found to decrease after an incubation period of 2 months. At the end of the incubation period the control had an OC level of 3.6 ± 0.29 and 6.56 ± 0.21% OM, whereas the soil amended with the biochar had levels ranging 3.6 ± 0.29 to 6.69 ± 0.04% OC and 6.56 ± 0.21 to 11.53 ± 0.41% OM which is increased by 45.6% OC and 43.1% OM. The highest OC and OM levels were recorded in the soil amended with 15t/ha of coffee husk biochar produced at 500°C temperature after 2 month incubation period. There were significant differences in the soil total nitrogen and total P after application of biochar. The highest increase was recorded in the soil amended with 15t/ha of coffee husk biochar produced at 500°C temperature. The observed increase in TN and total P could be due to some amount of decomposition which might have occurred when biochar is added to soil (Liang et al., 2006), which could induce net immobilization of inorganic N and organic P already present in the soil solution.

Table 3. The effect of biochar application on the soil OC, OM, TN and Total-P.

Biochar materials	Rate of biochar (t/ha)	%OC	%OM	%TN	Total-P
Control	0	3.64 ± 0.29^{g}	6.56 ± 0.21^{g}	0.32 ± 0.03^{g}	96.5 ± 1.3^{e}
	5	6.14 ± 0.06^{cd}	10.58 ± 0.37^{cd}	0.53 ± 0.016^{cc}	136.2 ± 2.6^{e}
CHB350	10	6.11 ± 0.04^{cd}	10.53 ± 0.37^{cd}	0.53 ± 0.0^{cd}	271.6 ± 3.25^{d}
	15	6.18 ± 0.06^{cb}	10.65 ± 0.37^{cb}	0.53 ± 0.01^{cb}	550 ± 4.32^{b}
CHB500	5	6.28 ± 0.08^{cb}	10.82 ± 0.38^{cb}	0.54 ± 0.01^{cb}	139.2 ± 3.2^{e}
	10	6.45 ± 0.08^{ab}	11.12 ± 0.39^{ab}	0.56 ± 0.08^{ab}	555 ± 4.50^{b}
	15	6.69 ± 0.04^{a}	11.53 ± 0.41^{a}	0.58 ± 0.0^{a}	615 ± 4.70^{a}
	5	5.58 ± 0.12^{f}	9.63 ± 0.33^{f}	0.48 ± 0.01^{f}	134.6 ± 3.3^{e}
CCB350	10	5.79 ± 0.06^{ef}	9.99 ± 0.35^{ef}	0.49 ± 0.01^{ef}	350 ± 3.60^{f}
	15	$5.86 \pm 0.04d^{ef}$	10.1 ± 0.35^{def}	0.51 ± 0.0^{def}	449.5 ± 3.50^{g}
	5	5.98 ± 0.05^{cde}	10.32 ± 0.36^{cde}	0.52 ± 0.0^{cde}	136.2 ± 2.6^{e}
CCB500	10	5.98 ± 0.05^{cde}	10.32 ± 0.36^{cde}	0.52 ± 0.0^{cde}	365 ± 4.52^{cf}
	15	6.07 ± 0.04^{cde}	10.46 ± 0.37^{cde}	0.52 ± 0.0^{cde}	469 ± 4.55^{g}
Pv<0.05		<.0001	<.0001	<.0001	<.0001
LSD		0.30	0.52	0.026	0.76

OC, Organic carbon, OM, Organic matter; TN, Total nitrogen; LSD, least significant difference.

Effect Of Biochar Application On Cec And Exchangeable Cations

The effect of biochar addition on CEC and the contents of exchangeable cations in the acidic soil are presented in Table 4. As shown in the Table, CEC and exchangeable cations were found to increase upon amendment of the acidic soil with coffee husk and corn cob biochar. Statistical assessment of the significance of the observed increases was carried out using ANOVA and, both increases were found to be significant ($P < 0.01$). The untreated acidic soil had 24 me/100 g level before treatment, however, due to the incorporation of biochar the CEC level increased to a level ranging 24.95 ± 1.05-38.46 ± 1.07 me/100 g after 2 months of incubation period and it increased by 34.13% .The highest increase in CEC was recorded in the soil amended with coffee husk biochar produced at 500°C, applied at a rate of 15t/ha and this increased of CEC results the increment of basic cations. The observed increase in CEC due to the application of biochar could have resulted from the inherent characteristics of biochar feedstock. Biochar has high surface area, is highly porous, possesses organic materials of variable charge that have the potential to increase soil CEC and base saturation when added to soil (Glaser et al., 2002). Available evidences also suggest that, on a mass basis, the intrinsic CEC of biochar is consistently higher than that of whole soil, clays or soil organic matter (Sohi et al., 2009).

Table 4. The effect of Biochar application on CEC and exchangeable cations

Biochar materials	Rate of biochar (t/ha)	CEC (me/100 g)	Ca (cmol (+)/kg)	Mg (cmol (+)/kg)	K(cmol (+)/kg)	Na (cmol (+)/kg
Control	0	24.95 ± 1.05^{e}	12.57 ± 0.82^{e}	1.3 ± 0.11^{e}	0.85 ± 0.04^{g}	0.05 ± 0.0^{bc}
	5	34.55 ± 1.09^{bcd}	26.94 ± 0.85^{bcd}	3.58 ± 0.11^{bcd}	1.46 ± 0.03^{efg}	0.05 ± 0.0^{bc}
CHB350	10	35.46 ± 1.07^{bcd}	27.65 ± 0.84^{bcd}	3.68 ± 0.11^{bcd}	1.55 ± 0.03^{edf}	0.05 ± 0.0^{bc}
	15	35.60 ± 1.09^{abcd}	27.83 ± 0.85^{abcd}	3.7 ± 0.11^{abcd}	1.00 ± 0.15^{bcde}	0.07 ± 0.0^{abc}
	5	36.03 ± 1.03^{abcd}	28.1 ± 0.80^{abcd}	3.73 ± 0.10^{abcd}	2.15 ± 0.19^{bcd}	0.05 ± 0.01^{bc}
CHB500	10	37.31 ± 0.86^{ab}	29.09 ± 0.67^{ab}	3.87 ± 0.09^{ab}	2.96± 0.09^{a}	0.06 ± 0.0^{abc}
	15	38.46 ± 1.07^{a}	29.99 ± 0.84^{a}	3.98 ± 0.11^{a}	3.01 ± 0.04^{a}	0.09 ± 0.0^{a}
	5	33.33 ± 0.52^{d}	25.98 ± 0.40^{d}	3.45 ± 0.05^{d}	1.25 ± 0.22^{fg}	0.04 ± 0.01^{c}
CCB350	10	34.37 ± 1.21^{cd}	26.8 ± 0.95^{cd}	3.56 ± 0.13^{cd}	1.46 ± 0.04^{efg}	0.05 ± 0.02^{bc}
	15	35.27 ± 0.46^{bcd}	27.5 ± 0.36^{bcd}	3.66 ± 0.05^{bcd}	1.72 ± 0.29^{cdef}	0.08 ± 0.01^{a}
CCB500	5	34.7 ± 0.46^{bcd}	27.06 ± 0.36^{bcd}	3.6 ± 0.05^{bcd}	1.85 ± 0.34^{bcdef}	0.05 ± 0.02^{bc}
	10	35.03 ± 1.15^{bcd}	27.32 ± 0.90^{bcd}	3.63 ± 0.12^{bcd}	2.23 ± 0.27^{bc}	0.06 ± 0.01^{abc}
	15	36.99 ± 0.70^{abc}	28.84 ± 0.55^{abc}	3.83 ± 0.07^{abc}	2.5 ± 0.54^{ab}	0.07 ± 0.01^{ab}
Pv<0.05		<.0001	<.0001	<.0001	<.0001	<.0001
LSD		2.80	2.18	0.29	0.68	0.029

CEC, Cations exchange capacity; me, Milliequivalents; cmol, centimol; CHB350, coffee husk biochar at 350°C; CHB500, coffee husk biochar at 500°C; CCB350, corn cob biochar at 350°C; CCB500, corn cob biochar at 500°C. LSD, least significant difference.

Therefore, it is quite logical that soil treated with biochar had a highest CEC than the corresponding soil. Studies by Agusalim et al. (2010) have also revealed the increase in soil CEC after the application of biochar. Application of CCB500 at a rate of 15 t/ha on the acidic soil was found to increase the levels of exchangeable Ca and Mg significantly (P < 0.01) from12.57 ± 0.82-29.99 ± 0.84 and 1.3 ± 0.11-3.98 ± 0.11 me/100 g respectively which is increased by 58.08% Ca and 67.3% Mg. Application of CHB500, on the other hand, increased the values of exchangeable K from 0.85 ± 0.04-3.01 ± 0.04 me/100 g and Na 0.05 ± 0.0-0.09 ± 0.0 me/100 g which is increased exchangeable K by 71.5 and 44.4% Na. The observed increase in exchangeable cations in the biochar treated soils might be attributed to the ash content of the biochar. The ash content of biochar helps for the immediate release of the concluded mineral nutrients like Ca, Mg, K and Na for crop use (Niemeyer et al., 2005). The increase in soil pH and CEC, that reduce the activity of Fe and Al, could also contribute for the highest values of available phosphorous in soils treated with biochar. Van Zwieten et al. (2010) also reported the increase in available phosphorous after the application of biochar.

Effect Of Biochar Application On Available Phosphorous

The amount of available phosphorous in acidic soil was also significantly (P<0.01) increased by application of biochar (Figure 2). The untreated acidic soil had 3.64±0.34 mg/kg available phosphorus at first week (W1) of incubation period and 4.99±0.24 mg/kg after an incubation periods of two months. However, due to the incorporation of biochar the available P level increased to a level ranging 3.64±0.34 -23.21±0.07 mg/kg after an incubation period of 2 months and it increased by 84.3% available phosphorus. The highest values of available phosphorous recorded when coffee husk biochar produced at 500°C temperature was applied at a rate of 15 t/ha after two months of incubation periods and at pH values of 6.6. The observed increase in available phosphorus due to application of biochar could be due to the presence of phosphorous in the coffee husk and corn cob and the increase in the availability of P with time was because of microbially mediated mineralization of soil organic P to form inorganic P (Opala et al., 2012).

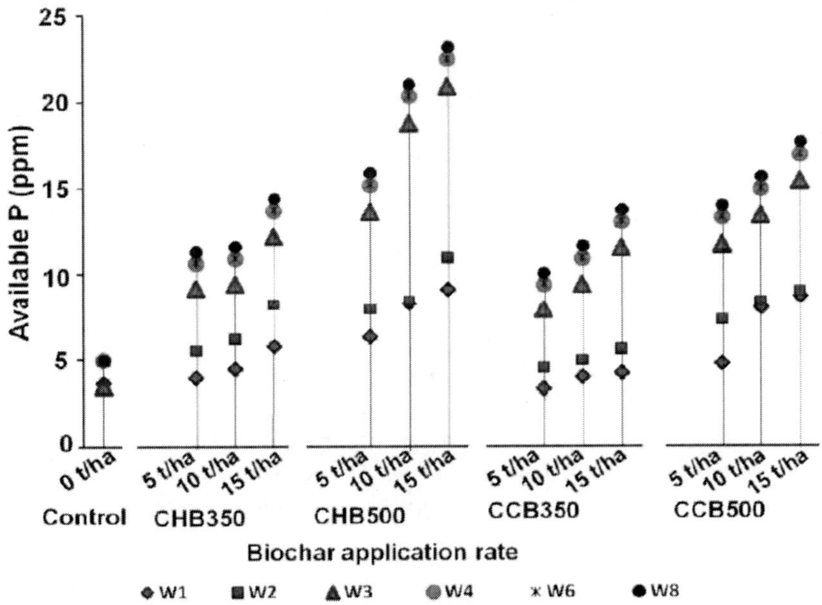

Figure 2. The effect of rate of application and incubation period of CHB350, CHBC500, CCB350 and CCB500, on soil available P

Significant differences were observed between soil available P and total levels of successive weeks of incubation of the biochar amended soil, that is, between available P levels of one week incubation and two weeks incubation, or between those of two weeks and three weeks of incubation period etc. (Figure 2). The biochar from both coffee husk and corn cob produced at 350°C and 500°C and applied at different levels on each treatment increased available P levels when compared with the control. The increase in available P with duration of incubation reported in this study is comparable to those reported by Laboski and Lamb (2003). The observed increase in available P with an increase in the duration of incubation was because of microbially mediated mineralization of soil organic P to form inorganic P.

CONCLUSION

The influence of coffee husk and corn cob biochar application on the level of available P of an acidic soil has been investigated. Findings of the study showed that, application of the biochar prepared from both feed stocks increased available P. The biochar application has also increased the pH and CEC of the soil. To enhance the bioavailability of P, alkaline biochars can be used on acidic soil and the finding of this study indicated that coffee husk residue biochar produced at 500°C highly increased available P as compared to corn cob reduced at the same temperatures. Corn cob residue biochar had the smallest effect on P availability but it could still improve soil quality parameters and increase available P. The effect of these biochar on soil P was in agreement with their characterization, suggesting that similar biochar evaluations might predict their effect on P availability and inform recommendations on biochar application to soils. Generally, in this study the application of both coffee husk and corn cob biochar produced at 350 and 500°C increases soil available P in all rating of application as compare to control (soil without application of biochar). Even though both feedstock biochar improved soil available P it is better to use coffee husk biochar produced at 500°C of pyrolyzing temperature applied at a rate of 15 t/ha. Moreover, further field researches are needed to evaluate the effect of biochar on the availability, fate and uptake of available p in soil.

CONFLICT OF INTERESTS

The authors have not declared any conflict of interests.

REFERENCES

1. Agusalim M, Wani HU, Syechfani MS (2010). Rice Husk Biochar for Rice Based Cropping System in Acid Soil: The Characteristics of Rice Husk Biochar and Its Influence on the Properties of Acid Sulfate Soils and Rice Growth in West Kalimantan. Indonesia. J. Agric. Sci. 2(1):39-47.
2. Arocena JM, Opio C (2003). Prescribed Fire- Induced Changes in Properties of Sub-Boreal Forest Soils. Geoderma. 113:1-16.
3. ASTM standard (2009). Standard test method for chemical analysis of wood charcoal. American Society for Testing and Materials (ASTM) International: Conshohocken, PA.Black CA (1965). Methods of soil analysis. Part I, American Society of Agronomy. Madison, Wisconsin, USA. 1572 p.Brady N, Weil R (2002). The Nature and Properties of Soils. 13thed.Chapman H (1965). Cation Exchange Capacity. In: Black CA (Eds.) Methods of Soil Analysis.Agronomy, Madison, Wisconsin. Am. Soc. Agro. Inc. 9:891-901.Chintala R, Mollinedo J, Schumacher TE,

Malo DD, Julson JL (2013).Effect of Biochar on Chemical Properties of Acidic Soil, Arch. Agron. Soil Sci. 60(3):393-404.

4. Gaskin JW, Steiner C, Harris K, Das KC, Bibens B (2008). Effect of Low Temperature Pyrolysis Conditions on Biochars for Agricultural Use, Trans. ASABE. 51:2061-2069.
5. Glaser B, Lehmann J, Zech W (2002). Ameliorating physical and chemical properties of highly weathered soils in the tropics with charcoal A review. Biol. Fertil. Soils 35:219-230.
6. Khanna PK, Raison RJ, Falkine RA (1994). Chemical properties of ash derived from Eucalyptuslitter and its effects on forest soils. For. Ecol. Manage. 66:107-125.
7. Laboski CAM, Lamb JA (2003). Changes in soil test phosphorus concentration after application of manure or fertilizer. Soil Sci. Soc. Am. J. 67(2):544-554.
8. Liang B, Lehmann J, Solomon D, Kinyangi J, Grossman J, O'Neill B, Skjemstad JO, Thesis J, Luizao FJ, Peterson J, Neves EG Black (2006). Carbon increases cation exchange capacity in soils. Soil Sci. Soc. Am. J. 70:1719-1730.
9. Niemeyer T, Niemeyer M, Mohamed A, Fottner S, Ha¨rdtle W (2005). Impact of prescribed burning on the nutrient balance of Heath Lands with particular reference to nitrogen and phosphorus. Appl. Veg. Sci. 8:183-192.
10. Novak JM, Busscher WJ, Laird DL, Ahmedna M, Watts DW, Niandou MAS (2009). Impact of biochar amendment on fertility of a southeastern Coastal Plain soil. Soil Sci. 174:105-112.
11. Ohno T, Amirbahman A (2010). Phosphorus availability in boreal forest soils: a geochemical and nutrient uptakemodeling approach. Geoderma. 155(2):46-54.
12. Ohno T, Fernandez JI, Hiradate S, and Sherman J F (2007). Effects of soil acidification and forest type on water soluble soil organic matter properties. Geoderma 140(2):176-187.
13. Opala PA, Okalebo JR, Othieno CO (2012). Effects of organic and inorganic materials on soil acidity and phosphorus availability in a soil incubation study. ISRN Agronomy, 20 Article ID597216. 10 p.
14. Ritchie GSP (1994). Role of dissolution and precipitation of minerals in controlling soluble aluminium in acidic soils,"Adv. Agron. 53:47-83
15. Rowell DL (1994). Soil science: Methods & Applications. Addison Wesley Longman Singapore Publishers (Pte) Ltd., England, UK. 350.
16. Scheuner ET, Makeschin F, Wells ED, Carter PQ (2004). Short-term impacts of arvesting and burning disturbances on physical and chemical characteristics of forest soils in western Newfoundland, Canada. European J. Forest Res. 123(4):321-330.
17. Sears G (1956). Determination of specific surface area of colloidal silica by titration with sodium hydroxide. Anal. Chem. 28(12):1981-1983.
18. Shaheen SM, Tsadilas CD, Eskridge KM (2009). Effect if Common Ions on Phosphorus Sorption and availability in Greek Alfisols with Different pH, Soil Sci. 174:21-26.
19. Sohi S, Lopez-Capel E, Krull E, Bol R (2009). Biochar, climate change and soil: A review to guide future research. CSIRO Land and Water Science Report.Tang C, Sparling GP, McLay CDA, Raphael C (1999). "Effect of short-term legume residue decomposition on soil acidity,"Aust. J. Soil Res. 37(3):561-573.
20. Van Reewijk LP (1992). Procedures for Soil Analysis (3rd Ed.). International Soil Refrence Center (ISRIC), Wageningen, the Netherlands.Van Zwieten L, Kimber S, Morris S, Chan KY, Downie A, Rust J, Joseph S, Cowie A (2010). Effects of biochar from slow pyrolysis of papermill waste on agronomic performance and soil fertility. Plant Soil. 327:235-246.
21. Walkley A, Black IA (1934). An examination of the Degtjareff method for determining soil organic matter and a proposed modification of the chromic acid titration method. Soil Sci. 37:29-38.
22. World Reference Base (2006). A Framework for International Classification, Correlation and Communication. Food and Agriculture Organization of the United Nations, Rome.

CHAPTER 4

Effect of Compost and Earthworm Production on Soil Properties, Growth and Dry Matter Yield of Maize in Crude Oil Degraded Soil

Nweke I. A.

Department of Soil Science, Chukwuemeka Odimegwu Ojukwu University, Anambra State, Nigeria.

ABSTRACT

The effect of compost and earthworm production on soil properties, growth and dry-matter yield of maize in crude oil degraded soil was studied. The treatment consisted of 250 g crude oil degraded soil (DS) and 50 g each of four different compost materials mixed differently with 250 g of degraded soil, and the compost were cassava peels (CP), cassava peels + poultry manure (PC), cassava peels + pig manure (GC) and cassava peels + pig manure + poultry manure (PGC) and 10 pieces of sub-adult earthworms (Eudrilus eugenae) were inoculated to each of the experimental pot after 11 days. The five treatments were replicated four times, data generated were subjected to analysis of variance test and treatment means were separated using least significant difference (LSD0.05). The results of the study indicated significant differences between the treatments in soil and agronomic parameters assessed. The application of compost and earthworm activities increased the plant height, leaf area, and number of leaves at 4 weeks after planting (WAP) and 6 WAP, and dry matter yield of maize. The shoot and root dry matter yield was observed to increase in the order PC > PGC > GC > CP > DS. Earthworm production in this study measured by the number survived and biomass weight showed CP and GC as the best culture for earthworm production in oil degraded soil. The degraded soil (DS) did not record any earthworm survival at harvest. The result of the soil analysis indicated less change in the textural class of the soil, and all the chemical parameters tested were enhanced by the compost and earthworm activities. Based on the results of growth rate and yield components of maize as well as soil chemical properties, the PC and PGC cultures having performed competitively better than the other treatments can be considered useful and adequate with the help of earthworm in reclaiming an oil degraded soil for crop production in a tropical environment like Nigeria.

Keywords: Compost, earthworm, maize, crude oil, soil properties

INTRODUCTION

Land is an essential part of earth crust which contains soil, rocks, minerals, water and natural vegetation. Land is in various shapes and forms, home for structures and mining activities. While soil is parts of land that man cultivate for the purpose of planting his crops and it is often the fine earth less than 2 mm in diameter.

Agricultural activities take place on land. So land is an important resource on which man depend for his livelihood as well as an important asset or legacy that he can bequeath to next generations. Therefore the way and manner man handles his environment with respect to land will determine to a greater extent his environmental comfort and healthiness and to lesser extent the life span of his next generation. For man live and derive his livelihood from the land. The advancement of science and technology, increase in urbanization and growth of human population has continuously accelerate the rates of consumption of mineral resource, the mining of this minerals either agricultural or non agricultural result to soil infertility, ecological imbalance and land degradation. Land degradation is the inability or non potentiality of the land to produce food, goods and services and maintain healthy ecosystem for over a long period of time. Soil is a living entity, with complex mixture of organic and inorganic materials that have been synthesized naturally through various actions and reactions, is a natural medium for plant growth as it provides crops with mineral nutrients and water, support, is an important reserve of water and mineral nutrients needed for crop production. So, a soil is degraded when it loses its capability to sustain crop production and other goods and services. Mbagwu (2003) asserted that soil degradation is the temporary or permanent lowering of the productive capacity of soil caused by over grazing, deforestation, inappropriate agricultural practices, over exploitation of fuel wood leading to desertification and other man-induced activities, such as: mining and oil drilling and refining, discharge of wastes and sludge, river diversion etc. Some of these man- induced activities often happen in the crude oil producing states of Nigeria, otherwise called Niger-Delta (south-south, Nigeria).

Soil degradation in the oil producing areas like the Niger-Delta region of Nigeria is mainly caused by oil pollution, oil drilling and spill, gas flaring and refining which affects the soil fertility status reducing the productivity of the soils. Kundu and Ghose (1997) observed that mining activities disrupt the landscapes and soil components such as soil horizons and structure, soil microbe population and nutrient cycles which are important for healthy ecosystem and therefore results in the destruction of vegetation and soil profile. When crude oil spills on soil the condition of the soil becomes unsatisfactory for plants, because there is insufficient aeration due to a decrease in the air filled pore space (Degong, 1990; Kostreba, 1999; Ufot and Akande, 2005). It leads to alteration of the physical and chemical properties of the soil such as structure, hydraulic conductivity, exchangeable cations and pH (Ogboghodo et al., 2001). Odu (1981) reiterated that oil pollution has less effect on soil physical properties than in chemical properties. Soil degradation due to oil exploration is caused by gas flaring pipelines leakage, oil waste dump and oil spills, drilling and refining (Odu, 1980; Okere et al., 2001). In oil producing regions of Nigeria, substantial amount of crude oil is spilled annually. Therefore the remediation or reclamation of such oil degraded soils is a process of improving the quality of the soil through some measures of biotic function that will help to return the productivity for agricultural purposes rather than abandoning the area. Tremblay and Levy (1993) and Mba et al. (2001) emphasized on the use of organic manure in soil remediation.

Compost is an organic matter that has been decomposed and use as organic fertilizer and soil amendment. It improves the physical and chemical properties of soil as well as biological activities that help to liberate nutrients for growth plant and ameliorate the structure of soil. Compost encourages plant growth capacity and help to withstand or resist tensile forces through its fibrous material (Goldsmith et al., 2011). Soil organisms and earthworms are organisms that assist the nature in maintaining nutrient flow from one system to another as well as minimize environmental degradation. Earthworms are natural bioreactors that proliferate along with other micro-organisms and provide required conditions for the biodegradation of wastes (Munnoli et al., 2010). They redesign the physical, chemical and biological structure of the soil environment and deposit casts rich in mineral nutrients on soil surface (Ansari, 2011; Verma and Shweta, 2011; Agburu, 2012). The remediation or reclamation of crude oil degraded soil like in the oil producing regions of Nigeria, according to Lone et al. (2008) and Kavamura and Esposito (2010), requires the establishment of stable nutrient cycles from plant growth and microbial process. This present study, therefore attempts to evaluate the effectiveness of cassava peel compost and earthworm production in liberating chemical nutrients and amelioration of oil degraded soil for optimum crop production.

MATERIALS AND METHODS

The experiment was carried out at the Faculty of Agriculture, Chukwuemeka Odimegwu Ojukwu University, Igbariam Campus, Anambra State. The area is located within latitude 6° 14'N and longitude 6° 45'E and 122 m above sea level in south eastern, Nigeria. The annual rainfall of the area varies from 1800 mm to about 2500 mm, the temperature of the area vary from 21 to over 25° while relative humidity is from 60 to over 85%.

Soil sample collection

Soil samples were collected from 0-30 cm depth at Asa in Ukwa-east local government area of Abia State, Nigeria where there is oil spillage on a vast land, due to breakdown in oil pipeline and due to oil exploration activities that take in this area. Asa is a community located between latitude 4° 45'N and 5°N and between longitude 7° E and 7° 15'E. The type of vegetation of the area was rain forest, with a mean annual temperature between 26.5 and 27°C and daily relative humidity of 70-85%. The major crops produced by farmers in the area are cassava, maize, yam and vegetable crops. The farmers in this community have abandoned those vast areas of land being degraded by the oil spillage do to nothing grows on it. In fact when the soil samples were being collected there was an oil films on the soil. The soil is heavily degraded. Soil samples were collected from five different locations within the affected area. These soil samples were bulk together inside a fiber jute bag, tied up and transported to Chukwuemeka Odimegwu Ojukwu University where it was allowed for 8 months before used for the potted experiment. The 8 months allowed before the soil samples can be used again for further studies was to assess the impact of the crude oil on the dry state of the degraded soil haven noted its effect on the fresh state. This is because, the unpublished work (though presented in one of the Faculty Lecture series), where by the crude oil degraded soil when freshly collected filled with oil films, was used in a potted experiment and guinea grass compost was used as treatment, earthworm introduced to the potted experiment and maize was used as test crop, no sign of germination was observed on the degraded soil throughout the period of the experiment and earthworm (*Eudrilus eugenae*) did not survive for 24 h on both the degraded and treated soils, though the guinea grass compost improve the germination and growth on the treated soils, the crops looked unhealthy and very tiny in structure. The physical and chemical properties of the soil before treatment application are presented in Table 1. Initial properties of the degraded soil before treatment application.

Parameter	Value
Sand	72.60%
Silt	2.40%
Clay	26%
Textural class	Sandy clay loam
pH_{H20}	5.05
Available P	20.10 $mgkg^{-1}$
Total N	0.098%
OC	0.65 gkg^{-1}
OM	1.11 gkg^{-1}
Ca	2.80 $Cmolkg^{-1}$
Mg	0.80 $Cmolkg^{-1}$
K	0.108 $Cmolkg^{-1}$
Na	0.078 $Cmolkg^{-1}$
EA	0.08 $Cmolkg^{-1}$
ECEC	3.87 $Cmolkg^{-1}$
BS	98%

Compost preparation, inoculation, allocation and experimental design

100 g of fresh cassava (*Manihot esculenta*) peels were air-dried for two weeks and then soaked in water for two weeks to ferment, in order to reduce the cyanide content and soften it for earthworm consumption. 50 g of poultry and pig manure each were moistened overnight. The fermented cassava peels and manures (poultry and pig manure) were composted in air tight polythene bags as follows; 100 g of cassava peels + 50 g poultry manure (PC); 100 g of cassava peels + 50 g pig manure (GC); 100 g of cassava peels alone (CP); 50 g of cassava peels + 25 g of poultry manure +

25 g of pig manure (PGC), the compost was harvested after 18 days and 50 g of each compost was then mixed thoroughly with 250 g of degraded soil. The treatments mixtures were as follows; PC, GC, PGC, CP, and DS (degraded soil without treatment). These treatments where replicated four times and then put into plastic pots of dimension 17 cm x 23 cm with perforations at the base, the perforations were closed with cotton wool to avoid excessive drainage. The experiment was laid out in an artificially prepared green house constructed with white tarpaulin and the experiment was properly blocked. This was to avoid external water inform of rainfall to interfere and to moderate the temperature. The degraded soil and the amendment mixtures were watered and left for 5 days before sowing of maize, the test crop. This was to give time for the transformation and further decomposition of the mixture. Ten (10) pieces of sub-adult earthworms (*E. eugenae*) sourced from costal area of River Niger, Onitsha Anambra state, Nigeria were introduced into each of the experimental pot after 11 days. Four seeds of maize orba super II sourced from Agricultural Development Programme (ADP) Awka, Anambra State were planted per pot and ten days after planting, supply was made for those that did not germinate. The choice of maize as test crop was based on the fact that it is one of the major crops grown by farmers in the oil degraded area. Secondly maize can be grown and studied in pot experiment. The treatment types were watered every other day at the rate of 25 ml per pot and monitored daily for maize germination, cast production activities of earthworm and compost degradation. The experiment lasted for 8 weeks.

Variables evaluated

The evaluated variable were: plant height, leaf area and number of leaves per plant were taken two weeks after planting and continued till the 6th week after planting (WAP). At the end of the 8th week the number of earthworms survived and their biomass was evaluated. The plants were carefully uprooted from the pots and the soil particles washed away. The roots were carefully cut-off from the shoot. Both the roots and shoots were oven dried for 24 hours for the dry matter yield. Soil samples were collected from respective pots and the variables analyzed were: pH using a digital pH meter with solid to liquid ratio of 1:2.5, exchangeable cation Ca^{2+} and Mg^{2+} were extracted with ammonium acetate by Perkin Elmer atomic absorption spectrophotometer (Tel and Rao, 1982), while Na^{+} and K^{+} were determined on flame photometer. Organic matter was determined by the method of Nelson and Sommers (1982) and total nitrogen was by a semi-micro Kjedahl procedure as described by Bremner and Mulvancy (1982). Available P was determined by Bray II method. Exchangeable acidity was determined by the titrimetric method of Mclean (1982), while the effective cation exchange capacity (ECEC) was determined by the summation method;

$$ECEC = TEB + TEA \quad (1)$$

Base saturation (BS %) was calculated using this formula;

$$BS\% = [TEA / ECEC] \times 100 \quad (2)$$

Statistical analysis

Data generated from the study was analyzed using analysis of variance (ANOVA) tested on randomized complete block design (RCBD) according to Steel and Torrie (1980) and treatment means were separated using least significant difference (LSD) at 5% probability level.

RESULTS

Daily examination of the experiment

The daily examination of the experimental cultures showed that maize seeds started to germinate on the treated soils four days after planting, but it took more than five days to record maize germination in the untreated soils. Within five days after earthworm introduction, cast which is one of the major evidence of earthworm activity in a habitable ecology or environment was observed in treated soils and more than 7 days later in untreated soils. Though germination and growth was observed in untreated (degraded) soil, the maize plant was very tiny and looking unhealthy and most of the plants do not survive to maturity and harvest. While the maize plants observed in treated soils looked healthy, robust and fresh. Secondly none of the earthworm introduced to the degraded soil (DS) survived to harvest and the cast found on the DS soils were very scanty and insignificant. The degradation of compost materials and earthworm activities were more on the PC and GC, followed by PGC treated soils as against CP and DS soils. Thus, this influenced the kind of results obtained from the study summarized in Tables 1 to 4.

Initial properties of the degraded soil

The properties of the studied soil presented in Table 1 show that the soil is a sandy clay loam soil, with low contents of organic matter (1.11%) and carbon (0.65%).

Table 2. Effect of compost and earthworm production on the growth, shoot and root dry matter yield of maize in a degraded soil.

Treatment	Maize height (cm)			Number of leaves			Leaf area (cm^2)			Shoot dry matter yield (gkg^{-1})	Root dried matter yield (gkg^{-1})
	2WAP	4WAP	6WAP	2WAP	4WAP	6WAP	2WAP	4WAP	6WAP		
DS	6.68^a	9.81^b	13.03^c	3.25^b	4.50^b	6.38^b	21.51^c	34.68^b	43.78^c	0.43^c	0.03^b
CP	6.59^a	11.34^b	12.50^c	3.63^b	5.00^b	6.13^b	19.92^c	23.35^b	35.50^c	0.59^c	0.38^b
PC	7.90^a	24.83^a	48.19^a	3.13^b	7.38^a	10.13^a	19.50^c	150.0^a	292.94^a	77.87^a	19.09^a
GC	10.73^a	15.51^b	26.78^b	4.13^b	7.13^a	9.25^a	37.43^c	66.95^b	107.5^b	24.65^b	10.82^a
PGC	12.86^a	26.53^a	47.43^a	4.38^b	7.75^a	10.13^a	38.51^c	144.3^a	266.92^a	74.30^a	15.13^a
LSD 0.05	NS	6.64	10.28	NS	1.82	1.57	NS	66.13	56.90	18.24	11.37

DS= Degraded soil; CP=Cassava peel compost; PC= Cassava peel + poultry manure compost; GC= Pig manure + cassava peel compost; PGC= Poultry manure + Pig manure + cassava peel compost; LSD= Least significant difference; NS= Not- significant; WAP= Weeks after planting. Means on the same column with the same letter do not differ significantly ($P<0.05$).

Table 3. Effect of compost materials on the biomass (g/pot) and survival of earthworm (*Eudrilus eugenae*) in a degraded soil at harvest.

Treatment	Number of earthworm survived/pot	Earthworm biomass at harvest (g/pot)
DS	0.00	0.00
CP	14.75^b	6.00^a
PC	7.25^b	1.18^c
GC	36.50^a	3.50^b
PGC	8.00^b	1.48^c
LSD 0.05	11.33	2.35

DS= Degraded soil; CP=Cassava peel compost; PC= Cassava peel + poultry manure compost; GC= Pig manure + cassava peel compost; PGC= Poultry manure + Pig manure + cassava peel compost; LSD= Least significant difference; NS= Not- significant. Means on the same column with the same letter do not differ significantly ($P<0.05$).

The soil is acidic and contain low level of nutrient elements, except in P that have high available content (20.10 $mgkg^{-1}$) compared to other nutrient elements, the studied soil is considered to be poor in these essential plant nutrient elements, according to the ratings and pH range classification of Landon (1991). Therefore crude oil degradation is an added impediment to the soil as this will adversely affect the productivity of the soil.

Effect of compost and earthworm production on the growth and dry matter yield of maize

The growth parameters presented on Table 2, showed non-significant differences among the treatments applied in 2 weeks after planting (WAP), however the highest maize height, number of leaves and leaf area was recorded in PGC, and the next in rank was the GC. The 4 and 6 WAP results on the Table 2 showed remarkable significant ($P < 0.05$) differences among the treatments. The order of increase in maize height in 4 WAP and 6 WAP were PGC > PC > GC > CP > DS and PGC > PC > GC > DS > CP respectively. The value of maize height recorded in PGC and PC at 4 and 6 WAP were statistical similar, but significantly better than the other treatments. The number of leaves and leaf area in 4 and 6 WAP follow the trend of maize height results with the highest value obtained in PGC. The analysis of the results recorded in this trial revealed that PC and PGC were at par in all the growth parameters assessed in this study, but competitively and significantly ($P < 0.05$) performed better than the other treatments. The result of the shoot and root dried matter yield showed that PC and PGC performed much better than the other amendments. The highest value of shoot and root dried matter yield was obtained from PC treated pots. The DS performed very poorly followed by the CP when compared to the values obtained from the other treatments. The analysis of the result on Table 2 equally showed that the values of maize height, number of leaves and leaf area increased as the weeks after planting increased from 2 to 6 WAP.

Effect of compost materials and maize production on the biomass (g/pot) and number of earthworm (*Eudrilus eugenae*) survived in a degraded soil at harvest

The result presented in Table 3 indicated significant ($P < 0.05$) difference among the treatments. The zero (0) value recorded in DS pots showed that no earthworm survived to harvest in DS, which may suggest un-habitable medium for earthworm production. The highest number of earthworm survived was recorded in GC treated pots, though this does not translate to the recorded biomass (g/pot), as the earthworm harvested from the treated GC pot at harvest were mainly young and very little sub adult earthworms. In contrast to GC, the more of sub-adult and matured earthworms was observed in CP treated pots which translated to the kind of biomass result obtained from the CP pots. Nonetheless, the result obtained may have indicated that GC and CP maybe considered as the best medium for earthworm production under the crude oil degraded soil.

Effect of compost and earthworm production on the soil chemical properties

The result of soil chemical parameters presented on Table 4 showed that there was significant ($P < 0.05$) difference among the treatments except for the result obtained for nitrogen (N) and base saturation (BS). The highest value recorded for these two parameters (N and BS) was observed in PC treated pots. For N, the value obtained from DS was even found to be higher in rank to PC than the rest of other treatments. The order of increase in BS were PC > GC > PGC > CP > DS. The pH of the treatments measured in water, indicated that the pH of the degraded soil was much enhanced. The pH changed from acidic medium of DS through neutral to alkaline medium observed in PC treated pots. The organic matter (OM) content of the treated soils were very significantly ($P<0.05$) improved. The percentage increase with respect to GC and PC were 66.54 and 31.06% respectively. The order of increase in OM content were GC > PGC > CP > PC > DS. The available P content of the treated soils was not enhanced except for PC pots, which showed rapid increase in the available P content. The P content obtained from DS and CP were statistically similar but better than the values obtain from PGC treated pots. The exchangeable Ca, Mg, K and Na vary significantly ($P < 0.05$) among the various amendments. The result showed that the compost and earthworm production did not much influence the values of the exchangeable parameters of the degraded soil. The exchangeable Ca only increased in CP and GC treated soils. The values obtained from PGC and PC was statistically similar and was not better than the degraded soil (DS).

The compost and earthworm production greatly enhanced the exchangeable Mg of GC treated pots, with respect to the rest of the other treatments which are not in any way better than the degraded soil. Similar to Mg, the exchangeable K of GC was improved by 41.67% by the application of compost and earthworm production. The rest of the other treatments were not enhanced. The exchangeable Na of PGC treated pots was increased in comparison to the values obtained from the other treated pots and the percentage improvement was 9.09%. The exchangeable acidity (EA) and effective cation exchange capacity (ECEC) values of the treatments were significant ($P < 0.05$). The EA and ECEC values of the degraded soil were higher than the values obtained from the other treatments. The other of decrease in EA and ECEC were PC < GC, PGC < CP < DS and PGC < PC < GC < CP < DS respectively. The percentage decrease in value of EA in PC relative to DS value was 1300%, while that of ECEC with regard to PGC value was 54.89%.

The result of the soil physical properties presented on Table 5 showed that application of compost and earthworm production influenced the physical parameter of the soil, more especially the silt content of the degraded soil. Hence the textural class of the soil was changed from sandy clay loam, through loamy sand to sandy loam, which has distinct characteristics on the nutrient content of the soils.

Table 4. Effect of compost and earthworm production on the soil chemical properties.

Treatment	pH of H_2O	N (%)	OM (%)	BS (%)	P ($mgkg^{-1}$)	Ca^{2+} ($Cmolkg^{-1}$)	Mg^{2+} ($Cmolkg^{-1}$)	K^+ ($Cmolkg^{-1}$)	Na^+ ($Cmolkg^{-1}$)	EA ($Cmolkg^{-1}$)	ECEC ($Cmolkg^{-1}$)
DS	5.79^d	0.32^a	0.91^e	85^b	18.00^b	4.00^{bc}	1.63^b	0.07^b	0.10^{abc}	1.12^a	7.28^a
CP	6.76^c	0.28^a	2.02^c	95^b	15.50^b	4.80^a	1.60b	0.06^{bc}	0.06^d	0.32^b	6.84^a
PC	8.34^a	0.56^a	1.32^d	99^b	30.60^a	3.60^{bcd}	1.35^c	0.04^d	0.08^{abcd}	0.08^d	5.42^b
GC	7.02^b	0.24^a	2.72^a	98^b	12.40^c	4.20^b	2.00^a	0.12^a	0.09^{abc}	0.16^c	6.57^a
PGC	7.62^b	0.12^a	2.58^b	97^b	9.80^c	3.20^d	1.20^d	0.05^{cd}	0.11^a	0.16^c	4.70^b
LSD 0.05	0.74	NS	0.10	NS	2.68	0.52	0.08	0.02	0.03	0.07	0.99

DS= Degraded soil; CP=Cassava peel compost; PC= Cassava peel + poultry manure compost; GC= Pig manure + cassava peel compost; PGC= Poultry manure + Pig manure + cassava peel compost; LSD= Least significant difference; NS= Not- significant; ECEC = Effective cation exchange capacity; EA = Exchangeable acidity, OM= Organic matter, BS= Base saturation. Means on the same column with the same letter do not differ significantly ($P<0.05$).

Table 5. Effect of compost and earthworm production on the texture of the soil.

Treatment	Sand %	Silt %	Clay%	Textural class
DS	70.60^a	1.40^d	28.00^a	Sandy Clay Loam SCL
CP	71.60^a	2.40^c	26.00^b	Sandy Clay loam SCL
PC	76.60^a	8.40^b	15.00^d	Loamy Sand LS
GC	70.60^a	8.40^b	21.00^c	Sandy Clay loam SCL
PGC	75.60^a	9.40^a	15.00^d	Sandy loam SL
LSD0.05	NS	0.67	1.35	

DS= Degraded soil; CP=Cassava peel compost; PC= Cassava peel + poultry manure compost; GC= Pig manure + cassava peel compost; PGC= Poultry manure + Pig manure + cassava peel compost; LSD= Least significant difference; NS= Not-significant. Means on the same column with the same letter do not differ significantly ($P<0.05$).

DISCUSSION

The result of this study indicated that cassava peel compost is capable of improving the productivity of crude oil degraded soil by improving the growth and yield components of maize as well as increasing the production of earthworm (*E. eugenae*). The germination and growth of maize, as well as very short lived activities of earthworm observed on the degraded soil, was due to 8 months allowed for the soil before it was used for the potted experiment. Therefore based on the unpublished work and present study experiences, leaving oil degraded soil for a while or soil removal alone can even be a good management strategy for an oil degraded land. For instance Amakiri and Onofegara (1983) and Rowell (2003) observed that crude oil degraded land appeared to recover or becomes useful for plant growth after years because of the ability of the soil to maintain an equilibrium in carbon nitrogen ratio. Secondary soil removal of crude oil degraded soil for the purpose of ameliorating the soil for agricultural purposes was an adopted technique in the Middle East as far back as early nineteenth century, according to Pinstrup and Lorch (2001) to reclaim an oil degraded environment. Kostreba (1991) in his own studies maintained that the soil removal should be followed with amendment with fertilizer and a mixture of compost manure and wood ships to improve water holding capacity and to provide micro organisms with sufficient carbon and nutrients. While Tremblay and Levy (1993) and Mbah et al. (2001) emphasized on the use of animal manure in soil remediation.

Effect of compost and earthworm production on the growth and yield components of maize

The significant differences observed in growth and yield parameters assessed in this study could be due to differences in the nutrient content of the compost applied and probable to preference in compost diet by the earthworm species (*E. eugenae*). The non-significant value observed in the 2WAP might be that the compost has not been decomposed enough to release nutrients that will influence the maize growth in 2WAP. The result of 4 to 6WAP of the growth parameters and yield component parameters indicated that there were more release of nutrients which reflected on the values obtained, showing that the compost and earthworm have stimulated the soil organisms to release phyto hormones that stimulate nutrient absorption and plant growth. Goldsmith et al. (2011) observed increases in plant growth parameters when they used compost as soil amendment. Paterson (2003) in his study noted that increases in plant growth, flowering and crop yields are as a result of earthworms activities in plant litter degradation, as they increase soil microbial population that produce plant growth hormones.

The high growth rate and yield observed can as well be due to the formation of humic substances resulted from the degradation of the compost materials by the earthworm; humic substances are noted to stimulate root respiration, formation and growth. Atiyeh et al. (2002), observed that the effect of humic substances result in increased efficiency of the plant rooting system, that in turn improves the upper growth of plants such as shoots, leaves, flowers and fruit yield, while Bahman et al. (2004), stated that application of compost provides nutrients and liming effects for the growing plant.

The shoot and root dry matter yield of maize result was a clear evidence of the impact of the compost and the earthworm on the maize yield. The compost materials PC and PGC were however found to yield better in root and shoot dry matter yield and even best in growth parameters compared to the other treatments. This probable might be due to faster degradation of the compost mixtures (PC and PGC) by the earthworm preferences relative to other composts materials used in this study. This resulted to more nutrient release in the condition and form that the maize plant can absorb them or might as well be due to that root exudates of maize have energized the activities of micro-organisms that invariable help to source and provide plant nutrients for the maize plants in treated pots, but more in PC and PGC treated pots as a result of good atmosphere created by the earthworm and compost. Lavelle and Martin (1992) showed that the interaction between earthworms and soil microbes increase the provision of plant nutrients and stimulate plant growth. Samaranayake and Wijekoon (2010) using soil mixtures in potted experiment observed that the growth rate, shoot dry matter yield of maize was increased by the introduction of earthworms.

The increase in growth and yield of maize observed in this study could as well be regard as the effect of combination of different interacting factors such as accumulations of earthworm casts rich in plant nutrients, compost degradation, soil aeration due to earthworm burrowing within the maize root zone of soil and many other factors as was reported by Fragoso et al. (1996). The productivity of maize depends on nutrients requirement particularly that of N, P and K (Arunkumar, 2007) and compost contribute more in building up the phosphorous status of the soil (Sawar et al., 2008) all of which believed to have contributed in the yield and growth result obtained from the treated pots especially for PC and PGC pots. The growth and yield result as well showed that the CP compost is not good compost for effective rehabilitation and reactivation of oil degraded soil and that cassava peels required the mixture of animal manure for effective utilization, nutrient release and reclamation of impaired soil for crop production.

The remarkable difference in the growth and yield values obtained from degraded soils and treated soils, showed the capability of the compost and earthworm production in bioremediation and rehabilitating of poorly degraded soil. Earthworms are ecosystem engineers that influence soil properties (Agburu, 2012) and help to turn wastes into resources and at the same time minimizes pollution (Edward, 1998). There cast have been reported to be more biologically active and richer in micro flora, growth promoting rhizobacteria than their surrounding indigested soils (Scheu, 1997; Parthasarathi et al., 2007; Sinha et al., 2010). They are major components that regulate nutrient cycling processes in many ecosystems (Edward and Bohlen, 1996).

With the activities of this organism, the physical, chemical and biological compositions of the soil tend to increase couple with the active degradation of the compost of which the effect was reflected in the nature of the results obtained. Ratty and Hutha (2004) found beneficial effect of earthworm on soil structural development, nutrient cycling and productivity when introduced to a reclaimed mining sites and peat. The result of maize growth and yield tend to prove that oil degraded soil can effectively be bioremediated, rehabilitated and reclaimed for efficient crop production earlier than expected using the combination of compost and earthworm.

Effect of compost materials and maize production on the biomass and survival of *Eudrilus eugenae*

The result of this study showed that earthworm (*E. Eugenae*) can successfully thrive or cultured in an oil degraded soil when assisted or supplied with organic materials as a source of feed. The significant difference in the number of earthworm that survived to harvest and weight could be attributed to the preference in the compost diet by the earthworm. Toutain et al. (1982) reported that earthworm show a preference for particular parts of a plant, while Nweke (2013) in his own study observed that earthworm (*E. eugenea*) not only show preference to grass diet but also show preferences to grass with or without animal manure. Hence he found out that Bracharia grass produced more compost than Bracharia mixed with pig manure. This probable might be what resulted in the biomass value obtained in CP (Cassava peel compost without animal manure), where sub-adult and matured earthworms was harvested at maturity suggesting best medium for earthworm production as against young and very little sub-adults earthworms harvested in the other treatments. This is why the number of earthworm survived in GC did not translate to weight obtained because earthworm harvested at maturity in GC were mainly young ones and few sub-adult. The significant difference in number of earthworm that survived and their biomass can as well be due to interaction between maize root, microbial processes and earthworm. Healthy and intact roots release a wide variety of chemicals into the surrounding soil, a process termed exudation (Rovira, 1969), some of these chemicals exuded by the plant roots act as signaling molecules or toxic allele chemicals (Grayston et al., 1998; Bertin et al., 2003). So probable maize roots might have released these toxic allele chemicals which are harmful to the earthworm activities coupled with lethal effect of crude oil and these acting together invariable reduced their production in this study. Though earthworm performed better in CP it does not translate or influence much the growth and yield parameters of the maize plant, probable the nutrient released are not in the form and condition that the maize plant can absorb them, because the result of soil chemical parameters presented in Table 4 indicated that the soil treated with CP are rich in soil nutrients. The zero (0) recorded for number and biomass of earthworm in DS showed that no earthworm survived to harvest. Though there were earthworm activities in the degraded soil (DS), it did not last beyond 5 weeks, probable there maybe still some toxic hydrocarbon materials which by virtue of watering the experiment, might have become harmful to the earthworm, hence limitation in their productivity and eventual death. Though the earthworm did not survive in DS to harvest its effect in growth, yield and soil chemical properties visa vies the initial soil properties in Table 1 is remarkable, probable due to its cast that have enriched the DS soil, before their death. For all intents and purposes, it can be deduced that for culturing earthworm (*E. eugenae*) using cassava peel as source of diet in crude oil degraded soil, it may be good not to add animal manure.

Effect of compost and earthworm production on the soil chemical and physical properties

The result of the soil chemical properties showed that production of the earthworm and compost enriched the chemical properties of the degraded soil, with the exception of %N and percentage base saturation (%BS) all the parameters tested in this study were significantly difference at $P<0.05$. The pH results of the treated soils in comparison to the degraded soils (DS) showed a favorable pH level, because at pH range of 6-7 most plant nutrients are readily available for plant uptake, while pH above 7 reduces the chances of the plants to absorb trace elements such as Fe, Mn, B etc (White, 1979; Greenland, 1981; Muller and Donahue, 1992; Tisdale et al., 1993). This was observable in the increase in growth and yield of maize obtained from the treated pots compared to the degraded soil (DS). The degraded soil (DS), result probable might have indicated the solubility of some trace elements such as Mn, Cu, cobalt and Zn which are toxic to the productivity of crops, some micro-organism proliferation and earthworm activities as well as less NO_3^- production. Hence the nature of the results obtained and death of earthworms inoculated. The OM content of the treated soils where highly increased relative to the DS soils which might be as a result of compost application thus the significant effect on the nutrient value of the plants consumed that translated into the growth and yield observed in this trial. Organic matter content is the store house of plant chemical nutrients. The available P, N and pH where increased more especially in PC treated pots compared to the other treatments and was readily made available for the maize plants, for PC pots gave the highest value in most of the growth parameters and in shoot and root dry matter yield of the maize plant among the other treatments. This result could also be explained on the ability of earthworm to modify characters of organic materials and productivity of the resultant product in the soil mixture. The high content of available P observed in the DS soils could be that traces of P are released from Al^{3t}, while low content of P recorded in most of the other treatments can as well be attributed to phosphorous (P) fixation by sesquioxides usually observed in highly weathered humid tropical soil like the studied soil. Therefore, result obtained from DS soils do signify that though a soil may be fertile or rich in plant nutrient but may not promote crop growth and yield due to nutrients fixation, soil pathogens, nutrient toxicity, soil pH and environmental factors like oil spillage.

The exchangeable cations (Ca^{2+}; Mg^{2+}; K^+; Na^+) of the soil were influenced by the compost application and earthworm production, though the influence is more in exchangeable Ca and Mg. The data obtained from the study also showed that exchange complexes were occupied mainly by Ca and Mg. The content of K is below the range given by Dutta (2005) for good crop production. The high record of Ca and Mg is a good omen as many physiological and biochemical processes in plants are due to the presence of Ca and Mg such as flower senescence and flower abscission (Glenn et al., 1998), adjustment of ethylene responses in plants (Zhang et al., 2002), fruit ripening (Ferguson, 1984), etc where due to calcium (Ca). Magnesium (Mg) activates more enzymes than any other mineral nutrient, making a significant contribution to plant growth and development (Epstein and Bloom, 2004).

The exchangeable acidity (EA) of the treated soils where greatly influenced compared to the acidity level of the degraded soil (DS), this signifies detoxification, nutrient availability and higher crop performance observed in the treated soils. The level of acidity dictated in DS, showed the impact of crude oil degradation as it lowers the soil pH thereby causing acidity. At low pH phosphate combines with aluminum and iron to form compounds (Al_3P0_4, $FeP0_4$) that are not readily available to plants, most micro-nutrients toxic to the growth of plants becomes soluble and available at about soil pH 5.5 and nitrogen fixation activities will be low. These activities could have been the reflection of the values obtained in DS soils in all the parameters tested in this study. The effective cation exchange capacity (ECEC) is the sum total of exchangeable bases (Ca^{2+}; Mg^{2+}; k^+; Na^+) plus the exchangeable acidity (EA, that is, $H^+ + Al^{3+}$), for soils in the humid tropical areas like the studied soil that are strongly weathered and are predominantly of variable charges and such variable colloid include Fe and Al-oxides, the ECEC measured give the more realistic CEC values which represents the potential fertility of the soil studied. The ECEC values obtained could be attributed to cations level especially Ca and Mg levels in the soils.

The soil physical properties were influenced especially the silt and the clay content of the soil. Odu (1980) reported that oil degradation has less effect on soil physical properties and that the texture of the soil degraded by oil has less significant changes. While SPDC (1998) argued that the structure of crude oil degraded soil may be weakened initially, but later improved when the hydrocarbons are incorporated by bacteria.

CONCLUSION

The result of the present study clearly showed that earthworm can be cultured in an oil degraded soil with organic materials as diet, the growth rate, shoot and root dry matter yield of maize as well as the soil parameters tested were greatly influenced by the compost and earthworm production. This signifies that an oil degraded soil can be reclaimed faster with the combination of earthworm and compost. From the data generated in this study, it is suggested that oil producing communities, where crude oil degradation due to oil drilling and oil spillage, gas flaring are paramount should imbibe the use of organic materials in combination with earthworm to reduce to the barest minimum scarcity of food, land abandonment, high cost of agricultural products, migration to urban centers, unemployment and improve generally environmental health of the community.

CONFLICT OF INTERESTS

The author has not declared any conflict of interest.

REFERENCES

1. Agburu EO (2012). The effect of earthworm on properties of soils
2. Amakiri JO, Onofegara FA (1983). Effect of crude oil polluted on the growth of maize, okra, and pepper. Petrochem. J. 2(4):199-205.Ansari AA (2011). Worms powered environmental biotechnology in organic waste management Int. J. Soil Sci. 6:25-30Arunkumar A (2007). Effect of different levels of NPLC on growth and yield parameter of sweet corn Karnaka J. Agric. Sci. 20(1):41-43.Atiyeh R, Lee S, Edwards C, Arancon Q, Metzger J (2002). The influence of humic acids derived from earthworm processed organic wastes on plant growth. Bioresour. Technol. 84(1):7-14.
3. Bahman E, Daniel G, John EG (2004). Residual effects of manure and compost application on corn production and soil properties, Agron. J. 96:442-447.
4. Bertin Yang X, Waston LA (2003). The role of root exudates and allele chemicals in the rhizophere. Plant Soil 256:67-83.
5. Bremner JM, Mulvancy CS (1982). Nitrogen total: In method of soil analysis part 2, American Society of Agronomy, Madison, WI. pp. 595-624.Degong E (1990). The effect of crude oil spill on cereals, Environ. Pollut. 22:187-196.
6. Dutta AC (2005). Botany for degree students, Oxford, New Delhi. pp. 220-233.Edward CA (1998). The use of earthworm in the break down and management of organic wastes: In Earthworm ecology, CA Edwards (ed), CRC, Press Boco Roton, FL pp. 327-254.Edward CA, Bohlen PJ (1996). Biology and Ecology of Earthworms, 3rd edn Chapman and Hall, London. 426pEpstein E, Bloom AJ (2004). Mineral nutrition of plants, principles and perspectives 2nd edn Sinauer associates

Sunderland MA ISBN 0-87893-472-4.Fragoso C, Brown GG, Patron JC, Blanchard F, Lavella P, Pashanasi B, Kuma TC (1996). Agricultural intensification, soil biodiversity in tropics: The role of earthworms. Appl. Soil Ecol. 6:17-35.

7. Ferguson B (1984). Calcium in plant senescence and fruit ripening. Plant Cell Environ. 7:477-489.
8. Glenn GM, Read ASN, Poovaiah BW (1988). Effect of calcium on cell structure, protein phosphorylation and protein profiles in senescing apples. Plant Cell Physiol. 29:565-572.Goldsmith W, Silva M, Fischenich C (2011). Determining optimal degree of soil compaction for balancing mechanical stability and plant growth capacity. ERDCTN- EMRRP-SR 26.Grayston SJ, Wang S, Campell CD, Edward AC (1998). Selective influence of plant species on microbial diversity in the rhizosphere soil, Biol. Biochem. 30:369-378.
9. Greenland DJ (1981). Characterization of soils. Oxford University Press, NewYork. 5p.Kavamura VN, Esposite E (2010). Biotechnology strategies applied to the decontamination of soil polluted with heavy metals. Adv. Biotechnol. 28:61-69.
10. Kostreba L (1999). The remediation technique of crude oil polluted soil using organic materials for crop production: In proceedings of oil spill conference, American Petroleum Institute Washington DC. pp. 40-41.Kundu NK, Ghose MK (1997). Soil profile characteristics in Rajmahal coalfield area. Indian J Soil Water Conserv. 25(2):110-113.Lavelle P, Martin (1992). Small scale and large scale effects of endogeic earthworms on soil organic matter dynamics in soils of the humid tropics. Soil Biol. Biochem. 24(12):1491-1498.
11. Lone MI, He Z, Stoffela PJ, Yang X (2008). Phyto-remediation of heavy metal polluted soils and water progress and perspectives. J. Zhejiang Univ Sci. B. 9:210-220.
12. Mbah CN, Ezeugwu G, Anikwe MAN (2001). Effect of poultry droppings on soil physical properties and heavy metals uptake by maize plant. J. Agric. Food Technol. Environ. 1(1):119-122.Mbagwu JSC (2003). Aggregate stability and soil degradation in tropics, paper presented at 2003 College of Soil Physics, ICTP Trieste Italy.Mclean EO (1982). Soil pH determination: In Pase DL (ed). Methods of soil analysis part 2, Chemical Analysis Agronomy Series, No. 9, ASA, SSA, Madison WI, USA.Muller WR, Donahue RL (1992). Soils: An Introduction to soils and plant growth, Prentice Hall India, New Delhi-11001. pp. 226-420.Munnoli PM, Jaime A Teixeira da silva, Saroj Bhosle (2010). Dynamics of the soil – earthworm-plant relationship. A review. Dynam. Soil Dynam. Plant 4(1):1-21.Nelson DW, Sommers LE (1982). Total organic carbon and organic matter: In Page CA, Miller RH, Keeny DR (eds). Method of Soil Analysis part 2, 2nd edn. Agronomy Monograph No. 9, ASA and SSSA, Madison, WI. pp. 539-579.Nweke IA (2013). Plant nutrient release composition in vermicompost as influenced by Eudrilus eugenae using different organic diets. J. Ecol. Natur. Environ. 5(11):346-351.
13. Odu CTI (1980). Degradation and weathering of crude oil under tropical conditions: In proceedings of a seminar on the petroleum industry. Nig. Environ. V1:143-152Odu CTI (1981). Petroleum industry and Nigeria Environment: In proceedings of a seminar on the petroleum industry and Nigeria environment held in Port Harcourt, River State, 1986.Ogboghodo AE, Osemwota IO, Choker JU (2001). An assessment of the effect of crude oil pollution on soil properties, germination and growth of Maize (Zea mays L.) Two crude oil types Forcados light (UQCC) and Escravos light (Chevron): In Proceedings of Annual Conference, SSSN, held at University of Calabar, Nigeria, March, 2001. pp. 299-300.Okere JU, Akamigbo FOR, Ufot UO (2001). The effect of local mining on soils around Enugu, Nigeria, 27th annual Conference of Soil Science Society of Nigeria (SSSN). pp. 291-299.Parthasarathi K, Ranganthan LS, Anandi V, Zeyer J (2007). Diversity of micro flora in the gut, and casts of tropical composting earthworms reared on different substrates. J. Environ. Bal. 28:87-97.Paterson E (2003). Importance of rhizodeposition in the coupling of plant and microbial productivity. Eur. J. Soil Sci. 54(4):741-750.

14. Pinstrup AP, Lorch FP (2001). Prospective on overcoming hunger, poverty and environmental degradation International Food Policy Research Institute, Washington DC. pp. 310-315.Ratty M, Hutha V (2004). Earthworm communities in birch stand with different origin in central Finland. Pedobiologia 48(1):283-291.
15. Rovira AD (1969). Plant root exudates. Bot. Rev. 35(1):35-57.
16. Rowell MJ (2003). The effect of crude oil spills on soils in the reclamation of agricultural soil after oil spills. University of Alberta Press, Edmonton. pp. 32-33.Samarayake JWK, Wijekoon S (2010). Effect of selected earthworms on soil fertility, plant growth and vermicomposting. Trop. Agric. Res. Extens. 13(2):33-40.Sawar G, Hussain N, Schmeisky H, Muhammad S, Ibrahim M, Safdar E (2008). Improvement of soil physical and chemical properties with compost application in rice-wheat cropping system. Pak. J. Bot. 40:275-282.Scheu S (1997). Microbial activity and nutrient dynamics in earthworm cast (Lumbricidae). Biol. Fertile Soils 50:230-234.Sinha RK, Aganwai S, Chauhan K, Valani D (2010). The wonders of earthworms and its vermicompost in farm production: Charles Darwin's friends of farmers' with potentials to replace destructive chemical fertilizers from agriculture. Agric. Sci. 1:76-94.
17. Shell Petroleum Development Company (1998). Environmental devastation, SPDC Publishers, Warri, Nigeria.Steel RG, Torrie JH (1980). Procedure of statistics. A biometrical Approach 2nd edn. McGraw-Hall, New York. pp. 633-690.Tel DA, Rao R (1982). Automated and semi-automated method for soil plant analysis, Manual Series No.7, IITA, Ibadan.Tisdale SL, Nelson WL, Beaton JD, Halin JL (1993). Soil fertility and fertilizers 5th edn Prentice Hall Inc. Upper Saddle River, New Jersey 07458, USA. pp. 45-561.Toutain F, Villemin G, Albrecht A (1982). Ultra structural study of biodegradation processes, 11 Enchytraeids-leaf litter models. Pedobiologia 23:145-156.Tremblay LK, Levy EM (1993). Bioremediation application of slow release fertilizer on low energy share lines: In Proceedings of oil spill conference, American petroleum Institute, Wessington, D. C. pp. 499-500.Ufot UO, Akande O (2005). Soil survey of Mkpat Enin FADAMA soils, Akwa Ibom State of Nigeria for food productions: In proceedings of the 30th Annual Conference of the SSSN, University of Agriculture Makurdi, Nigeria. pp. 18-20.Verma D, Shweta (2011). Earthworm resources of western Himalayan region. Indian Int. J. Soil Sci. 6:124-133.
18. White RE (1979). Introduction to the principles and practices of soil science, Blackwell Scientific publication, Oxford London p. 417.Zang BG, Rouland C, Lattand G, Lavelle P (1993). Activities of digestive enzymes in gut of the tropical earthworm Pontoscolex corethrus. Eur. J. Soil Biol. 29(1):7-11.

CHAPTER 5

Soil Quality Mapping Studies Using Nematodes as Bioindicators

Lakshmy Kamala Sthanu[1*], Jaya Divakaran Sarasamma[2], Radhakrishnan Thanu Iyer[3], Mohandas Chellappan[4]

[1]*Centre for Environment and Development, Thozhuvancode, Vattiyoorkavu P.O., Thiruvananthapuram, India;*

[2]*Department of Environmental Sciences, University of Kerala, Kariavattom Campus, Thiruvananthapuram, India;*

[3]*Division of Geo Informatics, Indian Institute of Information Technology and Management, Kerala, Techno Park Campus, Thiruvananthapuram, India;*

[4]*Division of Crop Protection, Central Tuber Crops Research Institute, Sreekaryam, Thiruvananthapuram, India.*

ABSTRACT

Soil quality is one of the most important factors in sustaining the global biosphere and developing sustainable agricultural practices. Land use and management practices greatly impact the direction and degree of soil quality changes in time and space. Understanding the effects of land use and management practices on soil quality and its indicators has been identified as one of the most important goals for modern soil science. Soil quality mapping study represents a method for assessing and mapping soil quality changes in time and space in small units. For the present study, changes in the physical, chemical parameters and nematode density of the soils in the rural and urban areas of Thiruvananthapuram district, Kerala, were determined. The soil samples were collected from seven different categories of contaminated soils namely coastal area, sewage disposal area, industrial area, road-side area, agricultural area, market area and gasoline station area, and also from two control stations in rural and urban areas. The soil physico-chemical parameters and nematode density were determined. Geostatistics combined with GIS was applied to analyze the spatial variability of soil physico-chemical characteristics and nematode density. This soil quality mapping study provides a basis for identifying tension zones and serves as a triggering mechanism for implementation of soil contamination mitigating strategies.

Keywords: Soil Quality; Mapping Studies; Nematode Density; Spatial Variability; Contaminated Soils

1. INTRODUCTION

Industrialization and urbanization have a strong impact on the environment, and its biotic and abiotic components. Many toxins added to the soils by different anthropogenic activities can build up to concentrations that become serious threats to plant and animal health (Alloway, 1996) [1]. The presence or absence of indicator species or indicator community reflects environmental conditions. Bio monitors provide a direct measurement of a biological effect rather than inferring values using soil extractions (Tarazona et al., 2005) [2]. The study of regional variations and the anthropogenic contamination by metals of soils is very important for environmental planning and monitoring in urban areas. Mapping allows for more efficient approach to remediation and monitoring of soil contaminants (Fortin and Dale, 2005) [3]. Geographical Information System (GIS) is used for the mapping of the contaminants in the study area soils. Based on the GIS analysis, the highest risk sites are identified in which majority of the remediation and monitoring should take place. The major objective of the present study was to monitor the soil quality of selected rural and urban areas in Thiruvananthapuram District, Kerala, using nematodes as bioindicators, and also aimed to provide base line information on the soil health status of rural and urban areas of the capital district of Kerala.

2. MATERIALS AND METHODS

2.1. Study Area

The rural and urban areas in Thiruvananthapuram District, Kerala were selected for the present study (http: www. censusindia.net/2001census results). Thiruvananthapuram District is located at 8°26'25" N latitude and 76°55'25"E longitude. Thiruvananthapuram, the southernmost District in the State of Kerala is the study area. Its southern most extremity, Kaliaikkavila, is only 54 kms away from Kanyakumari (Cape Comarin), the land's end of India. The district stretches 78 kms along the shores of the Arabian Sea on the west, Kollam district lies on the north with Thirunelveli and Kanyakumari districts of Tamil Nadu on the east and the south respectively. The district can be divided into three geographical regions; highlands, midlands and lowlands. The large forest reserves favorably affect the climate and induce rain. Cold weather is experienced in the mountain ranges whereas lower down, the weather is bracing and in the plains, it is generally hot. Though the mean maximum temperature is only around 900F, it is oppressive in the moisture-laden atmosphere of the plains. Humidity is high and rises to about 90 percent during the south-west monsoon. The average rainfall is around 150 cm per annum. It is significant that the district gets rainfall both from the south-west and the north-east monsoons. According to the final figures of 2011census (Census of India, 2011), the population of the district is 3,307,284. The Taluks in Thiruvananthapuram District are Neyyattinkara, Thiruvananthapuram, Nedumangad and Chirayinkeezhu. The district has three major rivers, several freshwater lakes and more than 300 ponds. The eastern region is forested, northern regions are mostly under rubber cultivation and the remaining areas have mixed dry land crops of coconut, plantain, tapioca, etc. Built up areas and rice fields complete the land use. Based on the land use pattern, Thiruvananthapuram District has been classified into different categories namely, coastal areas, sewage disposal areas, industrial areas, road-side areas, agricultural areas, market areas, gasoline station areas, benign environment, etc. These areas will be receiving different types of wastes based on the source of contamination. The contaminated sites due to different anthropogenic activities were identified in both rural and urban areas of Thiruvananthapuram District after conducting a reconnaissance survey. Sixteen sampling stations in the rural and urban areas of the Thiruvananthapuram District were selected. These include seven different categories of contaminated soils selected in the rural and urban areas namely coastal area, sewage disposal area, industrial area, roadside area, agricultural area, market area and gasoline station area. Control soil samples were collected from comparatively benign environments in both rural and urban areas which were selected as control stations (**Table 1**).

2.2. Soil Collection

Soil samples (0 - 15 cm depth) from the sampling stations were collected during the pre-monsoon, monsoon and post-monsoon seasons of the study period (April 2009 to January 2010). Composite samples were taken by mixing the samples collected from five different sites of each station. Soil sampling and analysis of soil physico-chemical characteristics were carried out according to the standard procedures (Gupta, 1999) [4]. The concentration of the heavy metals viz. lead, chromium, manganese, copper and zinc in the acid digested soil samples were determined (Saxena, 1994) [5] using an Atomic Absorption Spectrophotometer (Model, GBC 932 AA). Karl Pearson's Product Moment Correlation is used for finding significant correlation between nematode densities with all other soil quality parameters in the three different seasons (SPSS v - 16 for WINDOWS).

The Geographical coordinates of the sampling stations were collected using Hand-held GPS (Garmin eTrex Vista HCx). The Coordinate values were entered into a spreadsheet with properly assigning IDs for each location. Another spreadsheet was prepared with physico-chemical characteristics and nematode density. In this spreadsheet, the same IDs were entered against respective stations. Both spreadsheets were added in GIS software (Arc GIS version 10) as tables. The table with Geographic coordinate values were plotted as point layer using "Display XY" method and subsequently exported as a shape file with WGS84 as reference ellipsoid. The physico-chemical characteristics and nematode density table was joined with the attribute table of the point layer using ID as the foreign key using JOIN method in GIS software (Korte, 2001) [6]. The resultant layer comprised the geometry of stations as points and all the results as associated attribute table (**Figure 1**). The soil physico-chemical characteristics were analyzed and mapping of significantly correlated soil physico-chemical characteristics with nematode density was plotted.

3. RESULTS AND DISCUSSION

3.1. Mapping of Physico-Chemical Characteristics and Nematode Density in Soil

The results of physical characteristics of the soils in the study stations are given in Tables 2(a) and (b) respectively and the soil chemical characteristics analyzed are given in Tables 3(a) and (b) respectively. The results of heavy metal (Pb, Zn, Mn, Cu, Cr) content in soil are given in Tables 4(a) and (b) respectively. Correlation studies of soil physico-chemical characteristics with nematode density were done.

3.2. Nematode Sampling, Extraction and Density Calculation

One kg soil samples from 0 - 15 cm depth were collected. For soil nematode identification studies, 100 cc (approx. 150 gms) of soil samples were washed and the soil solutions were passed through 60 mesh (250 μ) and 350 mesh (45 μ) sieves (ASTM Endecott Sieves) respectively. The residues from 350 mesh sieve was washed and collected into a beaker and allowed to settle for one hour. Then decanted the supernatant and poured the rest on a wet tissue paper placed on metal gauze in a Petri dish containing water. After two days, collected the nematode containing solution, allowed to settle for decanting and added equal quantity of boiled 4% formalin to kill and fix it. 10 ml of this solution was placed in a nematode Perspex counting dish (Cobb, 1917) [7] and were observed under a Stereomicroscope (Model, Leica LED3000 NVI) and their abundance was calculated (Southwood and Henderson, 2000) [8] during the three seasons as shown in Figures 2(a)-(c). The soil sample solution was made up to 100 ml and stirred well. Pipetted 10 ml of this soil solution from three different depths of this 100 ml soil sample solution kept in a beaker and placed in Doncaster's circular mounting tray (Goodey, 1963) [9]. Nematode density is expressed in Number per 100 ml soil.

Table 1. Portrait of sampling stations in the study area.

Rural Stations			Urban Stations		
Name	Location		Name	Location	
	Latitude	Longitude		Latitude	Longitude
Vizhinjam (Station 1)	8°33'51"	76°53'25"	Shangumugham (Station 9)	8°27'50"	76°55'55"
Vilappilssala (Station 2)	8°30'18"	76°53'45"	Valiathura (Station 10)	8°33'51"	76°53'25"
Kochuveli (Station 3)	8°30'18"	76°53'45"	Peroorkkada (Station 11)	8°31'35"	76°58'03"
Mangalapuram (Station 4)	8°30'18"	76°53'45"	Thampanoor (Station 12)	8°29'15"	76°57'12"
Palode (Station 5)	8°28'53"	76°57'08"	Sreekaryam (Station 13)	8°30'18"	76°53'45"
Balaramapuram (Station 6)	8°27'06"	76°57'21"	Chalai (Station 14)	8°28'52"	76°57'08"
Pallichal (Station 7)	8°27'06"	76°57'31"	Karamana (Station 15)	8°28'56"	76°57'36"
Anappara (**Control Station** 8)	8°28'53"	76°57'08"	Ambalathara (**Control Station** 16)	8°27'06"	76°57'21"

S1 - S8: Rural Sampling Stations; S9 - S16: Urban Sampling Stations.

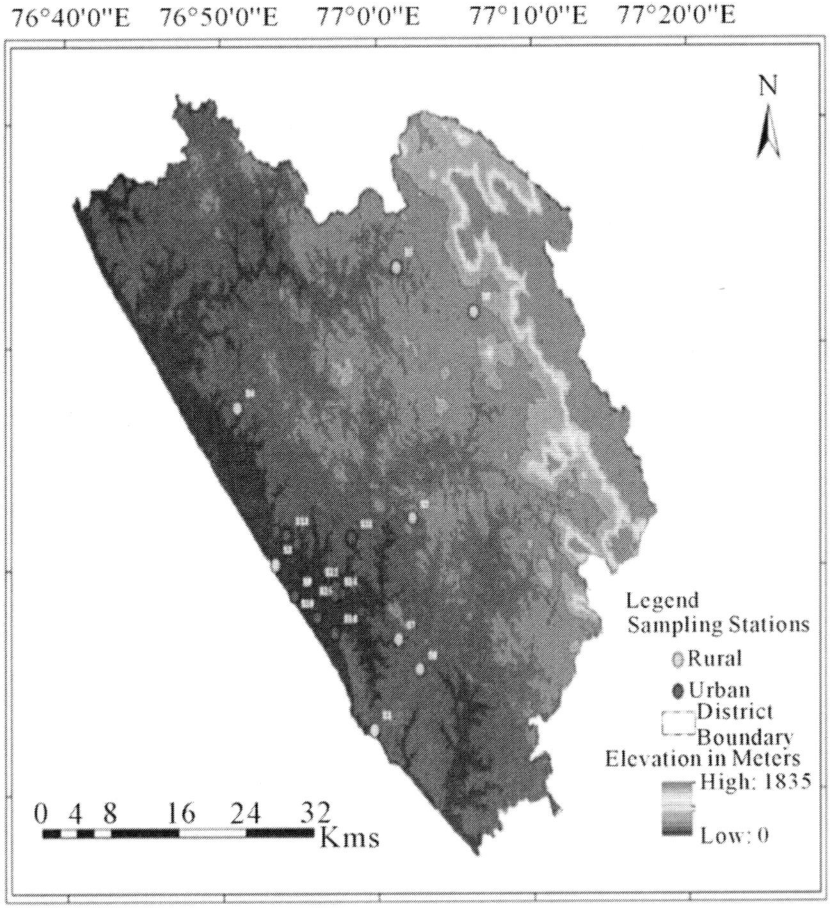

Figure 1. Base map showing the sampling stations in the rural and urban areasof.

Electrical conductivity, water holding capacity, moisture content, organic matter, total nitrogen content, phosphorus content, potassium and copper contents showed positive correlation with nematode density in soil. pH, temperature, pore space, bulk density, chloride, lead, zinc, manganese and chromium contents showed negative correlation with nematode density in soil samples collected from both rural and urban area as shown in Tables 5(a) and (b) respectively. Of the different physicochemical characteristics studied, water holding capacity, moisture content, organic matter content, pH, zinc and manganese contents only showed significant level of correlation with nematode density. Mapping of significantly correlated soil

Table 2. (a) Physical characteristics of rural and urban soils; (b) Physical characteristics of rural and urban soils.

(a)

St.No.	Chloride (mg/g)			Moisture content (%)			Organic matter (g %)		
	Pre-monsoon season	Monsoon season	Post-monsoon season	Pre-monsoon season	Monsoon season	Post-monsoo n season	Pre-monsoon season	Monsoon season	Post-monsoon season
S1	0.34**	0.338**	0.339**	1.74**	1.81**	1.76**	0.517**	0.591**	0.569**
S2	0.046**	0.046**	0.045**	12.54**	12.6**	11.8**	3.034**	3.839**	3.836**
S3	0.053**	0.052**	0.052**	5.46**	5.55**	5.54**	1.569**	1.772**	1.765**
S4	0.045**	0.045	0.044**	4.78**	4.9**	4.9**	1.397**	1.626**	1.562**
S5	0.046**	0.046**	0.046**	9.42**	9.62**	9.62**	2.569**	3.696**	3.6**
S6	0.046**	0.047**	0.046**	15.17**	15.25**	15.11**	5.655**	5.769**	4.448**
S7	0.046**	0.044**	0.046**	2.38**	2.4**	2.36**	1.155**	1.579**	1.155**
S8	0.046	0.046	0.045	8.55	8.61	8.59	1.69	2.081	2.512
S9	0.41**	0.406**	0.407**	0.19**	0.22**	0.22**	0.345**	0.367**	0.335**
S10	0.047**	0.047**	0.046**	11.85**	11.9**	11.5**	2.672**	2.838**	3.836**
S11	0.046	0.046	0.045	6.79**	6.82**	6.76**	1.569**	1.94**	1.867**
S12	0.044**	0.044**	0.043**	2.7**	2.9**	2.9**	1.328**	1.688**	1.455**
S13	0.046**	0.046**	0.046	8.67**	8.69**	8.6	2.207**	2.491**	2.255**
S14	0.046**	0.047**	0.046**	13.3**	14**	13.6**	3.5**	3.881**	3.87**
S15	0.046**	0.045**	0.046**	2.35**	2.36**	2.29**	0.69**	0.909**	0.905**
S16	0.046	0.046	0.046**	8.06	8.1	8	1.69	2.081	2.41

(b)

St.No.	o			Total Phosphorus (g %)			Total Potassium (g %)		
	Pre-monsoon season	Monsoon season	Post-monsoon season	Pre-monsoon season	Monsoon season	Post-monsoon season	Pre-monsoon season	Monsoon season	Post-monsoon season
S1	24.5**	42**	12.6**	4.42**	3.75**	3.62**	66.52**	64.67**	25.6**
S2	63.1**	105**	182.7**	52.46**	89.82**	75.31**	189.73**	147.41**	106.1**
S3	44.8**	70**	132.3**	6.21**	30.56**	14.41**	125.45**	145.54**	99.33**
S4	43.5**	63**	119.7**	5.31**	27.68**	13.15**	106.25**	143.53**	90.3**
S5	59.6**	91**	157.5**	43.57**	64.11**	69.21**	163.39**	184.6**	100.7**
S6	70**	168**	252**	159.82**	106.43**	76.36**	189.73**	187.32**	155**
S7	30.1**	56**	107.1**	4.45**	7.86**	5.68**	83.48**	128.46**	60.2**
S8	49.2	77	144.9	25.76	44.64	41.65	133.04	168.86	102.88
S9	20.3**	42**	40.8**	3.53**	0.71**	0.78**	34.38	17.08**	25.59**
S10	61**	98**	182.7**	52.46**	66.07**	50.2**	174.55**	398.44**	301**
S11	47**	70	138.6**	10.67**	33.21**	16.41**	127.23	64.51**	102.34**
S12	35.4**	56**	107.1**	5.31**	14.64**	7.78**	85.27	43.53**	84.28**
S13	49.6**	77**	157.5**	40.09**	63.04**	67.95**	144.2**	178.35**	155**
S14	64.4**	154**	195.3**	159.82**	92.14**	75.52**	189.73**	434.82**	415.5**
S15	29.6**	49**	103.83**	4.45**	4.11**	2.73**	66.52	26.56**	52.68**
S16	47.6	70	144.9	12.46	41.96	27.24	130.8	85.38	106.86

*significant at $p < 0.05$; **significant at $p < 0.01$ level.

Table 3. (a) Chemical characteristics of rural and urban soils; (b) Chemical characteristics of rural and urban soils.

(a)

St.No.	pH			Conductivity (mho/cm)			Temperature (°C)		
	Pre-monsoon season	Monsoon season	Post-monsoon season	Pre-monsoon season	Monsoon season	Post-monsoon season	Pre-Monsoon season	Monsoon season	Post monsoon season
S1	7**	7.7**	7.1**	0.07**	0.049**	0.032**	32.6**	30**	31.5**
S2	5**	5.7**	5**	1.52	0.88**	0.463**	30	29.7	31
S3	5.7**	6.4**	6.3**	0.26**	0.15**	0.094**	32.5**	30**	31.5**
S4	6.1**	7.1**	6.5**	0.26**	0.082**	0.057**	31.6**	30**	31.48*
S5	5.3*	6*	5.5*	1.09**	0.245**	0.326**	29.9**	29**	31**
S6	4.7**	5**	4.9**	3.14**	1.35**	1.343**	31.1**	30**	31.4**
S7	6.2**	7.3**	6.7**	0.14**	0.072**	0.048**	33**	30**	31.4**
S8	5.4	6.2	5.9	0.44	0.225	0.211	29.5	29.8	31
S9	8.7**	8.3**	7.3**	0.06**	0.042**	0.028**	32.6**	30**	31.5**
S10	5.7	5.7**	5.3**	1.33**	0.528**	0.451**	32**	30**	31.3**
S11	5.7	6.3	6.2	0.27**	0.174**	0.099**	32.2**	30**	31.1
S12	6.1**	7.3**	6.7	0.25**	0.073**	0.049**	31.3**	30**	31.4**
S13	5.3**	6.2	5.7**	0.52**	0.225**	0.214**	32**	29.9	31.2**
S14	4.8**	5.3**	4.9**	1.6**	1.305**	0.865**	31*	30**	31.3**
S15	6.9**	7.5**	7*	0.07**	0.064**	0.04**	33**	30**	31.4**
S16	5.7	6.2	6	0.33	0.197	0.118	31.1	29.8	31

(b)

St.No.	Pore Space (%)			Water holding capacity (%)			Bulk density (g/cc)		
	Pre-monsoon season	Monsoon season	Post-monsoon season	Pre-monsoon season	Monsoon season	Post-monsoon season	Pre-monsoon season	Monsoon season	Post-monsoon season
S1	57.8**	57.8**	57.8**	0.04**	0.05**	0.03**	1.56**	1.58**	1.57**
S2	30.19**	30.2**	30*	1.33**	1.34**	1.32**	0.71**	0.72**	0.71**
S3	38.32**	38.2**	38.1**	0.15	0.17**	0.15**	1.14**	1.14**	1.2
S4	40.12**	40.13**	40.1**	0.15**	0.15**	0.14**	1.23**	1.24**	1.23**
S5	30.28**	30.3**	30.2**	0.79**	0.75**	0.74**	0.5**	0.9**	0.9**
S6	25.43**	25.4**	25.3**	1.84**	1.87**	1.86**	0.2**	0.3**	0.3**
S7	45.25**	45.3**	44.9**	0.13	0.13**	0.11**	1.26**	1.25**	1.25**
S8	32.45	32	32	0.27	0.27	0.25	1.09	1.1	1.11
S9	58.87**	58.86**	58.84**	0.02**	0.03**	0.03**	1.58**	1.6**	1.6**
S10	30.19**	30.2**	30.1**	1.23**	1.25**	1.23**	0.8**	0.8**	0.78**
S11	35.45**	35.5**	35.2**	0.18**	0.2	0.2	1.11**	1.12	1.15**
S12	42.02**	42.1**	41.9**	0.13**	0.14**	0.12**	1.24**	1.25**	1.23**
S13	31.33**	31.34**	31.3**	0.54**	0.54**	0.51**	0.93**	1	0.97**
S14	29.26**	30**	30**	1.84**	1.84**	1.83**	0.3**	0.5**	0.46**
S15	48.3**	48.3**	48.19**	0.11**	0.12**	0.11**	1.28**	1.28**	1.28**
S16	35.09	35.1	34.8	0.25	0.26	0.23	1.11	1.1	1.11

*Significant at $p < 0.05$; **Significant at $p < 0.01$ level.

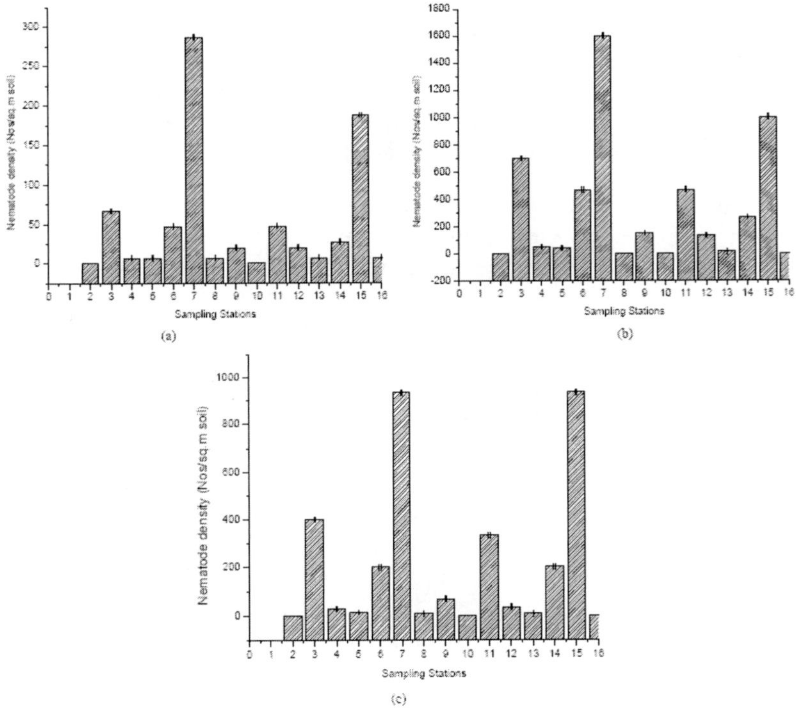

Figure 2. Nematode abundance in the Soils of Study stations during different seasons. (a) Pre-monsoon season; (b) Monsoon season; (c) Post-monsoon season.

physico-chemical characteristics with nematode density was plotted. The significance level of correlation coefficient is given in **Table 6**.

3.3. Mapping of Significantly Positively Correlated Physico-Chemical Characteristics and Nematode Density in Soil

Water holding capacity of soils in the selected stations of the study area showed significant and positive correlation with nematode density. Smiles (1988) [10] in his studies concluded that the rate of soil water movement controls many biological activities such as wilting and germination of plants and hatching of nematode cysts and spores. The spatial variability of water holding capacity and nematode density during the pre-monsoon, monsoon and post monsoon seasons were plotted and the results are shown in **Figure 3**. Water holding capacity showed very high positive correlation with nematode density during the pre monsoon season in all the stations.

But in the monsoon season due to the dilution factor, this positive correlation was very high in most of the study stations except in the stations S1, S2, S3, S4 and S16 where this positive correlation was only high. In the post-monsoon season, this positive correlation showed the same pattern as that in the monsoon season except for the stations—S3 and S4 where the positive correlation was reverted back to very high.

Moisture content in the soils of study stations showed a significant and positive correlation with nematode density. Townshend and Webber (1971) [11] showed that the movement and survival of nematodes was maximum in soils with maximum moisture retention. The spatial variability of moisture content and nematode density during the pre-monsoon, monsoon and post-monsoon seasons were plotted and the results are shown in **Figure 4**. Moisture content showed high positive correlation with nematode density during the pre monsoon season in most of the study stations except in the stations, S11, S12, S13, S14 and S15 where this positive correlation was very high. But in the monsoon season due to the dilution factor, this positive correlation was very high all the stations. In the post monsoon season, this positive correla tion showed the same pattern as that in the monsoon season except for the stations, S7 and S8 where the positive correlation was high.

Table 4. (a) Heavy metal content in rural and urban soils; (b) Heavy metal content in rural and urban soils.

(b)

St.No.	Cu (mg/kg)			Cr (mg/kg)		
	Pre-monsoon season	Monsoon season	Post-monsoon season	Pre-monsoon season	Monsoon season	Post-monsoon season
S1	19.2	18.7	20.5	0.2	BDL	BDL
S2	90.4	70.6	88.5	0.3	0.2	0.26
S3	79.6	71.9	77.9	0.4	0.2	0.33
S4	78.2	74.7	75.0	0.4	0.4	0.4
S5	79.6	76.2	77.5	0.2	BDL	0.57
S6	80.3	98.4	99.8	0.2	0.3	0.28
S7	95.2	90.5	93.6	0.7	0.6	0.8
S8	90.2	88.1	89.2	BDL	BDL	0.32
S9	20.2	18.9	20.0	0.2	BDL	BDL
S10	92.6	89.8	90.6	0.4	0.3	0.28
S11	86.8	85.4	85.6	0.4	0.2	0.38
S12	82.2	81.2	82.0	0.3	0.3	0.38
S13	87.2	83.7	86.5	0.3	BDL	0.6
S14	95.6	93.3	94.3	0.3	0.3	0.26
S15	98.3	84.1	87.4	0.8	0.6	0.82
S16	88.2	83.3	85.3	0.20	BDL	0.31

BDL-Below Detected Level.

(a)

St.No.	Pb (mg/kg)			Zn (mg/kg)			Mn (mg/kg)		
	Pre-monsoon season	Monsoon season	Post-monson season	Pre-monsoon season	Monsoon season	Post-monsoon season	Pre-monsoon season	Monsoon season	Post-monsoon season
S1	BDL	BDL	BDL	43.2	73.1	40.3	140	100.2	130.3
S2	0.6	0.3	0.44	70.2	352.1	245.4	105.2	310.5	310.6
S3	0.6	0.7	0.56	220.4	100.2	310.1	78.6	69.7	78.3
S4	0.9	0.6	0.76	510	220.9	535.4	401	225.6	512.6
S5	BDL	BDL	BDL	70.4	52.3	92.3	88.6	78.4	89.3
S6	0.6	0.6	0.4	90.1	70	186.5	111.4	142.5	127.9
S7	0.9	0.9	0.86	340.2	470.1	752	540.2	398.6	567
S8	BDL	BDL	BDL	75.4	40.4	72	70.3	63.2	70.2
S9	BDL	BDL	BDL	52.3	86	42	157.4	109.5	510
S10	0.8	0.3	0.4	69	314	252	146.4	152.3	152.1
S11	0.6	0.7	0.61	320	112.3	312.3	65.4	310.5	361.9
S12	0.6	0.6	0.7	360	216.9	502.9	512	213.9	509.4
S13	BDL	BDL	BDL	220	67.5	89.5	88.6	86.5	86.5
S14	0.4	0.5	0.38	98.1	72.7	182.9	115	152.3	138
S15	1	0.8	0.9	360.3	508	760.4	569.8	396.5	538
S16	BDL	BDL	BDL	74.4	52.3	77.1	78.6	68.6	65.8

Table 5. (a) Correlation of nematode density with physico-chemical characteristics in rural area; (b) Correlation of nematode density with physico-chemical in urban area.

(a)

Parameters	Pre monsoon	Monsoon	Post monsoon
pH	−0.680(**)	−0.841(**)	−0.820(**)
Conductivity	0.974(**)	0.967(**)	0.992(**)
Temperature	−0.223	−0.202	0.000
Pore space	−0.612(**)	−0.692(**)	−0.662(**)
Water holding capacity	0.845(**)	0.972(**)	0.965(**)
Bulk density	−0.817(**)	−0.933(**)	−0.923(**)
Moisture content	0.798(**)	0.892(**)	0.875(**)
Chloride	−0.245	−0.288	−0.277
Organic matter	0.958(**)	0.967(**)	0.804(**)
Total nitrogen	0.719(**)	0.980(**)	0.821(**)
Phosphorus	0.990(**)	0.907(**)	0.766(**)
Potassium	0.680(**)	0.567(**)	0.497(**)
Lead	−0.563(**)	−0.436(*)	−0.774(**)
Zinc	−0.311	−0.183	−0.254
Manganese	−0.275	−0.020	−0.237
Copper	0.333	0.450(**)	0.458(**)
Chromium	−0.464(**)	−0.405(*)	−0.434(**)

(b)

Parameters	Pre monsoon	Monsoon	Post monsoon
pH	−0.307	−0.796(**)	−0.819(**)
Conductivity	0.856(**)	0.989(**)	0.990(**)
Temperature	−0.616(**)	−0.016	−0.059
Pore space	−0.551(**)	−0.641(**)	−0.558(**)
Water holding capacity	0.915(**)	0.978(**)	0.961(**)
Bulk density	−0.893(**)	−0.902(**)	−0.886(**)
Moisture content	0.741(**)	0.875(**)	0.806(**)
Chloride	−0.266	−0.305	−0.256
Organic matter	0.828(**)	0.860(**)	0.756(**)
Total nitrogen	0.711(**)	0.991(**)	0.700(**)
Phosphorus	0.988(**)	0.903(**)	0.814(**)
Potassium	0.693(**)	0.929(**)	0.948(**)
Lead	−0.570(**)	−0.528(**)	−0.767(**)
Zinc	−0.354(*)	−0.254	−0.262
Manganese	−0.313(*)	−0.274	−0.537(**)
Copper	0.342(*)	0.438(**)	0.377(*)
Chromium	−0.138	−0.261	−0.455(**)

Table 6. Significance level of correlation coefficient.

Range	Significance
0.00 - 0.200	Negligible
0.200 - 0.400	Low
0.400 - 0.600	Moderate
0.600 - 0.800	Substantial
0.800 - 1.00	Very high

Organic matter content in the soils of study stations showed a significant and positive correlation with nematode density. This is in agreement with the studies by Gorres et al. (1998) [12] which showed that the soils rich in organic matter were reported with the highest nematode density.

The spatial variability of organic matter content and nematode density during the pre monsoon, monsoon and post monsoon seasons were plotted and the results are shown in **Figure 5**. Organic matter content showed very high positive correlation with nematode density during the pre monsoon and monsoon seasons in all the stations. In the post monsoon season, this positive correlation showed the same pattern as that in the pre monsoon and monsoon seasons except for the stations, S7 and S9 where the positive correlation was high.

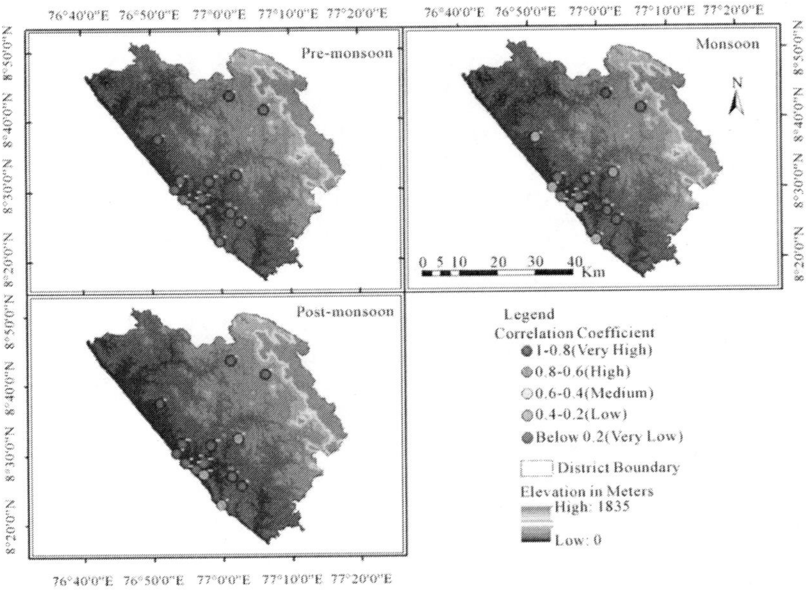

Figure 3. Mapping of water holding capacity and nematode density.

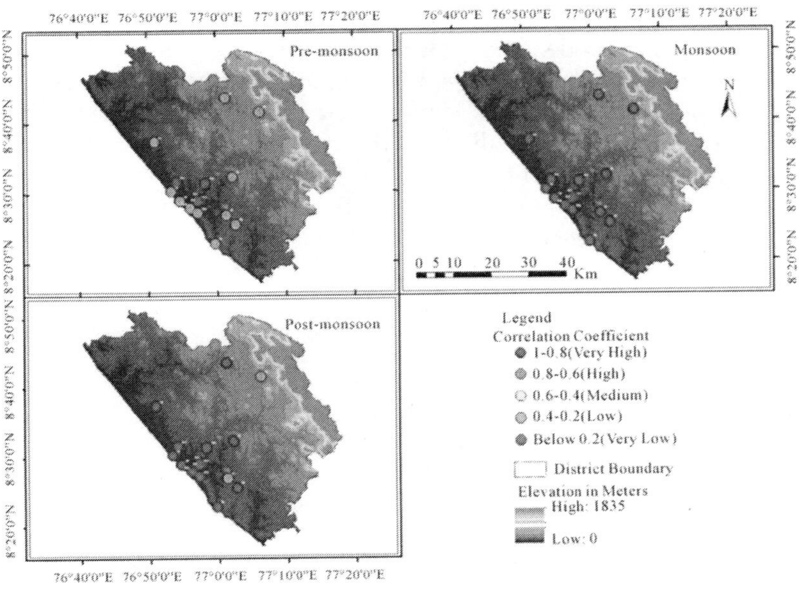

Figure 4. Mapping of moisture content and nematode density.

3.4. Mapping of Significantly Negatively Correlated Physico-Chemical Characteristics and Nematode Density in Soil

There is a significant and negative correlation between soil pH and nematode density. This is in agreement with the studies conducted by Murialdo et al. (2002) [13] which showed that when the soil pH was high, nematode density was low. The spatial variability of pH content in soil and nematode density during the pre monsoon, monsoon and post monsoon seasons were plotted and the results are shown in **Figure 6**. Soil pH showed very high negative correlation with nematode density during the pre monsoon season in stations, S13, S11, S12, S14 and S15. But in the monsoon season due to the dilution factor, this negative correlation was reduced. In the post monsoon season, the correlation showed the same pattern as that in the monsoon season.

Zinc content in the soils of study stations showed a significant and negative correlation with nematode density. Studies by Shao et al. (2008) [14] showed that nematode densities were negatively correlated with zinc concentrations as their population densities decreased with increased zinc concentrations. The spatial variability of zinc content and nematode density during the pre monsoon, monsoon and post monsoon seasons were plotted and the results are shown in **Figure 7**. Zinc content showed high negative correlation with nematode density during the pre monsoon season in the stations, S10, S11, S12 and S13. Due to the dilution factor in the monsoon season, this negative correlation was reduced in all the stations. But this negative correlation was very high in the stations, S14 and S15 in the monsoon season. The effect of raised water holding capacity and moisture content in the monsoon season influenced the soils in the station, S14 resulting in very high negative correlation. In the station, S15 the retention of zinc content in soil will be more due to the dilution factor resulting in the high negative correlation with nematode density. But in the post monsoon season due to the less water content in these soils, the negative correlation will be lesser.

Manganese content in the soils of study stations showed a significant and negative correlation with nematode density. Studies by Norton and Hoffmann (1974) [15] had shown that a negative correlation was observed between manganese content and egg/cyst ratio of nematodes which in turn will be reflected in their population densities. The spatial variability of manganese content in the soils and nematode density during the pre monsoon, monsoon and post monsoon seasons were plotted and the results are shown in **Figure 8**. Manganese content in soils showed the same negative correlation pattern with nematode density in both the pre monsoon and monsoon seasons. But in the post monsoon season, this negative correlation was very high in the station, S14 as these clayey soils were with less water content at this season as compared to that in the monsoon season and will augment the manganese content in soils. This is in agreement with the studies conducted by Rajmohan and Elango (2005) [16] which concluded that soils with high clay content will retain more manganese especially in the post monsoon season. In stations, S4, S6 and S9, the negative correlation was high. In station, S6 with high clay content in soils, the manganese retention will be more in this season. As S6 is a rural market area and S14 is an urban market area, the manganese contamination will be more in the urban market area due to different anthropogenic activities compared to the corresponding rural area. So station, S6 showed only high level of correlation with nematode density in the post monsoon season. In station, S4 which is a road-side area, the manganese deposition will be more in the post monsoon season. Previous studies by Singh and Singh (2006) [17] had shown that the manganese deposition in soils will be more in the post monsoon season in the road-side soils in India. Station, S9 which is an urban coastal area, this negative correlation will be more due to the hyper saline nature of these soils. This is in agreement with the studies by Alongi et al. (2005) [18] which showed that hyper saline soils will retain more manganese content in soils during the post monsoon season.

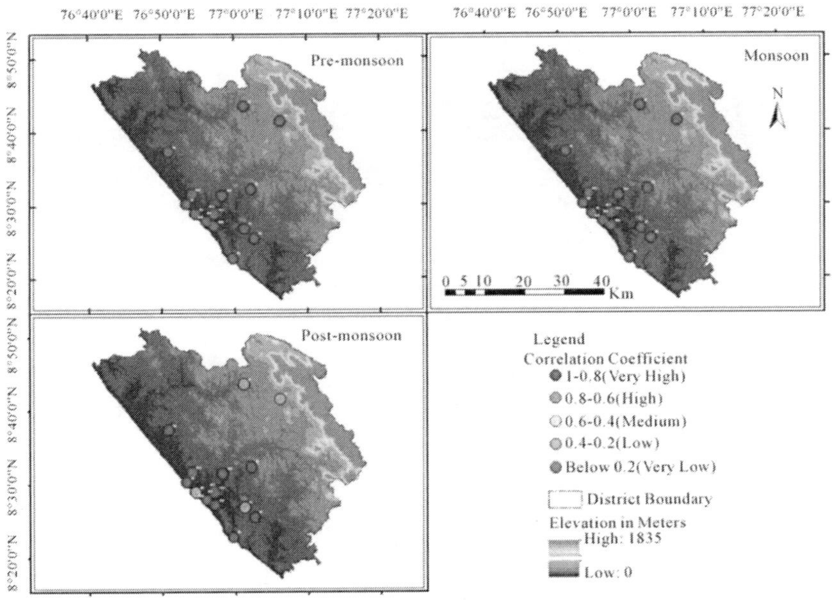

Figure 5. Mapping of organic matter content and nematode density.

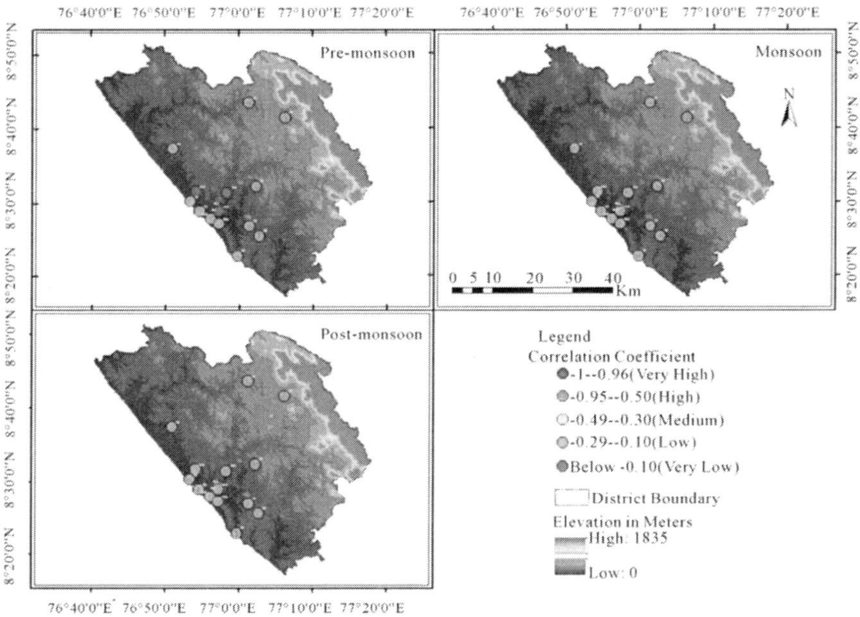

Figure 6. Mapping of pH and nematode density.

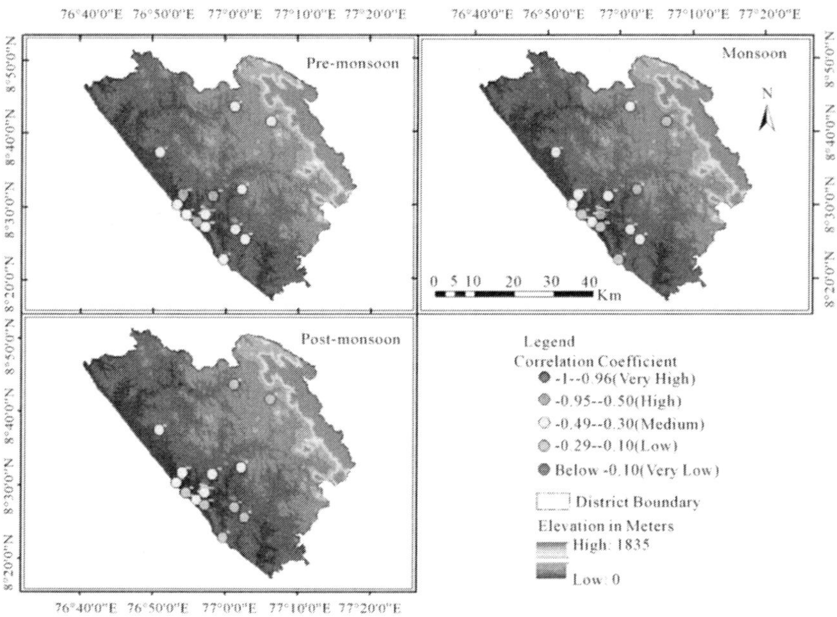

Figure 7. Mapping of zinc content and nematode density.

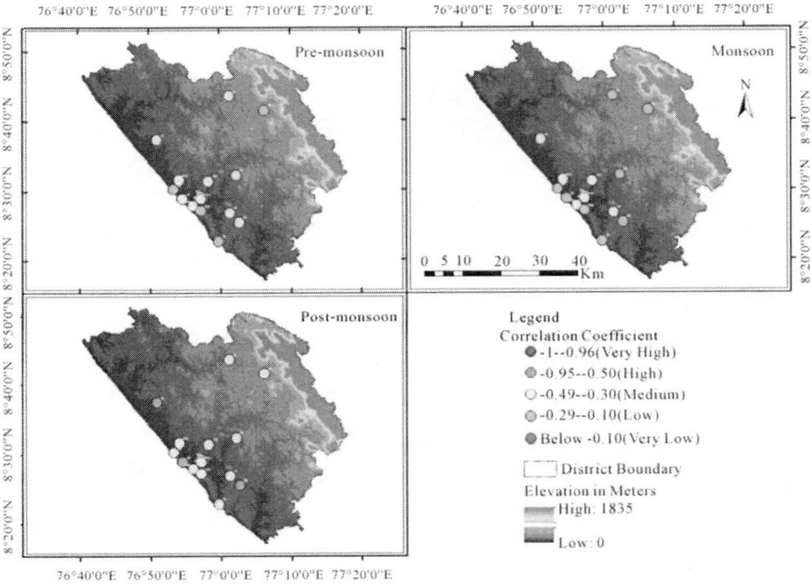

Figure 8. Mapping of manganese content and nematode density.

4. CONCLUSION

The soil physico-chemical characteristics which showed significant (positive and negative) correlations with nematode density were spatially plotted using the Arc GIS 10 to show their spatial variability. Water holding capacity, moisture content and organic matter content showed a significant and positive correlation with nematode density. PH, zinc content and manganese content showed a significant and negative correlation with nematode density. These soil mapping plots showed variations in the monsoon season compared to the pre monsoon and post monsoon seasons due to the dilution factor. In the study area, the organic matter content of urban soils with industrial activities and gasoline service stations was found to be significantly less compared to that of the rural area soils with the same land use patterns. Also the urban area soils were prone to more heavy metal contamination than the rural area soils especially in the gasoline station and roadside areas of Thiruvananthapuram District. This will reduce the soil microbial activity and soil health which in turn will be reflected in the crop production. This soil quality mapping provides a basis for identifying such tension zones and serves as a triggering mechanism for implementation of mitigating strategies.

5. ACKNOWLEDGEMENTS

Lakshmy Kamala Sthanu, Jaya Divakaran Sarasamma, Radhakrishnan Thanu Iyer, Mohandas Chellappan The authors sincerely thank Prof. M.S. Rajasree, Director, Indian Institute of Information Technology, TechnoPark, Thiruvananthapuram for providing facilities to conduct the spatial studies. Also extend sincere thanks to Mr. Vishnu, Project Fellow, Indian Institute of Information Technology, Thiruvananthapuram for the help rendered in doing the spatial analysis.

REFERENCES

1. B. J. Alloway, "Pollution-Causes, Effects and Control," 3rd Edition, The Royal Society of Chemistry Publishers, Cambridge, 1996.
2. J. V. Tarazona, M. D. Fernández and M. M. Vega, "Regulation of Contaminated Soils in Spain e a New Legal Instrument," Journal of Soils Sediments, Vol. 5, No. 3, 2005, pp. 121-124.
3. M. J. Fortin and M. Dale, "Spatial Analysis—A Guide for Ecologists," The Edinburgh Building, Published by Cambridge University, Cambridge, 2005.
4. P. K. Gupta, "Soil, Plant, Water and Fertiliser Analysis," Agro Botanica Publishers and Distributors, New Delhi, 1999, p. 438.
5. N. N. Saxena, "Environmental Analysis of Water, Soil and Air," Agro Botanical Publishers, 1994.
6. G. B. Korte, "The GIS Book," 5th Edition, SAN 694- 0269, On Word Press Publishers, 2001.
7. N. A. Cobb, "Notes on Nematology," Controlled Science Nematology, Vol. 5, No. 5, 1917, pp. 117-128.
8. T. R. E. Southwood and P. A. Henderson Ecological Methods," Published by Blackwell Science Ltd., London, 2000, pp. 218-240.
9. J. B. Goodey, "Laboratory Methods for Work with Plant and Soil Nematodes, Technical Bulletin No.2, Ministry of Agriculture, Fisheries and Food," Nematology Department, Rothamsted Experimental Station, Harpenden, Herts, London, 1963, pp. 21-22.
10. D. E. Smiles, "Aspects of the Physical Environment of Soil Organisms," Biology and Fertility of Soils, Vol. 6, No. 3, 1988, pp. 204-215.
11. J. L. Townshend and L. R. Webber, "Movement of Pratylenchuspenetrans and the Moisture Characteristics of Three Ontario Soils," Nematologica, Vol. 17, No. 1, 1971, pp. 47-57.
12. J. F. Gorres, M. J. Dichiaro, J. B. Lyons and J. A. Amador, "Spatial and Temporal Patterns of Soil Biological Activity in a Forest and an Old Field," Soil Biological Biochemistry, Vol. 30, No. 2, 1998, pp. 219-230.
13. S. E. Murialdo, M. G. Fuertes, J. F. Gonza'lez and E. Chavez, "Nematodes as Indicators of Wetland Pollution," Journal of Environmental Biology, Vol. 23, No. 4, 2002, pp. 423-428.
14. Y. Shao, W. Zhang, J. Shen, L. Zhou, H. Xia, W. Shu, H. Ferris and S. Fu, "Nematodes as Indicators of Soil Recovery in Tailings of a Lead/Zinc Mine," Soil Biological Biochemistry, Vol. 40, No. 2, 2008, pp. 2040-2046.

15. D. C. Norton and J. K. Hoffmann, "Distribution of Selected Plant Parasitic Nematodes Relative to Vegetation and Edaphic Factors," Journal of Nematology, Vol. 6, No. 1, 1974, pp. 81-86.
16. N. Rajmohan and L. Elango, "Distribution of Iron, Manganese, Zinc and Atrazine in Groundwater in Parts of Palar and Cheyyar River Basins, South India," Journal of Environmental and Monitoring Assessment, Vol. 107, No. 1-3, 2005, pp. 115-131.
17. A. K. Singh and M. Singh, "Lead Decline in the Indian Environment Resulting from the Petrol-Lead Phase-Out Programme," Science Total Environment, Vol. 368, No. 2-3, 2006, pp. 686-694.
18. D. M. Alongi, A. L. Ramanathan, L. Kannan, F. Tirendi, L. A. Trott and M. Bala Krishna Prasad, "Influence of Human-Induced Disturbance on Benthic Microbial Metabolism in the Pichavaram Mangroves, Vellar-Coleroon Estuarine Complex, India," Marine Biology, Vol. 147, No. 4, 2005, pp. 1033-1044.

CHAPTER 6

Phosphorus Adsorption of Some Brazilian Soils in Relations to Selected Soil Properties

Valdinar Ferreira Melo[1*], Sandra Cátia Pereira Uchôa[1], Zachary N. Senwo[2*], Ronilson José Pedroso Amorim[3]

[1]*Department of Soil and Agricultural Engineering, Federal University of Roraima, Boa Vista, Brazil*

[2]*Department of Biological & Environmental Sciences, Alabama A&M University, Huntsville, USA*

[3]*Agronomy, Federal University of Roraima, Boa Vista, Brazil*

ABSTRACT

A major nutritional problem to crops grown in highly weathered Brazilian soils is phosphorus (P) deficiencies linked to their low availability and the capacity of the soils to fix P in insoluble forms. Our studies examined factors that might influence P behavior in soils of the Amazon region. This study was conducted to evaluate the maximum phosphate adsorption capacity (MPAC) of the soils developed from mafic rocks (diabase), their parent materials and other factors resulting in the formation of eutrophic soils having A chernozemic horizon associated with Red Nitosols (Alfisol) and Red Latosols (Oxisol) of the Amazonian environment. The MPAC was determined in triplicates as a function of the remnant P values. The different concentrations used to determine the MPAC allowed maximum adsorption values to be reached for all soils. The Latosol (Oxisol) and Nitosol (Alfisol) soils presented higher phosphate adsorption values that were attributed to the oxidic mineralogy and high clay texture while the Chernosol (Mollisol) soils presented the lowest pho- sphate adsorption values.

Keywords: Amazon Soil, Eutrophic Soils, Oxidic Soils, Phosphate, Alfisol, Oxisol

1. INTRODUCTION

Due to the climatic variations impacting the Amazon region of Brazil, the Northeast area of Roraima has transitioned from a semi-arid climatic condition to one with high precipitation [1] . The increase in precipitation has greatly affected the soil's biogeochemical weathering attributes, erosion and deposition processes, thus contributing to the formation of deep, highly weathered soils dominant in kaolinite and oxidic mineralogy [2] . Some areas of the Amazon have soils originated from mafic rocks (diabase), resulting in the formation of Eurtophic Red Nitosol (Haplustalf), Dystrophic Yellow Red Latosol (Xanthic Haplustox), Eutropic Haplic Tb Cambisols, with A chernozem, Orthic black Chernosol (Mollisol) and vertic Orthic Ebanic Cheronosol (Mollisol) soil series [3] .

Studies have shown that the soils consist typically of kaolinitic mineralogy and low fertility, with the predominant classes being Yellow Latosols (Oxisol) and Yellow Argisols (Ultisol) [4] -[6] . The indigenous areas of Roraima in general are located where these soils are found and reflect certain aspects of the sustainability and quality of life of the indigenes cultivating the soils for agricultural production. In the Raposa Serra do Sol Indigenous Reserve (approximately 1,747,464 ha); the proportion of soils (approximately 19,900 ha) considered as fertile for sustainable agricultural production is situated on gentle undulating relief and includes diabase soils of the "Pedra Preta" Sill [7] .

A major nutritional problem to crops grown in these soils is P deficiency linked to low available P content and capacity to fix fertilizer P in insoluble form. Evaluating the P availability in these soils is important because soil P adsorption capacity directly influences plant response to phosphate fertilization of soils. Phosphorus fixation in soils, widely designated as adsorption, is a slow process that can take years to reach a balance and may lead to decrease in P availability. In most Brazilian soils, studies have shown that the major factors influencing soil P adsorption are the clay fractions, mineralogy, amorphous colloid content, pH, exchangeable aluminum and organic matter [8] -[10] .

Phosphorus adsorption depends on the nature and quality of sites available on the mineral surfaces and is therefore affected by high clay contents within the same mineralogy [11] . Studies by Ker et al. [9] indicate that the mineralogy, crystal size, and specific surfaces of soils will be more important than the quantity of clay in determining soil's P adsorption capacities [12] -[14] . Bedin et al. [14] state that the presence of large proportions of sesquioxides on clays influences phosphate adsorption and precipitation with iron and aluminum. The objective of this study was to evaluate the maximum P adsorption capacity (MPAC) of the soils developed from mafic rocks (diabase) in northeastern Roraima and its relationships with certain physical, chemical and mineralogical attributes.

2. MATERIALS AND METHODS

The soils used in this study were sampled from a toposequence and consisted of five soil profiles, developed from mafic rocks that form one of the largest mafic rock bodies in North Amazon's "Pedra Preta" Sill. The Maloca do Flechal study areas (Figure 1) is inhabited by the Macuxi Indians and situated in the northeast of the state of Roraima, Brazil. The climate average about 1200 mm in rainfall, with sunshine duration of 12 hours/day. The region's relief is characterized as undulated with step savannah vegetation and patches of seasonal forest on the East-West transect line.

The fieldwork consisted of opening trenches in a toposequence, collecting the soil samples in each profile soil and describing them according to Santos et al. [15] . Soils were classified based on the Brazilian Soil Classification System manual published by Embrapa [16] and analyzed for chemical properties (Table 1) as described in a soil analysis manual published by Embrapa [17] . The profiles were classified as Eutrophic Red Nitosol (Alfisol)- NVe, Tb Eutrophic Haplic Cambisol (Inceptisol)-CXbe; Orthic Ebanic Chernosol (Mollisol)-MEo; vertic Orthic Ebanic Chernosol (Mollisol)-MEov and Distgrophic Yellow Red Latosol (Oxisol)-LVAdf.

The Fe associated with the low crystal minerals in the surface horizons was extracted with ammonium oxalate [18] , and the free Fe extracted using dithionite-citrate bicarbonate (DCB) [19] . The Fe content was determined using induced plasma emission spectrometry while the Fe associated with the crystalline minerals was calculated by the difference between the Fe extracted by DCB (Fe_d) and the Fe extracted with oxalate (Fe_o). The mineralogy was determined by X-ray diffraction of the clay and silt fraction.

From the natural, Fe-free clay fraction, oriented slides were prepared and irradiated in an X-ray diffractometer at sweeping angle (2θ) intervals between 2° and 40° and a goniometer speed of 2° in 2θ/min, using CuKα radiation with Ni filter [20] . Powdered silt slides were analyzed under the same conditions while the diffraction patterns were interpreted according to Chen [21] .

Phosphorus (P) was evaluated as described by Alvarez and Fonseca [22] using 5 g of fine ground soils mixed with a solution of $CaCl_2$ (0.01 mol·L^{-1}) containing 50 mg·L^{-1} P and shaken for an hour then filtered and the P concentration using Murphy and Riley [23] procedure. The Langmuir isotherm were obtained by adding 25 mL $CaCl_2$ (10 mol·L^{-1}) solution containing P in doses ranging from 0 to 60 mg·L^{-1}. After 24 hours agitation, the samples were centrifuged and the P in the equilibrium solution was quantified colorimetrically [23] .

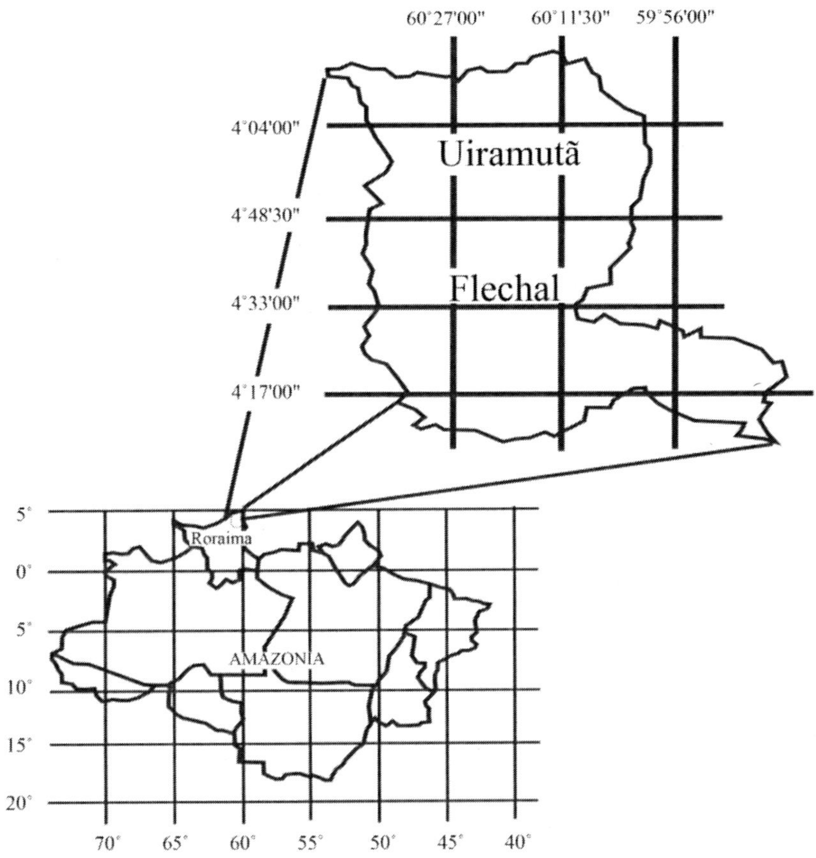

Figure 1. The maloca do flechal study areas.

Table 1. Chemical characteristics of soils studied.

Soil	Hor.	pH	Ca^{2+}	Mg^{2+}	Na^{+}	K^{+}	Al^{3+}	H + Al	S	CEC	V	SOM	P
			------	------	------	$cmol_c \cdot kg^{-1}$	------	------	------	------	(%)	$g \cdot kg^{-1}$	$mg \cdot kg^{-1}$
NVe	Ap	5.7	2.63	2.48	0.13	0.10	0.05	6.87	5.34	12.21	43	24.6	0.96
	Bnit	5.9	1.03	0.91	0.03	0.01	-	2.20	1.98	4.18	65	9.3	0.72
MEo	Ap	6.1	6.87	3.16	0.32	0.15	-	4.34	10.50	14.84	71	23.8	2.50
	Bnit	6.2	5.80	2.62	0.04	0.05	0.05	3.57	8.51	12.08	71	14.2	1.01
CXbe	Ap	6.8	8.69	2.79	0.11	0.05	0.05	2.75	11.64	14.39	81	15.3	1.33
	Bi_2	7.3	9.21	0.00	0.12	0.05	-	1.54	9.38	10.92	86	9.5	1.44
MEov	Ap	7.0	9.31	0.04	0.09	0.05	-	2.53	9.35	11.88	79	20.6	1.16
	Bi_2	7.4	12.7	0.06	0.11	0.05	-	1.43	12.92	14.35	90	3.7	1.39
CXbe	Ap	6.5	6.75	0.37	0.07	0.04	-	3.24	7.23	10.63	69	19.3	1.22
	Bi_2	7.8	7.31	0.00	0.09	0.03	-	0.71	7.43	8.14	91	3.7	1.08
LVAd	A	5.4	0.00	0.49	0.14	0.11	0.50	6.60	0.74	7.34	10	24.6	1.48
	Bw	4.3	0.00	0.50	0.04	0.04	0.50	8.52	0.52	9.04	06	22.5	0.53

Hor. = Horizon. 1) pH in water. 2) S = sum of bases. CEC = Cation exchange capacity at pH 7.0. V = percentage of saturation by bases; SOM = Soil organic matter. P = Phosphorus extractable by Mehlich-1. NVe-Red Nitosol. MEo-Orthic Ebanic Chernosol. Cxbe-Tb Eutrophic Haplic Cambisol. MEov-Vertic Orthic Ebanic Chernosol, LVAd-Distrophic Red Yellow Latosol.

The data were fitted to the non-linear Langmuir model, relating the concentration of the adsorbed element per adsorption unit (soil) and the concentration of the element in the equilibrium solution (supernatant), and the maximum adsorption capacity calculated. The sorption values of each soil were plotted according to the Langmuir isotherm: $C/(x/m) = (1/Kb) + (C/b)$. Where C, is equilibrium P concentration ($mg \cdot L^{-1}$), x/m is the amount of sorbed P ($mg \cdot kg^{-1}$), m and b, are the constants related to P sorption maximum ($mg \cdot kg^{-1}$) and K is the bonding energy ($L \cdot mg^{-1}$), respectively. The plot of C/(x/m) versus C should give a straight line from which 1/b (slope) and K (slope or intercept) can be calculated [24] . The Pearson analysis of simple linear correlation using the Statistic program 7.0 between MPAC and some soil characteristics was carried out, and the correlations greater than 95% were considered significant ($\alpha < 0.05$).

3. RESULTS AND DISCUSSION

The Fe contents determined from successive extractions using DCB (2), ranged from 29.4 (MEov) to 258.7 $g \cdot kg^{-1}$ (LVAdf). The values seemed high and atypical of Roraima [5] [6] and Amazonia soils [25] [26] , given that the predominant soils in these environments have iron contents averaging less than 73 $g \cdot kg^{-1}$. The LVAdf stood out among the soils studied with lower Fe_o/Fe_dratio indicating the predominance of high crystalline iron oxide.

The soils mineralogical compositions (Table 2) showed the presence of high mineral activities. Illite was common in all the soil horizons but smectite was restricted to soils with the A Chernozem horizon (CXbe and MEov).

Kaolinite also occurred generally in most of the profiles, indicating intense weathering with strong lixiviation of bases, low fertility and lower (CEC) especially for LVAdf. The mineralogy of the 2:1 group might have resulted from the geology and climatic variations in the Amazon over geological time. The northeast of Roraima has passed through a recent sub-period of semi-arid climate to the current conditions of greater precipitations [27] [28] . However, this mineralogy contrasts with the data obtained by Vale Júnior [5] and Melo et al. [6] for other Roraima soils.

The P values varied greatly among the soil profiles and horizons. The highest values were observed in the surface horizons (Table 3), ranging from 19.8 (LVAdf) to 50.2 $mg{\cdot}L^{-1}$ (MEo). For the surface horizons, the values ranged from 6.6 $mg{\cdot}L^{-1}$ (LVAdf) to 46.9 (MEov) and this could be attributed to the type of clay mineralogy present in the soils. According to Alvarez et al. [29] , the soils and horizons analyzed presented low P retention, except for LVAdf, which could be attributed to the highly negative charge on the smectites surfaces of the clays. The high P values in most of the soils could be explained by the striking presence of 2:1 minerals. Alves and Lavorenti [30] worked with Latosols in the state of São Paulo, Brazil and reported the lack of correla tions between P and goethite and P and kaolinite. High iron-sesquioxides contents are often considered as the determining factor of P adsorption in soils [31] ; however, the low crystalline forms should fix more P than the crystalline forms [30] . The differences in behavior observed for LVAdf was more related to the presence of gibbsite than to the presence of the low crystalline forms, considering the low Fe_o/Fe_d ratio presented by this soil (Table 4).

Table 2. Mineralogical composition of the clay, silt and sand fractions of soils studied.

Soil	Hor.	Clay	Silt	Sand
NVe	Ap	Ct, Gt	Qz, Ct, Bt, Fs, Ilm	Qz, Mi
	Bnit	Ct, Il, Gt	Qz, Ct, Bt, Fs, Ilm, Pg	Qz, Mi,
MEo	Ap	Ct, Il, Gt	Qz, Bt, Fs, Px	Qz, Mi
	Bnit	Ct, Il, Gt	Qz, Ct, Mi	Qz, Mi
CXbe	Ap	Ct, Il/Mt, Es, Gt	Qz, Fs, Ct, Px, Af	Qz, Fs, Pg, Mi
	Bi_2	Ct, Il, Es	Pg, Fs, Qs, Ct, Mi, Af	Pg, Qz, Fs, Mi, Mg, Cn
MEov	Ap	Ct, Il, Es	Fs, Pg, Qz, Ct, Mi, Af	Qz, Pg, Af, Fs
	Bi_2	Es, Ct, Il	Ac, Cl, Mi, Fs	Qz, Pg, Fs, Cn
CXbe	Ap	Ct, Il, Gt	Ct, Qz, Zl, Pg	Qz, Fs, Ru
	Bi_2	Ct, Il, Gt	Ct, Qz, Hb, Bt	Qz
LVAd	Ap	Gb, Ct, Il, Gt	Qz, Ct, Gb, Ru, Hm	Qz
	Bw_1	Ct, Gb, Il	Qz, Ct, Gb	Qz

Qz-Quartz, Ct-Kaolinite, Es-Smectite, Fs-Feldspar, Mi-Mica, Pg-Plagioclase, Af-Amphiboles, Px-Piroxene, Il-Illite, Bt-Biotite, Ac-Actinolite, Cl-Chlorite, Zl-Zeolite, Hb-Hornblendite, Ru-Rutile, Gb-Gibbsite, Gt-Goethite; Hm-Hematite, Mg-Manganite, Cn-Corundum

Table 3. Maximum phosphate adsorption capacity for the soils studied.

Soil	Horizon	Clay g·kg^{-1}	P-remaining mg·L^{-1}	MAPC mg·kg^{-1}
NVe	Ap	520	39.8	439.0
	Bnit.	640	4.2	1280.0
MEo	Ap	280	50.2	353.0
	Bnit.	370	36.2	475.0
CXbe	Ap	320	43.3	409.0
	Bi$_2$	470	29.5	551.0
MEov	Ap	290	47.9	370.0
	Bi$_2$	290	46.9	378.0
CXbe	Ap	350	40.7	431.0
	Bi$_2$	380	37.7	460.0
LVAdf	Ap	650	19.8	701.0
	Bw$_1$	700	6.6	1111.0

Nve-Red nitosol, MEo-Orthic ebanic chernosol, Cxbe-Tb eutrophic haplic cambisol, MEov-vertic orthic ebanic chernosol LVAdf-distrophic red latosol.

Table 4. Soil iron and silicon dioxide contents of soils.

Soil	Hor.	Fe_d	Fe_o	Fe_o/Fe_d	Fe_c	SiO_2
				g·kg^{-1}		
NVe	Ap	58.3	3.5	0.06	54.8	1.41
	Bnit	75.5	3.2	0.04	72.3	1.13
MEo	Ap	106.0	7.7	0.07	98.3	1.26
	Bnit	95.1	5.6	0.06	86.9	1.90
CXbe	Ap	52.0	6.4	0.12	45.6	2.01
	Bi$_2$	51.1	4.9	0.10	46.2	2.65
MEov	Ap	34.9	7.7	0.22	27.2	2.01
	Bi$_2$	29.4	1.6	0.05	27.8	1.37
CXbe	Ap	77.1	7.2	0.09	69.9	1.61
	Bi$_2$	64.8	2.5	0.04	62.3	2.03
LVAdf	A	180.3	1.8	0.01	178.9	0.73
	Bw	258.7	1.0	0.004	257.7	0.88

Nve-Red nitosol, MEo-Orthic ebanic chernosol, Cxbe-Tb eutrophic haplic cambisol, MEov-vertico orthic ebanic chernosol, LVAdf-distrophic red yellow latosol. Fe_d-Dithionite-extracted Fe. Fe_o-Oxalate-extracted iron. Fe_c-crystalline Fe.

The MPAC values (Table 3) ranged from 353 to 701 mg·kg^{-1} in the surface horizons and from 378 to 1280 mg·kg^{-1} for the sub-surface horizons. Except for the values obtained in the MEov profile and the MEo surface layer, the rest of the soils were in the range of MPAC values as indicated by Novais and Smyth [32] for Cerrado Latosols. In Amazon soils, Lima et al. [26] reported MPAC values ranging from 210 to 2170 mg·kg^{-1} in Oxisol, while Singh et al. [33] obtained values between 160 and 980 mg·kg^{-1} P. For central Amazon, Falcão and Silva [34] reported MPAC ranging from 298 to

888 mg·kg^{-1} for Ultisols. The MPAC values were lower than those normally found in highly weathered soils. In basalt-derived soils, Oxisol and Alfisol, Bognola [35] found MPAC values between 1620 and 2740 mg·kg^{-1}, while Valladares et al. [36] obtained values between 526 and 1.667 mg·kg^{-1} for Ultisols. Soils originating from this material have higher MPAC because of the higher clay content and occurrence of oxides such as hematite and goethite [37] [38] . Despite the parental materials, climate and relief contributed to the soil formation with high smectite contents in undulated relief, highly weathered rich Fe and Al oxides contents. The structure, chemical compositions, exchangeable ion type and small crystal size of smectite clays are responsible for the unique properties, including a large chemically active surface area, a high cation exchange capacity and low anion exchange capacity.

In the high clay soils (Table 3), phosphate adsorption was more pronounced. The Yellow Red Latosol (Oxisol) presented the greatest MPAC in the two horizons, which can be explained by the high clay content, high Fe_d and low Fe_c contents (Table 4). The Nitosol also presented high MPAC, greater than LVAdf, in the subsurface horizon, which can be due to the high clay content. Vertic Orthic Ebanic Chernosol had the least adsorbed P in the two horizons, which can be explained by the low clay contents and low Fe_d and Fe_c values. Generally the soils showed increase in P adsorption with increase in soil depth, which can be attributed to increase in the clay contents from horizon A to horizon B. The high phosphate adsorption in the Latosol and Nitosol in the Amazon soils can be attributed to the low Fe contents predominate in this environment [6] .

The clay, Fe_d and Fe_c contents correlated positively and significantly with the MPAC, suggesting that these attributes contributed to an increase in the MPAC in the horizons and profiles studied (Table 5). Given the strict correlations between MPAC and clay, Fe_d and Fe_c contents, it can be postulated that the presence of goethite in all the soils was a major determinant for the MPAC and not kaolinite soils. Bahia Filho et al. [39] tested the effects of the mineralogical components on MPAC and observed that goethite (Gt) was mainly responsible, con tributing 6% of the total MPAC. The high affinity of geothite soils for P is likely due to the easy access of surface phosphate anions. The $H_2PO_4^-$ occupies site of the hydroxides (OH^-) previously coordinated to Fe^{3+}, forming much more stable surface complexes.

Table 5. Simple correlations between maximum phosphate adsorption capacity of soils and some soils properties.

	A Horizon	B Horizon
Clay	0.79^{*}	0.94^{**}
T	-0.87^{*}	-0.79^{*}
C	$0.38^{n.s.}$	$0.54^{n.s.}$
Fe_d	0.83^{*}	0.93^{**}
Fe_o	-0.87^{*}	-0.62^{*}
Fe_c	0.84^{*}	0.94^{**}

T-Clay. Fe_d-Dithionite-extracted Fe. Fe_o-Oxalate-extracted iron. Fe_c-crystalline Fe. n.s.-not significant. *mean of $0.01 < probability < 0.05$; **significant at $probability < 0.01$.

The correlation between the soil organic C and MPAC was not significant for the soils studied. This might be because organic C can influence P adsorption positively in numerous ways [40] , negatively [41] or no effect [42] . The latter has been observed in Cerrado ecosystem (Savanna) soils where the C stock increases linearly with the soil clay and silt contents, demonstrating that total C does not increase the P adsorption, but enhances the richness of the minerals that retain P.

The CEC values and Fe_o correlated significantly and negatively, respectively, with MPAC. Indeed P adsorption tends to diminish with increase in CEC in the soils, because the negative charges repel the phosphate ion. On the other hand, negative correlation of MPAC with Fe_o is in line with reported studies. Hernández and Meurer [43] studied three forms of soils in Uruguay, an Argiudoll, a Hapludert, and a Natraqualf, and observed positive correlations between P adsorption and low Fe crystalline forms (Fe_o).

4. CONCLUSION

Valdinar Ferreira Melo, Sandra Cátia Pereira Uchôa, Zachary N. Senwo*, Ronilson José Pedroso Amorim The Oxisol and the Alfisol soils presented the highest phosphate adsorption values, due to Fe and Al oxides rich minerals. The Molissol soils showed the lowest adsorption values due to the high presence of smectite clay. The soils that showed positive correlations with MPAC were in the order: A Horizon-Fe_c > Fe_d > clay and B Horizon-clay = Fe_c > Fe_d. The soils of the northern border of the Amazon exhibit P adsorption behavior different from other areas of the Amazon. Sustaining agricultural production in these soils will require management approaches that will enhance P availability. An adequate pool of labile P in these soils could be enhanced with inorganic P fertilizations and other organic residues.

REFERENCES

1. Schaefer, C.E.R. and Dalrymple, J. (1996) Pedogenesis and Relict Properties of Soils with Columnar Structure from Roraima, North Amazonia. Geoderma, 71, 1-17.
2. Benedetti, U.G., Vale Jr., J.F., Schaefer, C.E.G.R., Melo, V.F. and Uchôa, S.C.P. (2011) Genesis, Chemistry and Mineralogy of Soils Derived from Plio-Plestocene Sediments and from Volcanic Rocks in Roraima—North Amazonia. Brazilian Journal of Soil Science, 35, 299-312.
3. Melo, V.F., Schaefer, C.E.G.R. and Uchôa, S.C.P. (2010) Indian Land Use in the Raposa-Serra do Sol Reserve, Roraima, Amazonia, Brazil: Physical and Chemical Attributes of a Soil Catena Developed from Mafic Rocks under Shifting Cultivation. Catena, 80, 95-105.
4. Schaefer, C.E.G.R., Resende, S.B., Correia, G.F. and Lani, J.L. (1993) Chemical and Minerological Characteristics of Sodium-Afected Soils from Northeeastern Roraima. Brazilian Journal of Soil Science, 17, 431-438.
5. Vale Jr., J.F. (2000) Pedogenesis and Changing of Soil under Shifting Cultivation in Areas of Acid and Basic Volcanic Rocks in Northeastern Roraima. Ph.D. Thesis, Universidade Federal de Viçoa, Viçosa.
6. Melo, V.F., Schaefer, C.E.G.R., Fontes, L.E.F., Chagas, A.C., Lemos Jr., J.B. and Andrade, R.P. (2006) Physical Chemical and Mineralogical Characteristics of Soils from the Agricultural Colony of Apiaú (Roraima, Amazonia), under Different Land Uses and after Burning. Brazilian Journal of Soil Science, 30, 1039-1050.
7. Melo, V.F. (2002) Soil and Indicators of Agricultural Use in Roraima: Indigenous Areas of Maloca Flechal and Colonization of Apiaú. Ph.D. Thesis, Universidade Federal de Viçoa, Viçosa.
8. Brennan, R.F., Bolland, M.D.A., Jeffery, R.C. and Allen, D.G. (1994) Phosphorus Adsorption by a Range of Western Australian Soils Related to Soil Properties. Communications in Soil Science and Plant Analysis, 25, 2785-2795.
9. Motta, P.E.F., Curi, N., Siqueira, J.O., Van Raij, B., Furtini Neto, A.E. and Lima, J.M. (2002) Adsorption and Forms of Phosphorus in Latosols: Influence of Mineralogy and Use. Brazilian Journal of Soil Science, 26, 349-359.
10. Moreira, F.L.M., Mota, F.O.B., Clemente, C.A., Azevedo, B.M. and Bomfim, G.V. (2006) Phosphorus Adsorption in Ceará State Soils, Brazil. Revista Ciência Agronômica, 37, 7-12.
11. Bahia Filho, A.F.C., Braga, J.M., Resende, M. and Ribeiro, A.C. (1983) Relationship between Phosphorus Adsorption and Mineralogical Components of Latosols Clay Fraction of the Central Plateau. Brazilian Journal of Soil Science, 7, 221-226.
12. Torrent, J., Barron, V. and Schwertmann, U. (1990) Phosphate Adsorption and Desorption by Goethites Differing in Crystal Morphology. Soil Science Society of American Journal, 54, 1007-1012.
13. Ruan, H.D. and Gilkes, R.J. (1996) Kinetics of Phosphate Sorption and Desorption by Synthetic Aluminous Goethite before and after Thermal Transformation to Hematite. Clay Minerals, 31, 63-74.
14. Bedin, I., Furtini Neto, A.E., Resende, A.V., Faquin, V., Tokura, A.M. and Santos, J.Z.L. (2003) Phosphate Fertilizers and Soybean Production in Soils with Different Phosphate Buffer Capacities. Brazilian Journal of Soil Science, 27, 639-646.
15. Santos, R.D., Lemos, R.C., Santos, H.G., Ker, J.C. and Anjos, L.H. (2005) Description Handbook and Collecting of Soil in Field. 5 Edition, Sociedade Brasileira de Ciência do Solo, Viçosa, MG.
16. Embrapa (Empresa Brasileira de Pesquisa Agropecuaria) (2006) Brazilian System of Soil Classification. 2nd Edition, Embrapa Solos, Rio de Janeiro.
17. Embrapa (Empresa Brasileira de Pesquisa Agropecuaria) (1997) Manual of Soil Analysis Methods. 2nd Edition, National Soil Research Center, Rio de Janeiro.

18. Mckeague, J.A. and Day, J.H. (1966) Dithionite and Oxalate Extractable Fe and Al as Aids in Differentiating Various Classes of Soils. Canadian Journal of Soil Science, 46, 13-22.
19. Mehra, O.P. and Jackson, M.L. (1960) Iron Oxide Removal from Soils and Clay by a Dithionite-Citrate System Buffered with Sodium Bicarbonate. Clays and Clay Minerals, 7, 317-327.
20. Whittig, L.D. and Allardice, W.R. (1986) X-Ray Diffraction Techniques. In: Klute, A., Ed., Methods of Soil Analysis, Part 1: Physical and Mineralogical Methods, American Society of Agronomy, Madison, 331-362.
21. Chen, P.-Y. (1977) Table of Key by Lines in X-Ray Power Diffraction Patterns of Minerals in Clays and Associated Rocks. Occasional Paper 21, Department of Natural Resources, Geological Survey, Bloomington, 67 p.
22. Alvarez, V.V.H. and Fonseca, D.M. (1990) Definition of Doses of Phosphorus to Determine the Maximum Adsorption Capacity for Phosphate in Greenhouse Trials. Brazilian Journal of Soil Science, 14, 49-55.
23. Murphy, J. and Riley, J.P. (1962) A Modified Single Solution Method for the Determination of Phosphate in Natural Waters. Analytica Chimica Acta, 27, 31-36.
24. Bera, R., Seal, A., Bhattacharyya, P., Mukhopadhyay, K. and Giri, R. (2006) Phosphate Sorption Desorption Characteristics of Some Ferruginous Soils of Tropical Region in Eastern India. Environmental Geology, 51, 399-407.
25. Silva, J.R.T. (1999) Soils of Acre: Physical, Chemical and Mineralogical and Phosphate Adsorption. Thesis of Ph.D., Federal Universtiy of Viçosa, MG, Brazil.
26. Lima, H.N., Melo, J.W.V., Schaefer, C.E.G.R., Ker, J.C. and Lima, A.M.N. (2006) Mineralogy and Chemistry of Three Soils along a Toposequence from the Upper Solimões Basin, Western Amazonia. Brazilian Journal of Soil Science, 30, 59-68.
27. Schaefer, C.E.G.R. (1994) Soils and Paleosols from Northeast Roraima, North Amazonia: Geomorphology, Genesis and Landscape Evolution. Ph.D. Dissertation, University of Reading, Reading, Berkshire.
28. Schaefer, C.E.G.R. and Vale Júnior, J.F. (1997) Climate Changing and Landscape Evolution in Roraima: A Review from Cretaceous to Recent. In: Barbosa, R.I., Ferreira, E.J.G. and Castellón, E.G., Eds., Man, Environment and Ecology in the State of Roraima, INPA, Manaus, 231-293.
29. Alvarez, V.V.H., Novais, R.F., Dias, L.E. and Oliveira, J.A. (2000) Determination and Use of the Remaining Phosphorus. Newsletter, Sociedade Brasileira de Ciência do Solo, 25, 27-33.
30. Alves, M.E. and Lavorenti, A. (2004) Remaining Phosphorus and Sodium Fluoride pH in Soils with Different Clay Contents and Clay Mineralogies. Pesquisa Agropecuária Brasileira, 39, 241-246.
31. Uehara, G. and Gillman, G. (1981) The Mineralogy, Chemistry and Physics of Tropical Soils with Variable Charge Clays. Westview, Boulder.
32. Novais, R.F. and Smyth, T.J. (1999) Phosphorus in Soil and Plant in Tropical Conditions. Universidade Federal de Viçosa-MG, Viçosa, 399 p.
33. Singh, R., Moller, M.R.F. and Ferreira, W.A. (1983) Phosphorus Adsorption Kinetics in Amazonian Soils. Revista Brasileira de Ciência do Solo, 7, 227-231.
34. Falcão, N.P. and Silva, J.R.A. (2003) Phosphorus Adsorption Characteristics in Some Central Amazonian Soils. Acta Amazônica, 34, 337-342.
35. Bognola, I.A. (1995) Chemical, Physical and Mineralogical Soil Intermediaries between Latosols Brunos and Latosols Purple. Master's Thesis, Universidade Federal de Viçosa, Viçosa.
36. Valladares, G.S., Pereira, M.G. and Anjos, L.H.C. (2003) Phosphate Sorption in Low Activity Clay Soils. Bragantia, 62, 111-118.
37. Curi, N., Camargo, O.A., Guedes, G.A.A. and Silveira, J.V. (1988) Phosphorus Sorption in Latosols of Southeast and South of Brazil. In: Reunião de Classificação e Correlação de Solos e Interpretação de Aptidão Agrícola, 3rd Edition, Empresa Brasileira de Pesquisa Agropecuária, Rio de Janeiro, 150.

38. Fontes, M.P.F. and Weed, S.B. (1996) Phosphate Adsorption by Clays from Brazilian Oxisols: Relationships Whit Specific Surface Area and Mineralogy. Geoderma, 72, 37-51.
39. Bahia Filho, A.F.C., Vasconcellos, C.A., Santos, H.L., Mendes, J.F., Pitta, G.V.E. and Oliveira, A.C. (1982) Inorganic Phosphorus and Available Phosphorus Forms in a Dark Red Latosol, Fertilized with Different Phosphates. Brazilian Journal of Soil Science, 6, 99-104.
40. Parfitt, R.L. (1978) Anion Adsorption by Soils and Soil Materials. Advances in Agronomy, 30, 1-50.
41. Borggaard, O.K., Jorgensen, S.S., Moberg, J.P. and Raben-Lage, B. (1990) Influence of Organic Matter on Phosphate Adsorption by Aluminum and Iron Oxides in Sandy Soils. European Journal of Soil Science, 41, 443-449.
42. Zinn, Y.L., Lal, R. and Resck, D.V.S. (2005) Texture and Organic Carbon Relations Described by a Profile Pedotransfer Function for Brazilian Cerrado Soils. Geoderma, 127, 168-173.
43. Hernández, J. and Meurer, E.J. (1990) Phosphorus Sorption in Soils from Uruguay and Its Relationship with Iron Oxides. Brazilian Journal of Soil Science, 22, 223-230.

CHAPTER 7

Influence of Wood Charcoal from Chlorophera Excelsa on Soil Properties and Yield Components of Maize

Nweke, I. A.

Soil Science Department, Chukwuemeka Odumegwu Ojukwu University, Anambra State, Nigeria.

ABSTRACT

Wood charcoals were mainly used for smoking of fish, meat and frying of corns and many times dump off as wastes in southeast, Nigeria. Their nutrient recycles and release when used as soil amendment has not been considered and evaluated by farmers in the southeast, Nigeria. This study was therefore carried out to assess the efficiency of wood charcoal technology in improving soil productivity. Influence of wood charcoal from Chlorophera excelsa on soil properties and yield components of maize were evaluated. Four different rates of wood charcoal (6, 4, 2 and 0 tha-1 that received no treatment application) were used. The treatments were laid out in randomised complete block design (RCBD) and treatment means were compared using least significant difference (LSD). The findings from the study showed non-significant ($P<0.05$) differences in most of the growth and yield components of maize tested. Significant differences were recorded in all the soil properties assessed except for percentage nitrogen

(N). Higher values were observed in 4 tha-1 rates except for the values of pH, available phosphorus and percentage base saturation, where higher values were observed in 6 tha-1 rates of wood charcoal than the other rates of wood charcoal. The percentage increase in calcium content of the soil relative to the values obtained in 0 tha-1 rate was 13.04% (2 tha-1), 11.76% (4 tha-1) and 7.69% (6 tha-1). While the percentage increase in the value of Mg2+ obtained in 4 tha-1 plots relative to control (0 tha-1) plots was 28.5%. The data generated from both growth and yield component and soil properties showed that the values of parameters tested increased as the rate of wood charcoal applied increased, though this was not consistent in the growth and yield parameters and few cases in soil parameters. In comparison of the four different rates of the wood charcoal applied, 4 tha-1 rate seems to be an ideal than the other rates as it performed competitively better and were able to liberate plant nutrients more than the other rates of wood charcoal though these liberated nutrients were not reflected in most of the growth parameters assessed.

Keywords: Chlorophera excelsa , crop growth, soil properties, wood charcoal.

INTRODUCTION

The availability of soil mineral nutrients in non-fixed and toxic form is natural requirement for crop growth. Crop plants require them for healthygrowth and good yield. Agricultural soils show mineral deficiency problems merely after short period of cultivation, especially in the south eastern soils of Nigeria. They are very fragile and delicate due to nature and prevailing environmental conditions. Hence,they suffer rapid decline in mineral nutrients after short cultivation and these are soils that tend to harbour crop production. The soils of this area are usually influenced by high rainfall and temperature coupled with the climatic change that have influenced greatly the periodic cycle of the rainfall with heavy downpour. This situation invariably causes low nutrient content and soil organic matter mineralisation.

Also with accelerated decrease in cation exchange capacity (CEC) resulting to increase in acidity of the soils. Hence, agricultural sustainability in the area faces a large constraint. Farmers, therefore, in their quest to improve their crop yield then sought to supply additional plant nutrients by application of organic or inorganic chemical fertilizers so that yield of crops will no longer be limited by the amount of plant nutrients that the natural system can supply. Thus, both the organic manure and chemical fertilizers are common forms of soil amendment that are routinely used in agricultural soils. Though their impact on soil mineral nutrients differ remarkably when they are surface applied or incorporated into the soil. Also because of the high rainfall observed in the study area, the efficiency of applied mineral fertilizers is very low.

In addition, many farmers in the study area cannot afford the cost of regular application of chemical fertilizers. Hence, nutrient deficiency is very prevalent in many crop production systems of the area. The application of organic wastes or manure in its own case have frequently been shown to increase soil fertility, but have not fared better because, according to Tiessen et al. (1984), under tropical conditions organic matter is usually mineralised very rapidly and from the work of Fearnside (2000) only a small portion of the applied organic compounds will be stabilised in the soil in the long term but successively released to the atmosphere as carbon iv oxide (CO_2).

Therefore, under this circumstance, the use of more stable compounds such as carbonised materials or their extract becomes useful both in availability and cost for agricultural sustainability. The works by Glaser (1999) and Glaser et al. (2000), showed that carbonised materials from the incomplete combustion of organic material (charcoal, black C, pyrogenic C) are responsible for maintaining high levels of soil organic matter and available nutrients. The works of Radlein et al., (1996) equally showed that coal from geological deposit (coal) and from various specialised procedure were successfully used for soil amelioration as carbonised materials comprised a wide range of materials from partly charred materials to graphite and soot particles. Therefore, the objective of this study is to examine the influence of wood charcoal from C. excelsa (Iroko) on soil properties and yield components of maize.

Materials and Methods

The experiment was conducted at the Teaching and Research Farm of the Faculty of Agriculture, Chukwuemeka Odumegwu Ojukwu University, Anambra State, Nigeria. The experimental site is located within latitude 06°14'N and longitude 06°45'E. The total annual mean rainfall ranges between 1800 and 2000 mm while the temperature range during the rainy season is between 21° and 24°C and an estimated relative humidity range of 64 to 90%.

Land Preparation and Treatment Allocation

The experimental site was manually cleared and debris removed. The plots were then laid out in a randomised complete block design (RCBD) with four treatment materials and four replications to give 16 plots. The size of each plot measured 3 m × 4 m (12 m^2) with a distance of 1 m between the blocks and 0.5 m pathway between plots. The treatment consisted of the appropriate rate of wood charcoal made from *C. excelsa* (Iroko) which were applied to their respective plots. The treatment summaries are:

2 tha^{-1} wood charcoal equivalent to 2.4 kg/ha

4 tha^{-1} wood charcoal equivalent to 4.8 kg/ha

6 tha^{-1} wood charcoal equivalent to 7.2 kg/ha

0 tha^{-1} control (CO) that received no treatment

Each of these rates of wood charcoal were grounded and applied evenly on the plot and incorporated into the soil one week before sowing to allow mineralisation of nutrients in the treatment. Maize hybrid seeds (Oba super II) were planted two per hole at the spacing of 76 cm × 25 cm, which was done one week after the incorporation of the wood charcoal. The seedlings were thinned down to one plant per stand two weeks after germination, while empty stands were supplied. The experimental field was kept relatively weed free till harvest. Soil samples prior to treatment application was collected from different locations after field preparation and bulked together and analysed for the physical and chemical parameters of the soil (Table 1). At the end of the study, soil samples were collected from individual plots and used for the determination of physical and chemical parameters of the soil according to the procedure described by Black (1965). Five maize plants were randomly selected and used to measure plant height and leave area at 2, 4, 6 and 8 weeks after planting as well as length of cob/plant, numbers of cob/plant and cob's diameter.

Data collected from the study was tested on analysis of variance based on randomised complete block design (RCBD) according to Steel and Torrie (1980), while least significant difference (LSD) at 5% was used to compare treatment means.

Table 1. Physical and chemical properties of the studied soil prior to treatment application.

Parameter	Values
Coarse sand	46%
Fine sand	43%
Silt	7%
Clay	4%
Textural class	Sandy
pH H_2O	6.8
OC	0.71%
OM	1.23%
N	0.06%
Na^+	0.15 $cmolkg^{-1}$
K^+	0.19"
Ca^{2+}	0.13 "
Mg^{2+}	2.00 "
CEC	18.4 "
BS	82.29 "
Avail. P	5.60 $mgKg^{-1}$
H^+	0.80 $cmolkg^{-1}$

RESULTS

The initial soil properties of the study area in Table 1 showed that the soil is sandy with low contents of organic carbon (0.71%) and organic matter (1.23%) as well as low values in major plant nutrients tested. Hence, the soil of the experimental site was found to be deficient in the major plant nutrients.

The result of the growth parameter recorded in Table 3 showed that the rates of wood charcoal applied have no effect on the plant height and leaf area tested for 2 to 8 weeks after planting (WAP) except for plant height at 2 WAP. The plant height and leaf area result obtained equally showed that the plots that received different rates of wood charcoal did not perform better than the control plots except at 8WAP where the treated plots especially the 4tha^{-1} wood charcoal gave the highest plant height and leaf area followed by 6 tha^{-1} wood charcoal, but they were not significantly ($P<0.05$) different. The result of the length of cob/plant and cob's diameter showed non-significant ($P<0.05$) difference among the treatments. While number of cob/plant show significant difference among the treatments, with the values obtained from 2 and 4 tha^{-1} being statistically similar, but significantly ($P<0.05$) better than the 6 tha^{-1} plots. Critical observation of the data obtained for the growth parameters showed that the plots treated with 4 tha^{-1} wood charcoal performed better than the other treatments. Though its effectiveness were minimally observed in plant height at 2 and 4 WAP and leaf area at 4 and 6 WAP as well as cob's diameter.

Table 2. Nutrient content of the wood charcoal used for the study.

Parameter	Value
pH H_2O	10.4
OC	10.47%
OM	18.06%
N	0.084%
Na^+	0.19 cmolkg^{-1}
K^+	0.21 cmolkg^{-1}
Ca^{2+}	0.06 cmolkg^{-1}
Mg^{2+}	0.10 cmolkg^{-1}
Avail. P	0.07 mgkg^{-1}

Table 3. Effect of wood charcoal on the growth and yield component.

Treatment (tha^{-1})	2WAP		4WAP		6WAP		8WAP		Length of cob/plant (cm)	Numbers of cob/plant	Cob's diameter (cm)
	PH (cm)	LA (cm^2)	PH (cm)	LA (cm^2)	PH (cm)	LA (cm^2)	PH (cm)	LA (cm^2)			
0	8.16	6.42	21.34	33.65	36.17	59.73	48.92	70.94	5.82	1.9	3.96
2	7.25	5.67	19.08	24.34	30.17	41.44	43.25	60.45	5.70	2.1	3.87
4	7.17	6.50	21.60	31.92	35.00	54.42	50.42	84.36	5.82	2.1	3.62
6	7.42	6.00	19.75	29.54	32.42	46.15	48.92	76.64	5.17	1.5	3.43
LSD0.05	0.04	NS	NS	NS	NS	NS	NS	NS	NS	0.33	NS

WAP = Weeks after planting; PH = Plant height; LA = Leaf area; LSD = Least significant difference; NS = Not – significant.

The values obtained for soil chemical properties recorded in Table 4 indicated that the wood charcoal from the *C. excelsa* influenced the chemical properties of the soil. All the parameters tested show significant ($P<0.05$) difference among the treatments with the exception of total nitrogen result.

Table 4. Effect of wood charcoal on the chemical properties of the soil.

Treatment (t/ha)	pH H_2O	OM (%)	N (%)	P ($mgkg^{-1}$)	BS (%)	CEC ($cmolkg^{-1}$)	EA ($cmolkg^{-1}$)	Na^+ ($cmolkg^{-1}$)	K^+ ($cmolkg^{-1}$)	Ca^{2+} (%)	Mg^{2+} (%)
0	6.20	0.94	0.04	3.73	75	18.40	0.80	0.121	0.147	12.0	2.0
2	6.40	1.08	0.07	3.73	77	18.40	0.60	0.133	0.164	13.8	0.4
4	6.50	1.14	0.08	4.66	78	19.20	0.80	0.146	0.182	13.6	2.8
6	6.60	1.08	0.08	8.39	87	19.20	0.80	0.133	0.173	13.0	1.4
LSD0.05	0.27	0.02	NS	0.08	1.29	0.11	0.14	0.014	0.013	0.97	1.49

Table 5. Effect of wood charcoal on the particle size of the soil.

Treatment (t/ha)	Coarse sand (%)	Fine sand (%)	Clay (%)	Silt (%)	Textural class
0	45	44	4	7	Sandy
2	46	43	4	7	Sandy
4	43	46	4	7	Sandy
6	40	51	4	5	Sandy
LSD0.05	0.62	0.40	NS	0.66	

The result of the soil pH showed that the soil pH increased gradually as the rates of wood charcoal application increased and the order of increase were 6 > 4 > 2 > 0 tha^{-1}. The highest organic matter (OM) content was observed in 4 tha^{-1} plots compared with the other treated plots and was significantly ($P<0.05$) different with the other treatments. The values obtained in 2 and 6 tha^{-1} treated plots were statistically similar but significantly ($P<0.05$) better than the values obtained in 0 tha^{-1} plots. The result of available P and base saturation (BS) showed that the plots treated with 6 tha^{-1} gave the highest value and were significantly ($P<0.05$) different from the other treatments. The next in rank is 4 tha^{-1} treated plots of which its available P value is significantly better than the 2 and 0 tha^{-1} results. The base saturation (BS) values indicated that as the rate of wood charcoal application increased the value of BS increased and the order were 6 > 4 > 2 > 0 tha^{-1}. The result of cation exchange capacity (CEC) and exchangeable acidity (EA) recorded in Table 4 showed significant difference among the treatments. The values of CEC indicated that 4 and 6 tha^{-1} values were statistically similar, but significantly better than the values obtained in 2 and 0 tha^{-1}, while values of EA showed that the values obtained in 0, 4 and 6 tha^{-1} are statistically similar, but significantly better than the values obtained in 2 tha^{-1} rates of wood charcoal applied.

The result of Na^+ and K^+ showed significant ($P<0.05$) difference among the treatments. The highest value in Na^+ and K^+ was observed in 4 tha^{-1} treated plots of which its Na^+ value was significantly better than the other treatments. While its K^+ value was statistically similar with 6 tha^{-1} value but significantly better than the value obtained in 0 and 2 tha^{-1}. The exchangeable Ca^{2+} and Mg^{2+} result in Table 4 were significant ($P<0.05$) and the order of increase for Ca^{2+} were 2 > 4 > 6 > 0 tha^{-1} while Mg^{2+} order of increase were 4 > 0 > 6 > 2 tha^{-1}.

The percentage increase in Ca^{2+} relative to the values obtained in 0tha^{-1} was 13.44% (2 tha^{-1}), 11.76% (4 tha^{-1}) and 7.69% (6 tha^{-1}). The percentage increase in the value of Mg^{2+} obtained in 4tha^{-1} relative to the value obtained in control plots were 28.57% while its percentage decrease relative to 2 tha^{-1} plots were 600%. The soil chemical parameter results showed that the wood charcoal applied at 4tha^{-1} performed competitively better and were able to liberate mineral nutrients better than the other rates of wood charcoal applied, though it is not true for soil pH, available P, BS and Ca^{2+} contents of the soil studied. Nonetheless, the release of these chemical nutrients was not adequately reflected in the growth and yield parameters of the tested crop.

The result of the particle size analysis of the soil in (Table 5) indicated significant differences in coarse and fine sand particles and silt content of the soil, but showed non-significant differences as well as non-change (as the values are the same) in the clay content of the soil. Thus, with regard to the clay content of the soil the influence of the different rates of wood charcoal are the same and equal. The influence of 0, 2 and 4 tha^{-1} rates of wood charcoal on the silt content of the soil were equal and statistically similar, but significantly better than silt content of 6 tha^{-1} rate of wood charcoal. The influence of wood charcoal on the sand particles showed that 6 tha^{-1} rates had more sand particles than the other rates. While the rates of 0, 2 and 4 tha^{-1} the same values of sand particles on the average were observed.

DISCUSSION

The soil analysis result in Table 1 showed that the studied soil is very deficient in major mineral nutrients; the values are below the fertility classification of Ibedu et al. (1988) and Landon (1991). Hence, the soil is deficient in plant nutrients. The nutrient content in wood charcoal (Table 2) applied indicated that the organic matter, organic carbon, pH and some other nutrients tested are higher relative to quantity of the nutrients available in the soil before treatment application.

The effect of wood charcoal on the growth and yield component of maize was observed to be non-significant. This problem may be attributed to the amount present in the individual rates applied and their slow release as well as non-synchronisation of nutrient released within the short period of growth of the maize plant. The differences in values of these parameters tested, however, may be as a result of differences in plant nutrient in the rate of treatments applied.

However, non-significant values in growth and yield component of crops have been reported in the works of Nweke and Nsoanya (2013a,b), Nweke and Obasi (2013) and Nweke et al. (2014) following organic waste application in soil, the post-harvest soil analysis result of this study show that wood charcoal from *C. excelsa* when used as soil amendment has the capacity of increasing the pH level thereby reducing the acidity levels of the soil required for crop production as was observed in Table 4. This increase in soil pH level is good omen for agricultural production as low pH value

limits soil productivity, since it affects availability and uptake of nutrients by plants because at low pH, acid soils are normally flocculated (Haynes and Naidu 1998). The wood charcoal at rate of 4 tha^{-1} was observed to have being a good estimate of amount required significantly to improve soil properties and enhance growth and yield components of maize (Table 3).

Using cocoa pod ash and wood ash, by Ayeni et al. (2008) and Mbah et al., (2010) respectively, as soil amendment reported increased soil pH relative to non-ash treated soil. The OM and TN content of the treated plots were observed to be improved relative to the control plots. This improvement could be attributed to the higher level of OM and TN in wood charcoal applied than the soil. Organic matter according to Tisdale et al., (1993) and Brady and Weil (2006) play important roles in essential nutrient availability and soil improvement.

The influence of the wood charcoal on the soil available P, BS% and CEC values varied among the rates and appreciated relative to the control rate. The level of soil available P and other nutrients might have been influenced by the changes in soil pH due to application of wood charcoal since the availability of P and its solubility is pH dependent. In line with the findings of this study, Mbah et al. (2010) and Nweke and Nsoanya (2013b) reported significant increase in available P and plant nutrients in soils amended with wood ash and organic wastes respectively relative to control plots. Kayode and Agboola (1983) attributed the increased CEC in wood ash amended soil to increased cations Ca, K and Na as was observed in the present study. In support of the findings of this study, previous study like Ayeni et al. (2008), Adele et al. (2010), Nweke and Nsoanya (2013b) and Nweke et al. (2014) had shown that organic wastes increased soil OM, N, pH, P, CEC, exchangeable base and reduced soil exchangeable acidity (EA).

The post-harvest soil analysis showed that the rates of wood charcoal applied significantly influenced the particle size distribution but did not affect the textural class of the soil. There were more sand fraction which indicates good aeration and will not present problem with drainage and plant root penetration. There were no differences between the clay contents of the different rates of wood charcoal and the silt contents of 0,

2 and 4 tha^{-1} rates of wood charcoal. Nweke and Nsoanya (2012) observed no difference between the clay and silt content of soils of the same locality.

CONCLUSION

The findings from the study have shown that wood charcoal from *C. excelsa* is capable of improving soil properties and crop production. The values obtained from 4 and 6 tha^{-1} wood charcoal in most of the soil and plant parameters tested were statistically similar. This placed 4 tha^{-1} in this particular study as the optimum rate of wood charcoal application for soil fertility improvement and maize growth. The material is cheap, available and can be sourced by poor resource farmers and used for soil and crop improvement.

CONFLICT OF INTERESTS

The authors have not declared any conflict of interests.

REFERENCES

1. Adeleye EO, Ayeni LS, Ojeniyi SO (2010). Effect of poultry manure on soil physico-chemical properties, leaf nutrient contents and yield of yam (Dioscorea rotundata) on Alfisol in southwestern Nigeria. J. American Science 6 (10):509-516.
2. Ayeni LS, Adetunji MT, Oyeniyi SO (2008). Comparative nutrient release from cocoa pod ash, poultry manure and NPK 20:20:20 fertilizer and their nutrient combination incubation study Nig. J. Soil Sci. 18:114-123.
3. Black CA (1965). Method of soil analysis II chemical and microbiological properties. Am. Soc. Agron. Madison, Wisconsin. 157 p.
4. Brady NC, Weil RR (2006). Nature and properties of soil, 13th edition London, Prentice Hall. pp. 595-624.
5. Fearnside PM (2000). Global warming and tropical land use change, green house gas emissions from biomass burning, decomposition and soils in forest conversion, shifting cultivation and secondary vegetation, Climatic Change. 46:115-155.
6. Glaser B (1999). Ergenschaften und stabilch des Humuskrpers der indianerschwar zerden Amazon iens Bayreuther Bodenkunliche Berichte. 68.
7. Glaser B, Haumaier L, Guggenberger G, Zeck W (2001). The terra preta phenomenon a model for sustainable agriculture in the humid tropics naturwissenschaften 88:37-41.
8. Glaser B, Balachov E, Haumaier L, Guggenberger G, Zech W (2000). Black carbon in density, fractions of anthropogenic soils of the Brazilian Amazon region. Org. Geochem. 31:669-678.
9. Haynes RJ, Naidu RC (1998). Influence of lime fertilizer and manure application on soil organic matter content and soil physical conditions, a Rev., Nutrient Cycling in Agrosys. 51:123-137.

10. Ibedu MA, Unambra RPA, Udealor A (1988). Soil management systems in relation to farming systems development in the southeastern agricultural zone of Nigeria. Paper presented at the National Farming System Research Workshop, Jos, Plateau State, Nigeria. pp. 26 -29.
11. Kayode GO, Agboola AA (1983). Investigation on the use of macro-and micro-nutrients to improve maize yield in South Western Nigeria. Fertilizer Research 4(3):211-221.
12. Landon JR (1991). Booker tropical soil manual. A Handbook for Soil Survey and Agricultural Land Evaluation in the Tropics and Subtropics, Longman Scientific and Technical New York. P 33.
13. Mbah CN, Nwite JN, Njoku C, Nweke IA (2010). Response of maize (Zea mays L.) to different rates of wood ash application in acid ultisol in southeast Nigeria. Afr. J. Agric. Res. 5(7):580-583.
14. Nweke IA, Nsoanya LN (2012). Inventory of physico-chemical characteristics of soils of the Teaching and Research Farm, Faculty of Agriculture, Anambra State University, Igbariam Campus, Nigeria. In Proceedings of International Agricultural Conference 6th - 9th May, 2012, held at Anambra State University, Igbariam Campus, pp. 488-493.
15. Nweke IA, Nsoanya LN (2013a). Effect of Poultry Manure and Inorganic Fertiliser on the Performance of Maize (Zea Mays) and Selected Physical Properties of Soils of Igbarian Southeastern Nigeria. J. Agric. Rur. Dev. 16(1):1348-1353.
16. Nweke IA, Nsoanya LN (2013b) Effect of different rates of rice mill waste on soil chemical properties and grain yield of maize (Zea Mays). J. Agric. Rur. Dev. 16(1):1431-1436.
17. Nweke IA, Obasi MN (2013). Effect of different levels of pig manure on the growth and yield of okra (Abelmoschus esculentus Ush). In proceedings of the 47th annual conference of the Agricultural Society of Nigeria, Ibadan 2013. pp. 23-26.
18. Nweke IA, Okoli PSO, Enyioko CO (2014). Effect of different rates of poultry droppings and plant spacing on soil chemical properties and yield of cucumber. Elixir Agriculture 70(2014):23934-23940.
19. Radlein D, Piskorz JK, Maher P (1996). Method of producing slow-release nitrogenous organic fertilizer from biomass. European patent application 0716056AI.
20. Steel GD, Torrie JN (1980). Principles and Procedures of Statistics. A Biometrical Approach 2nd ed. McGraw Hill Book Co., New York. P. 633Tiessen H, Stewart JWB, Cole CV (1984). Pathways of phosphorous transformation in soils of differing pedogenesis. Soil Sci. Soc. Am. J. 48:853-858.
21. Tisdale SL, Nelson WL, Beaton JD, Havlin A (1993). Soil Fertility and Fertilizers. 5th edn. Macmillan Publishing Company New York USA.

CHAPTER 8

Suitability of Kabanyolo Soils for Fruit and Vegetable Production

Paul Okiror[1*], Julius Bunny Lejju[1], Joseph Bahati[2], Grace Kagoro Rugunda[1], Collins Inno Sebuuwufu[1], Patrick Mulindwa[1], Jolly Joe Ocan[3]

[1]*Faculty of Science, Mbarara University of Science and Technology, Mbarara, Uganda*

[2]*School of Forestry, Environmental and Geographical Sciences, Makerere University, Kampala, Uganda*

[3]*School of Engineering, Makerere University, Kampala, Uganda*

ABSTRACT

This paper presents results of soil analysis for samples extracted from Makerere University Agricultural Research Institute Kabanyolo. The study was aimed at ascertaining the suitability of MUARIK soils for fruit (pineapple, passion fruit and watermelon) and vegetable (cabbage, eggplant, tomato and pumpkin) growth, yield and maturity. Soil morphological, physical and chemical properties were probed by excavating a 1.86 meter soil profile pit and sampling soils in 9 equidistant locations at MUARIK. The soil morphology and physical properties were described using the Munsel colour chart and USDA classification system while chemical parameters were determined in Makerere University Soil Laboratory following procedures described by Okalebo et al. [1] . Three pedon layers (RSK-H1, RSK-H3 and RSK-H5) had gradual boundaries while RSK-H4 was continuous. However, RSK-H2 and RSK-H6 had abrupt boundary layers. The texture of the profiles was clayey (52%). The soils were acidic with a pH of 6.08 (sub soil) to 6.12 (top soil). On comparison with standard requirements for target fruits and vegetables, the site was deficient in most of the minerals including P (8.88 - 12.42 pmm), Ca (6.46 - 7.92 pmm), K (0.06 - 0.47 pmm), Mg (1.37 - 1.70 pmm), Na (0.17 - 0.23 pmm), N (0.18 - 0.19 pmm), and soil organic matter (4.53% - 4.81%). Sand and clay had the highest negative significant correlation ($r = -0.94$, $p < 0.01$) in the top soil. The highest positive significant correlation was observed between Ca and Mg both in the topsoil ($r = 0.96$, $p < 0.01$) and subsoil ($r = 0.99$, $p < 0.01$). MUARIK is suitable for growing cabbage and pumpkin. However, for improved growth, yields and maturity, the soils should be amended with manure (watermelon), lime (passion fruit and tomato) and sulphur (eggplant and pineapples).

Keywords: Fruits, Kabanyolo, Vegetables, Soil, Suitability

1. INTRODUCTION

Uganda's soils are classified into six (6) major types notably: Soils of high productivity (8% of the land area), Soils of medium productivity (14%), Soils of fair productivity (43%), Soils of low productivity (30%) while the Soils of negligible productivity constitute 5% of the land area. In addition, the country's main soil types are 18 divided into 7 groups based on their occurrence and agricultural productivity. These include: 1) the Uganda surfaces cover most areas south of Lake Kyoga. This group embraces five types of deep, sandy clay loams with medium to high productivity. 2) Tanganyika surfaces cover most areas north of Lake Kyoga, West Nile and some parts of the South Western tip of Uganda, embracing five types of sandy clay loam with low to medium productivity. 3) Karamoja surfaces cover the North Eastern part of the country and include two soil types of sandy clay loams and black clays with very low productivity. 4) Rift valley soils in the Western and Northern parts of the country, bordering on the Western Rift Valley, embracing two types of mainly sandy clay loams with alluvial parent rock of medium to high productivity. 5) Volcanic soils are dominant in Mt. Elgon, Northern Karamoja, and the extreme South Western tip of Uganda (Kabale and Kisoro) with medium to high productivity except in N. Karamoja where their productivity is reportedly low. 6) Alluvial soils are found outside the Rift Valley, mainly in Central Northern Uganda (Lango and Acholi) as well as West of Lake Victoria. The productivity of these sandy soils is very low. 7) The last group of soil types is in Northern Uganda and their productivity is low [2] .

Despite this broad soil classification, fruits and vegetables have specific soil requirements for growth, high yields and maturity. Eggplants, for example, can be grown with considerable success in fine and rich loam soils that are deep and well-drained. The soil pH should range from 5.5 to 6.0 for its better growth and development. As the crop remains in the field for a number of months therefore, the soils should be well prepared by being ploughed 4 - 5 times before transplanting the seedlings. When the field is well prepared and leveled, the beds of suitable size should be made in the field before transplanting. However, in most fields, mineral supplies to crops (N + P_2O_5 + K_2O) are always lower than the recommended quantities and imbalanced with excess of Nitrogen due to inappropriate use of urea [3] . The mineral bill is 100 (kg of N) – 100 (kg of P_2O_5) – 200 (kg of K_2O) for African eggplant. According to Nono-Womdim et al. [3] , eggplants require 1 - 3 applications of the growth fertilizer a week until early fruit formation whereby a development fertilizer will be applied.

Tomato grows well in sandy loam soil with well-drained clay subsoil. Light soils are good for early variety [3] . While clay loam or silt loam soils are well suited for heavy yield, tomato grows at pH 6.0 to 7.0 satisfactorily. The soil should be well prepared and leveled by ploughing the land 4 - 5 times. Similarly, pumpkins may be grown on a wide range of soil types provided the soil is free- draining. They are tolerant of fairly acid conditions and liming should not be necessary unless the pH is less than 5.5 [3] . Well-drained soils that warm rapidly are ideal for the production of the quick-maturing pumpkins. Friable loams with an abundance of organic matter and the capacity to retain moisture are ideal. A deep, friable seedbed should be prepared for improved pumpkin growth and yield.

As for leafy vegetables such as cabbages, a balanced ratio between nitrogen and potassium supplies is needed for maximum yield. Cabbages grow on a wide range of soil types ranging from light sand to heavier clays. Cabbages are divided into three main groups: ballhead (or roundhead), conical and the large drumhead types. Soils with high organic matter give the best yields. Soil pH should be in the range of 6.0 - 6.5 for cabbage growth and clubroot control [3] .

Although, soil tests are useful in determining the availability of nutrients and fertiliser requirements prior to fruit and vegetable cultivation [4] , the soil morphological and physico-chemical features of the Makerere University Agricultural Research Institute (MUARIK) remain largely unclear. Relatedly, much as Pomeroy [5] undertook a detailed survey of termite mounds in the central parts of Uganda, Wakiso district and MUARIK in particular, was left out. A study by Tenywa et al. [6] , focused on the resilience and degradation stress of soils at MUARIK. Additionally, Okwakol and Sekamatte [7] studied the soil macrofauna in Uganda but recommended regular assessment of soil features in agricultural landscapes.

It was therefore important to undertake soil tests at MUARIK as a precursor for propagation of priority fruits (pineapple, passion fruit and watermelon) and vegetables (cabbage, eggplant, tomato and pumpkin). Fungo et al. [8] assert that precision agriculture requires site specific modeling of soil properties. Soil analysis was conducted because soil conditions usually influence the growth of fruits and vegetables [8] [9] [10] . In addition, incidence of pests and diseases depend on soil and weather conditions of a site.

2. MATERIALS AND METHODS

2.1. Description of Study Area

The study was conducted at Makerere University Agricultural Research Institute Kabanyolo (Figure 1). MUARIK is located on spatial coordinates 0°27'60"N, 32°36'24" E at an altitudinal range of 1250 m to 1320 m above mean sea level [11] . The study site is within the administrative boundaries of Nangabo Sub County, Wakiso district and about 14 km north of Kampala, Uganda's capital city. Kabanyolo is part of the Lake Victoria basin that receives an average annual precipitation of 1218 mm and slightly drier periods in June to July and December to February. The average annual temperature is 21.5°C [12] .

Agricultural research and demonstrations are the main activities conducted at MUARIK by the Makerere University College of Agricultural and Environmen- tal Sciences and collaborating institutions. The institute conducts several agri culture and rural innovation research projects. However, as part of the Buganda surface, the predominant farming system around MUARIK is the banana-coffee system [13] . According to the map (Figure 1), MUARIK occupies 3 square miles of land. Yost and Eswaran [11] indicated the MUARIK soils to be formed on residuum and colluvium from quartzites, gneiss and basement complex rocks. On the side slopes, colluvium enriched with lateritic gravel is common. This study was aimed at ascertaining the current morphological, physical and chemical properties of soil that may influence fruit and vegetable growth, yield and maturity.

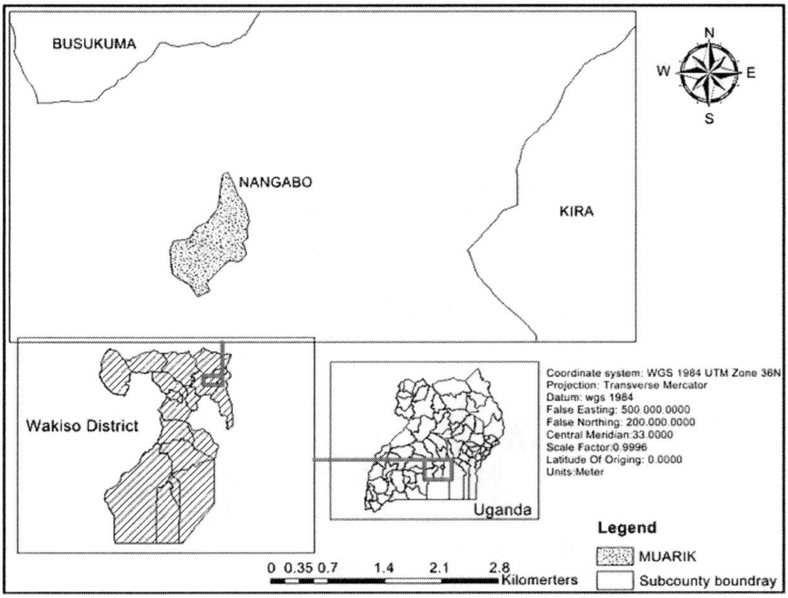

Figure 1. Map of Uganda showing the location of MUARIK.

2.2. Soil Sampling

The influence of soil depth of on soil properties at the experimental site was probed by excavating, describing and sampling soil in a 1.86 meter deep soil profile pit following Okia [14] . The pit was located at the centre of the experimental site. Surrounding the soil profile pit were sweet potato, passion fruit plants and a 1 year fallow. The soil profile was described in terms of depth, number of horizons, width of each horizon and boundary distinctness between horizons following Okia [14] , Fungo et al. [8] and Szyma?ski et al. [15] . A linear tape (Colt Germany 5 m) was used to measure the profile depth while distinctness of the horizon boundaries was described as abrupt (a), gradual (g), clear (c) and diffuse (d). The Munsell colour chart was used to provide standard descriptions and the colour name for each horizon. Okia [14] asserts that the Munsell colour system specifies colour based on three colour dimensions: hue (predominant spectral colour), value (lightness and darkness against a neutral gray scale) and chroma (purity or richness). This method of soil colour analysis is described in MCC [16] . Other scholars [15] [17] support the use of the Munsell colour soil charts to describe color of the soil material in moist state. In each horizon, about 500 g of soil was collected by chiselling, placed in plastic sample bags, labelled, packed in a sample box (Marina cooler 24S) and delivered to Makerere University Soil Laboratory for further processing and analysis. A similar approach was used by Okia [14] and Fungo et al. [8] .

In addition, to ascertain soil properties across the site and thus site suitability for the propagation of fruits and vegetables, three (3) patches within MUARIK were selected. The patches were selected in locations that represented different land uses and the elevations. Site conditions included a sweet potato field and the 1 year fallow. Three (3) sampling points that were equidistant from each other, separated by 50 and 30 m across and along the slope, were placed along the contour of each patch. Soil samples were taken at two (2) depths including 0 to 20 and 20 - 40 cm as done by Fungo et al. [8] . A total of 18 samples were therefore obtained from across the experimental site.

2.3. Laboratory Analysis of Soils

Soil samples were double labelled and transported to Makerere University Soil Science laboratory for drying, preparation and subsequent analyses. Parameters determined included soil organic matter (OM), pH, N, P, K, Ca, Mg and Na. Available Phosphorus (P) was determined by using Bray and Kurtz No. 1 method [18] . The soil was extracted by Brady 1 solution and the P determined by the calorimetric procedure using a spectrophotometer. Soil pH was ascertained using a pH meter as done by Rhoades [19] . Exchangeable K, Ca and Mg were measured by treating the soil samples with excess 1 M ammonium acetate solution. Later, the concentrations of exchangeable Na and K in the extract were measured by flame photometer and the concentration of Ca and Mg was measured by atomic absorption spectrophotometry [20] .

Considering the limitation of the field method for determination of soil texture, particle size analysis was performed in the laboratory using the pipette method. Since soil texture refers to the relative proportions of sand, silt, and clay particles in a mass of soil, this method applies Stokes' Law to measure settling rates of different soil particles in a given soil sample [1] . The proportions of sand, clay and silt in each soil sample were then applied to the soil textural class estimation chart (texture triangle) to derive the respective textural classes [17] . The textual classes were recorded as sand (s), loam (l), loamy sand (ls), clay loam (cl), sandy loam (sl), clay (c), sandy clay loam (scl), silt (si), sandy clay (sc), silt loam (sil) silty clay loam (sicl), and silty clay (sic) based on the United States Department of Agriculture (USDA) soil textural classes [21] following Okia [14] .

2.4. Data Analysis

Laboratory data on soil texture and minerals was entered in MS Excel to generate frequencies, means, standard errors of the means, coefficients of variation, percentages and correlation coefficients. Correlations were run to ascertain the strengths and directions of associations between soil physical and chemical parameters. The results were presented in tables and charts as done by Okia [14] , Fungo et al. [8] and Szyma?ski et al. [15] .

3. RESULTS

3.1. Soil Morphology

The field description of the morphology of the studied soil profiles is given in Table 1. Three pedon layers (RSK-H1, RSK-H3 and RSK-H5) had gradual boun- daries while RSK-H4 was continuous. However, the second pedon (RSK-H2) and the last (RSK-H6) had abrupt boundary layers. The first three horizons were soft to slightly hard with some roots. The last three horizons ranged from hard to very hard to hard consistence with few to no roots. The upper part of the studied pedons (RSK-H1 and RSK-H2) shows a dark yellowish brown (10YR 4/4) or dark brown (7.5YR 4/4) colour. The rest of the pedons were yellowish red with the colour codes of 5YR5/6 for RSK-H3 and RSK-H4 and 5YR5/6 in RSK- H5 and RSK-H6 (Table 1).

3.2. Physical Properties

The texture of the studied soil profiles, in the 1.86 m pit, was generally clay (Table 2). The clay fraction clearly dominated and ranged from 40% to 62% with an average of 52%. The upper part of the pedons (i.e. RSK-H1 and RSK-H2 horizons) contained a smaller clay fraction (i.e. 40% - 44%) in comparison with the lower part. The highest amount of clay (62%) was recorded in the 4thHorizon (RSK-H4). Amount of the sand fraction ranges from 35% to 52% and was almost evenly distributed throughout the studied profiles (except the RSK-H3) with an average of 43%. Silt fraction was the least ranging from 1% to 8%, with a mean of 5%. Based on the USDA textual classification, the top and the 5th pedons were sandy clay (sc) while the rest (RSK-H3, RSK-H4 and RSK-H6) were clay (c). When data was pooled, the overall textual classification is clay (Figure 2). Clay had an average of 52%, followed by sand at 43% and distant silt with 5%.

Table 1. Field description of the studied soils.

Site* Horizon**	Depth (cm)	Boundary Distinctness***	Roots	Consistency	Colour (moist)	Colour name
RSK-H1	0 - 26	g	+++	Soft	10 YR4/4	Dark yellowish brown
RSK-H2	26 - 42	a	++	Soft	7.5YR4/4	Dark brown
RSK-H3	42 - 80	g	+	Slightly hard	5YR5/6	Yellowish red
RSK-H4	80 - 112	c	Few	Hard	5YR5/6	Yellowish red
RSK-H5	112 - 165	g	Absence	Very hard	5YR5/8	Yellowish red
RSK-H6	165 - 190	a	Absence	Hard	5YR5/8	Yellowish red

*RSK = RELOAD Site Kabanyolo; **H1 to H6 = Soil horizons in the profile pit; *** Abrupt (a), gradual (g), clear (c) diffuse (d); + + + = Horizon has 50% root coverage; + + = Horizon has 25% root coverage; + = Horizon has 15% root coverage [8] [14] [15] [16].

3.3. Chemical Properties

Across the study site, the soils were generally acidic with a pH of 6.12 ± 0.15 in the top soil and 6.08 ± 0.11 in the subsoil layer (Table 3). The mean value of phosphorus was 12.42 ± 3.65 in the top soil compared to 8.88 ± 3.2 in the sub soil. Whereas Ca, K, Mg, Na, N, OM, Sand and Silt were 7.92 ± 0.76, 0.47 ± 0.13, 1.70 ± 0.17, 0.23 ± 0.05, 0.19 ± 0.01, 4.81 ± 0.35, 54.44 ± 2.64 and 7.22 ± 0.87 on average in the top soil, these parameters were 6.46 ± 0.74, 0.29 ± 0.06, 1.37 ± 0.16, 0.17 ± 0.02, 0.18 ± 0.02, 4.53 ± 0.15, 52.56 ± 1.68 and 6.67 ± 1.22 in the subsoil horizon respectively (Table 3). Phosphorus was the most variable in both the top and sub soil, considering the absolute value of the coefficient of variation of 88.2 and 108.1 respectively. Whereas pH was the least variable property in both the top soil (CoV = 7.2) and subsoil layer (CoV = 5.3). Besides clay, all the soil properties we investigated were higher in the top soil than the subsoil horizon. The average clay composition was 38.33 ± 2.66 in the top soil compared to 40.78 ± 2.7 in the sub layer (Table 3).

Table 2. Some physical properties of the investigated soil profiles.

Site-Horizon	Depth (cm)	% of sand	% of clay	% of silt	USDA textual class***
RSK-H1	0 - 26	52	40	8	sc
RSK-H2	26 - 42	50	44	6	sc
RSK-H3	42 - 80	35	62	3	c
RSK-H4	80 - 112	38	56	6	c
RSK-H5	112 - 165	45	54	1	sc
RSK-H6	165 - 190	38	56	6	c

***sandy clay (sc), clay (c).

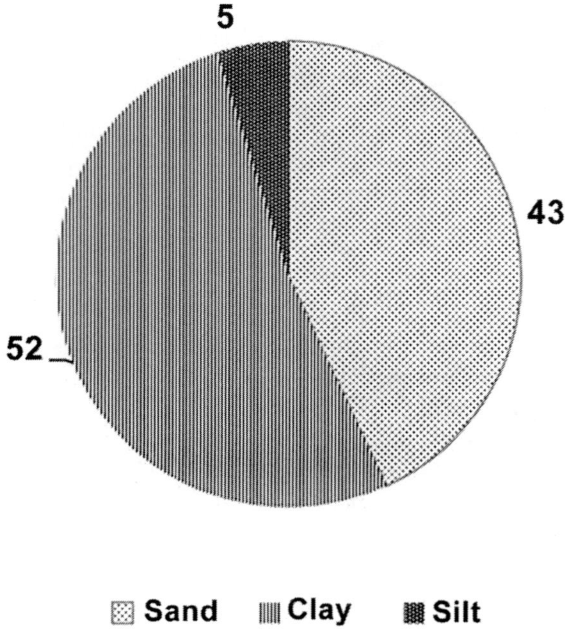

Figure 2. Soil texture.

Table 3. Descriptive statistics of the top and sub soils.

Statistic	pH	P (ppm)	Ca (ppm)	K (ppm)	Mg (ppm)	Na (ppm)	N (%)	OM (%)	Sand (%)	Silt (%)	Clay (%)
Top Soil											
Mean	6.12	12.42	7.92	0.47	1.70	0.23	0.19	4.81	54.44	7.22	38.33
Standard Error of the Mean	0.15	3.65	0.76	0.13	0.17	0.05	0.01	0.35	2.64	0.87	2.66
Coefficient of Variation	7.20	88.20	28.85	85.36	30.23	66.20	12.79	21.84	14.53	36.04	20.86
Sub Soil											
Mean	6.08	8.88	6.46	0.29	1.37	0.17	0.18	4.53	52.56	6.67	40.78
Standard Error of the Mean	0.11	3.20	0.74	0.06	0.16	0.02	0.02	0.15	1.68	1.22	2.13
Coefficient of Variation	5.28	108.07	34.14	59.72	33.98	41.72	30.87	10.06	9.58	54.77	15.67

A correlation matrix (Table 4) showed that sand and clay had the highest negative significant correlation ($r = -0.94$, $p < 0.01$) while Ca and Mg had the highest positive significant correlation ($r = 0.96$, $p < 0.01$) in the top soil. Soil pH was highly correlated negatively with sand in the subsoil ($r = -0.88$, $p < 0.01$) whereas Ca and Mg had the highest positive significant correlation ($r = 0.99$, $p < 0.01$) in the sub-soil layer. Phosphorus and silt did not show any relationship with soil pH in both the top ($p > 0.05$) and sub-soil layers ($p > 0.05$). In the topsoil, only silt did not show a significant relationship with soil P and Ca. In addition, Na was not significantly ($p > 0.05$) correlated with K and N in the top soil layer.

Having carefully reviewed the contents of Table 4, this paragraph should read as follows: in the subsoil, pH had no significant ($p > 0.05$) relationship with P or silt. In addition, P was not significantly ($p > 0.05$) influenced by Ca, OM or Silt. Other insignificant ($p > 0.05$) relationships in the sub soil were between Ca and Silt, K and Na, K and N, Na and N, and OM and Silt (Table 4).

Table 4. Spearman's Correlations of field based soil properties of top and sub soils.

Parameter	pH	P (ppm)	Ca (ppm)	K (ppm)	Mg (ppm)	Na (ppm)	N (%)	OM (%)	Sand (%)	Silt (%)
Top Soil										
pH										
P (ppm)	−0.03									
Ca (ppm)	0.79*	0.05								
K (ppm)	0.68**	−0.20**	0.68**							
Mg (ppm)	0.72**	0.05**	0.96**	0.69**						
Na (ppm)	0.57**	−0.33**	0.51**	0.93	0.45**					
N (%)	0.36**	0.56**	0.33**	0.40*	0.46**	0.17				
OM (%)	0.36**	−0.19*	0.32**	0.30**	0.26**	0.27**	−0.39**			
Sand (%)	0.32**	−0.04**	0.47**	0.73**	0.55**	0.60**	0.16**	0.52**		
Silt (%)	0.07	0.44	0.30	0.10**	0.38**	−0.02**	0.67**	−0.69*	−0.13**	
Clay (%)	−0.34**	−0.10**	−0.56**	−0.75**	−0.66**	−0.59**	−0.38**	−0.29**	−0.94**	−0.21**
Sub Soil										
pH										
P (ppm)	−0.19									
Ca (ppm)	0.20**	0.83								
K (ppm)	0.25**	0.49*	0.81**							
Mg (ppm)	0.31**	0.78*	0.99**	0.81**						
Na (ppm)	0.58**	−0.19*	0.16**	0.08	0.19**					
N (%)	0.38**	−0.12*	−0.19**	−0.31	−0.14**	−0.19				
OM (%)	0.09**	0.11	0.29*	0.32**	0.34**	−0.39**	0.21**			
Sand (%)	−0.88**	0.06**	−0.42**	−0.58**	−0.48**	−0.49**	−0.29**	−0.23**		
Silt (%)	−0.34	0.30	0.29	0.56**	0.26**	−0.27**	−0.30**	−0.04	0.06**	
Clay (%)	0.88**	−0.22**	0.16**	0.14**	0.23**	0.54**	0.40**	0.20**	−0.82**	−0.62**

Values with * and ** were significantly correlated at 0.05 and 0.01 alpha level, respectively.

Table 5. Comparison of MUARIK soil with requirements for fruits and vegetables.

Parameter	MUARIK soils	Comparisons						
		Requirements for fruits			Requirements for vegetables			
		Pineapples	Passion fruit	Watermelon	Cabbage	Tomato	Egg plant	Pumpkin
pH	6.08 - 6.12	*5	2	3	3	2	4	3
P (ppm)	8.88 - 12.42	2	2	2	2	2	2	2
Ca (ppm)	6.46 - 7.92	1	1	1	1	1	1	1
K (ppm)	0.29 - 0.47	1	1	1	1	1	1	1
Mg (ppm)	1.37 - 1.70	1	1	1	1	1	1	1
Na(ppm)	0.17 - 0.23	1	1	1	1	1	1	1
N (ppm)	0.18 - 0.19	1	1	1	1	1	1	1
Organic Matter (%)	4.53 - 4.81	1	1	1	1	1	1	1
Clay (%)	38.33 - 62.00	5	4	5	3	5	4	3

*1 = Very low; 2 = Low; 3 = Moderate; 4 = High; 5 = Very high for target fruits (Pineapple, Watermelon, Passion fruits) and vegetables such as Cabbage, Eggplant, Tomato and Pumpkin [22] [23] [24].

3.4. Comparison of MUARIK Soil with Requirements for Fruit and Vegetable Production

The comparison of key MUARIK soil parameters with the standard requirements for fruits and vegetable propagation is presented in Table 5. Soil organic matter and minerals including Ca, K, Mg, Na and N were generally lower than the requirement for fruits and vegetable propagation. The soil pH is slightly favorable for water melon, cabbage and pumpkins but unsuitable for pineapple, passion fruits, egg plant and tomato. The predominantly clayey (av. 52%) soil texture could tolerate pumpkins and cabbages but may not be suitable for the rest of the target fruits and vegetables (Table 5).

4. DISCUSSION

4.1. Soil Morphology

A half of the horizons (RSK-H1, RSK-H3 and RSK-H5) had gradual boundaries while RSK-H5 was continuous. However, the second pedon (RSK-H2) and the last (RSK-H6) had abrupt boundary layers (Table 1). Similar results were obtained by Okia [14] along the Balanites aegyptiaca belt in Uganda. Accordingly, abrupt boundaries between horizons are indicative of highly weathered soils. These boundaries also indicate that the soils have experienced some anthropogenic disturbance [14] .

The first three pedon layers were soft to slightly hard with some roots (Table 1). The last three horizons ranged from hard to very hard to hard consistence with few to no roots. These results concur with findings by Szyma?ski et al. [15] . Such morphology indicates that the pedons could have resulted from the translocation of colloids (clay minerals and iron oxides) and cyclic reduction-oxida- tion processes [25] [26] .

The upper part of the studied pedons (RSK-H1 and RSK-H2) had dark yellowish brown to dark brown colour. The rest of the pedons were yellowish red (Table 1). Earlier classification of MUARIK soil by Yost and Eswaran [11] indicated the surface layer to be dark reddish brown and the sub soil to be red neutral. The changes in soil colour over time could be attributed to the translocation of colloids and reduction-oxidation processed triggered by cultivation of research fields at MUARIK.

4.2. Physical Properties

The texture of the studied soil profiles was generally clay (Table 2 and Figure 2). According to Mugagga et al. [27] , the percentage of clay, in particular, gives a clear indication of the problem nature of the soils. A 10% clay threshold has been used as an indicator of the expansion potential, whilst 32% clay content exhibits extreme expansion potential [28] . Such clay content has implications for the shrink-swell properties of the soil. Yang et al. [29] and Wati et al. [30] agree that fine-textured clayey soils have small pores and liberate water gradually, which renders them susceptible to water logging because of their high water retention capacity. Such soils may not be suitable for some fruit and vegetable propagation such as water melon that require well drained soils. Similarly, pumpkins may be grown on a wide range of soil types provided the soil is free-draining. However, cabbage grows on a wide range of soil types ranging from light sand to heavier clays. Cabbage could therefore be recommended at MUARIK.

A report by Morton [22] showed that passion fruit vines could be grown on many soil types but light to heavy sandy loams, of medium texture are most suitable. Good drainage is also essential to minimize the incidence of collar rot. This implies that passion fruits may not present high yields at MUARIK that is predominately clay. Furthermore, watermelon grows best on soils with a sandy loamy texture. It requires good drainage at site and therefore clayey, damp, and water-logged soil must be avoided [23] . This revelation points to inadequacy of the MUARIK soils that are mainly clayey to support watermelon production.

According to DAF [24] , the best soils for pineapple production are non-com- pacted, well-aerated and free-draining loams, sandy loams and clay loams with no heavy clay or rock within one meter of the surface. Well drained soils are essential because poor drainage makes the plant more susceptible to root and heart rot diseases owing to a weak rooting system [24] . The fact that MUARIK has predominantly clayey soils makes the site unsuitable for pineapple growing except after remediation action is undertaken on the soil. In addition, eggplants can be grown with considerable success in fine and rich loam soils that are deep and well-drained [3] . Eggplant propagation at MUARIK should therefore be preceded by fertilizer application for better growth and development.

The upper part of the pedons (i.e. RSK-H1 and RSK-H2 horizons) contained a smaller clay fraction (i.e. 40% - 44%) in comparison with the lower part (Table 2). The highest amount of clay (62%) was recorded in the 4th Horizon (RSK-H4). Amount of the sand fraction ranges from 35% to 52% and is almost evenly distributed throughout the studied profiles (except the RSK-H3) with an average of 43%. MUARIK could therefore be favourable for tomatoes which are reported by Muzaale [31] to grow well in sandy loam soil with well-drained clay subsoil.

4.3. Chemical Properties

The MUARIK soils are generally acidic with a pH of 6.12 in the top soil and 6.08 in the subsoil layer (Table 3). Soil pH is the measure of the acidity or alkalinity of the soil. The results in Table 3 concur with Majaliwa [32] who reported MUARIK soils to be acidic. Previous studies have provided possible scenarios for low soil pH in a site. Fungo et al. [8] noted that frequent cultivation may result in rapid decomposition of organic matter and weakening of soil structure, which later results in lowering soil pH. Relatedly, Steenwerth et al. [10] found lower values of soil pH in the grassland than in cultivated soils, reportedly due to leaching in the grasslands. Mulumba [9] reported some cropping systems to pose acidifying risk to the soil due to the amount of materials removed at harvest, amount and type of fertilizers normally used.

Muzaale [31] indicated that tomato growers in Uganda require slightly acidic soils with a pH level of 6.0 to 7.0 for high quality yields. Since the soil tests present a pH of 6.08 - 6.12, it therefore means that the MUARIK soils could be suitable for tomato growing. Soil pH can be adjusted if necessary. If the pH is too low, lime is added to the soil. If it is too high, some sulphur is mixed to the soil [31] . However, these adjustments come at a cost to the farmer. In addition, pumpkins may be grown at MUARIK because they are tolerant of fairly acid conditions and liming should not be necessary unless the pH is less than 5.5.

Based on the pH results of 6.08 - 6.12 (Table 3), passion fruits and cabbages can be grown at MUARIK after reducing the acidity by lime application for better yields. This is so because passion fruits have been associated with soils with a pH of 6.5 to 7.5 (22). Cabbage requires soils with pH in the range of 6.0 - 6.5, making Kabanyolo a suitable site for production of cabbages [3] . In contrast, watermelons prefer a soil pH between 6.0 and 6.8 [23] making MUARIK is a suitable site. However, other parameters such as soil texture, soil organic matter at MAURIK fall below the required standards for watermelon propagation. Eggplants require a site with pH 5.5 to 6.0 for its better growth and development [3] . Similarly, a soil pH in the range of 4.5 - 5.6 is optimal for pineapple production [33] . It therefore implies that lime must be applied to lower the pH of MUARIK soil from 6.12 to the required range of 4.5 - 5.6 and 5.5 - 6.0, respectively, before pineapples and eggplants are introduced.

Phosphorus was 12.42 ± 3.65 ppm in the top soil compared to 8.88 ± 3.2 pmm in the sub soil (Table 3). However, according to Obreza and Rhoads [33] , the critical level of P is 10 $mg \cdot kg^{-1}$. This threshold shows that phosphorus is a deficient nutrient in the MUARIK sub soils. The lower level of P could be due to the large quantities drawn by plants [8] from the sub-soil layer. Perhaps P fixation is high as the soils in the area are highly weathered with potential of high content of aluminum oxides. The abrupt boundaries between horizons in the top and bottom pedons (Table 1) point to highly weathered soils. Phosphorus was also the most variable mineral in both the top and sub soil (Table 3). This could be attributed to the behavior of P in most soils. Fungo et al. [8] noted that P tends to move less than Ca and K because of various types of chemical reactions that may occur, rendering it insoluble. The quantity and availability of P ought to be improved prior to fruit and vegetable propagation at MUARIK.

Soil organic matter (OM), Ca, K, and Mg were 4.81 ± 0.35, 7.92 ± 0.76, 0.47 ± 0.13 and 1.70 ± 0.17, on average in the top soil, these parameters were 4.53 ± 0.15, 6.46 ± 0.74, 0.29 ± 0.06 and 1.37 ± 0.16 in the subsoil horizon respectively (Table 3). Similarly, Majaliwa [32] had earlier indicated MUARIK soils to have moderate to low fertility. According to Obreza and Rhoads [33] , the critical levels of Ca, K, and Mg are 250, 45 and33 $mg \cdot kg^{-1}$, respectively. By these standards, potassium is the most deficient nutrient in MUARIK. For high fruit and vegetable yields, fertilizer application may be required. According to a guide by Adekunle et al. [23] , the soil must be fertile with good organic matter content. Thus, for watermelon yields, the MUARIK soils ought to be amended with aged manure, seaweed, and/or compost before planting because watermelons are heavy feeders.

A correlation matrix (Table 4) shows sand and clay with the highest negative significant correlation ($r = -0.94$, $p < 0.01$). The highest positive significant correlation was observed between Ca and Mg both in the topsoil ($r = 0.96$, $p < 0.01$) and subsoil ($r = 0.99$, $p < 0.01$). The observed high and negative correlation between sand and silt is not uncommon because these two soil properties are complementary to each other [8] . In contrast, the high positive correlation between Ca and Mg can be attributed to the possibility of the two minerals having similar parent material mineralogy. Chadwik and Graham [34] assert that mafic mantle- derive rocks typically weather to a smectite and iron oxide-rich colloidal fraction with the simultaneous release of both Ca and Mg.

5. CONCLUSION

This study reveals that the upper part of the studied pedons had dark yellowish brown to dark brown colour while lower soil layers were yellowish red. The texture of the studied soil profiles was generally clay (52%). Furthermore, the soils were generally acidic with a pH of 6.08 to 6.12. The site was deficient in most of the minerals including P, Ca, K, Mg, Na, N, and soil organic matter (OM). Phosphorus was the most variable property in both the top and sub soil. Sand and clay had the highest negative significant correlation ($r = -0.94$, $p < 0.01$) in the top soil. Similarly, soil pH had the highest negative and significant relationship ($r = -0.88$, $p < 0.01$) with sand in the sub-soil layer. The highest positive significant correlation was observed between Ca and Mg both in the topsoil ($r = 0.96$, $p < 0.01$) and subsoil ($r = 0.99$, $p < 0.01$). Thus, MUARIK is suitable for cabbage and pumpkin because they grow on a wide range of soil types ranging from light sand to heavier soils. However, passion fruit and tomato should be grown at MUARIK after adjusting the soil pH with lime. Similarly, watermelon will require aged manure before planting because watermelons are heavy feeders and slight sulphur amendments are needed to accommodate eggplant and pinea- pple (Table 5).

ACKNOWLEDGEMENTS

Financial support was obtained from the German Federal Ministry of Education and Research under the Reduction of Post-Harvest Losses and Value Addition in East African Food Value Chains Project (RELOAD/A401UNCST2012). We are grateful for the support from the University of Kassel (Prof. Oliver Hensel and Michael Hesse) and management of the MUARIK especially Prof. Phinehas Tukamuhabwa, Prof. Jacob Godfrey Agea and Jimmy Wesiga. Mr. Edward Baka of the Makerere University Agriculture Soil Laboratory is appreciated for the assistance in the preparation and laboratory analysis of soil samples. Harryson Lyagoba, Esther Birungi Tendero and Phillip Kihumuro are commended for the support during soil sampling.

CITE THIS PAPER

Okiror, P., Lejju, J.B., Bahati, J., Rugunda, G.K., Sebuuwufu, C.I., Mulindwa, P. and Ocan, J.J. (2017) Suitability of Kabanyolo Soils for Fruit and Vegetable Production. Open Journal of Soil Science, 7, 19-33.

REFERENCES

1. Okalebo, J.R., Gathua, K.W. and Woomer, P.L. (2002) Laboratory Methods for Soil and Plant Analysis: A Working Manual. TSBF, Nairobi.
2. Mahadevan, P. and Parsons, D.J. (1970) Livestock. In: Jameson, J. D., Ed., Agriculture in Uganda.
3. Nono-Womdim, R., Ojiewo, C., Abang, M. and Oluoch, M.O., Eds. (2009) Good Agricultural Practices for African Indigenous Vegetables. Proceedings of a Technical Consultation Workshop in Arushia, Tanzania, 7-8 December 2009, International Society for Horticultural Science.
4. Lal, R. (1994) Sustainable Land Use Systems and Soil Resilience. In: Greenland, D.J. and Szabolcs, I., Eds., Soil Resilience and Sustainable Land Use, CAB International Publishers, 41-67.
5. Pomeroy, D.E. (1977) Distribution and Abundance of Large Termite Mounds in Uganda. Journal of Applied Ecology, 14, 456-475.
6. Tenywa, M.M., Lal, R., Majaliwa, M.J.G. and Lufafa, A. (1999) Characterization of the Stages of Soil Resilience to Degradative Stresses: Erosion. In: Stott, D.E., Mohtar, R.H. and Steinhardt, G.C., Eds., Sustaining the Global Farm, The 10th International Soil Conservation Organization Meeting, Purdue University and the USDA-ARS National Soil Erosion Research Laboratory, 24-29 May 1999, 606-610.
7. Okwakol, M.J.N. and Sekamatte, M.B. (2007) Soil Macrofauna Research in Ecosystems in Uganda. African Journal of Ecology, 45, 2-8.
8. Fungo, B., Grunwald, S., Tenywa, M.M., Vanlauwe, B. and Nkedi-Kizza, P. (2011) Lunnyu Soils in the Lake Victoria Basin of Uganda: Link to Toposequence and Soil type. African Journal of Environmental Science and Technology, 5, 15-24. [Citation Time(s):11]
9. Mulumba, L.N. (2004) Land Us Effect on Soil Quality and Productivity in the Lake Victorial Basin of Uganda. Ph.D. Thesis, The Ohio State University, USA.
10. Steenwerth, K.L., Jackson, L.E., Calderon, F.J., Stromberg, M.R. and Scow, K.M. (2002) Soil Community Composition and Land Use History in Cultivated and Grassland Ecosystems of Coastal California. Soil Biology and Biochemistry, 34, 1599-1611.
11. Yost, D. and Eswaran, H. (1990) Major Land Resource Areas of Uganda. World Soil Resources, Soil Conservation Service-USDA, Washington DC, 227 p.
12. Komutunga, E.T. and Musiitwa, F. (2001) Climate. In: Mukiibi, J.K, Ed., Agriculture in Uganda, Volume 1: General Information, Fountain Publishers, Kampala, 21-32.
13. Kisamba-Mugerwa, W. (2001) Social Background. In: Mukiibi, J.K, Ed., Agriculture in Uganda, Volume 1: General Information, Fountain Publishers, Kampala, 186-199.

14. Okia, C.A. (2010) Balanites aegyptiaca: A Resource for Improving Nutrition and Income of Dryland Communities in Uganda. PhD Thesis, University of Wales, Bangor, 310 p. [Citation Time(s):9]
15. Szymański, W., Skiba, M., Nikorych, V.A. and Kuligiewicz, A. (2014) Nature and Formation of Interlayer Fillings in Clay Minerals in Albeluvisols from the Carpathian Foothills, Poland. Geoderma, 235-236, 396-409.
16. MCC (1954) Munsel Colour Charts. Munsel Colour Company, Baltimore.
17. Cleland, T.M. (1921) A Practical Description of the Munsell Color System with Suggestions for Its Use. Strathmore Paper Company, Russell.
18. Bray, R.H. and Kurtz, L.T. (1945) Determination of Total, Organic, and Available Forms of Phosphorus in Soils. Soil Science, 59, 39-45.
19. Rhoades, J.D. (1982) Soil Testing. In: Page, A, Miller, R.M. and Keeney, D.R., Eds., Methods of Soil Analysis, Part 2, 2nd Edition, Am. Soc. Agron, Madison, 16-21.
20. Anderson, J.M. and Ingram, J.S.I. (1989) Tropical Soil Biology and Fertility. Handbook of Methods, CAB. International, Wallingford.
21. United States Department of Agriculture (1987) Soil Mechanics Level 1 Module 3 USDA Soil Textural Classification Study Guide. USDA Soil Conservation Service, Washington DC.
22. Morton, J. (1987) Passionfruit. In: Morton, J., Ed., Fruits of Warm Climates, Florida Flair Book, Miami, 320-328.
23. Adekunle, A.A., Fatunbi, A.O., Adisa, S. and Adeyemi, O.A. (2007) Growing Watermelon Commercially in Nigeria: An Illustrated Guide. USAID ICSNIGERIA and IITA. http://www.fao.org/sd/erp/toolkit/BOOKS/watermelon_illust_ guidebook.pdf
24. Department of Agriculture and Fisheries (2013) Land Requirements for Growing Pineapples. Queensland Government. https://www.daf.qld.gov.au/ plants/fruit-and-vegetables/fruit-and-nuts/pineapples/land-requirements-pineapple
25. Lindbo, D.L., Rhoton, F.E., Hudnall, W.H., Smeck, N.E., Bigham, J.M. and Tyler, D.D. (2000) Fragipan Degradation and Nodule Formation in Glossic Fragiudalfs of the Lower Mississippi River Valley. Soil Science Society of America Journal, 64, 1713-1722.
26. Szymański, W., Skiba, M. and Skiba, S. (2011) Fragipan Horizon Degradation and Bleached Tongues Formation in Albeluvisols of the Carpathian Foothills, Poland. Geoderma, 167-168, 340-350.
27. Mugagga, F., Kakembo, V. and Buyinza, M. (2011) A Characterisation of the Physical Properties of Soil and the Implications for Landslide Occurrence on the Slopes of Mount Elgon, Eastern Uganda. Natural Hazards, 52, 1113-1131.
28. Baynes, F.J. (2008) Anticipating Problem Soils on Linear Projects. Conference Proceedings on Problem Soils in South Africa, Midrand, 3-4 November 2008, 9-21.
29. Yang, H., Adler, R. and Huffman, G. (2007) Use of Satellite Remote Sensing in the Mapping of Global Landslide Susceptibility. Nat Hazards, 43, 245-256.
30. Wati, S.E., Hastuti, T., Wijojo, S. and Pinem, F. (2010) Landslide Susceptibility Mapping with Heuristic Approach in Mountainous Area. A Case Study in Tawangmangu Sub District, Central Java, Indonesia. International Archives of the Photogrammetry, Remote Sensing and Spatial Information Science, 38, 248-253.
31. Muzaale, F. (2014) Growing Good Quality Tomatoes: Farming Magazines, Daily Monitor, Kampala, Uganda.
32. Majaliwa, M.J.G. (1998) Effect Vegetation Cover and Biomass Development on Soil Loss from Maize Based Cropping Systems. MSc Dissertation, Makerere University, Kampala, 102 p.
33. Obreza, T.A. and Rhoads, F.M. (1988) Irrigated Corn Response to Soil Test Indices and Fertilizer Nitrogen, Phosphorus, Potassium and Magnesium. Soil Science Society of America Journal, 52, 701-706.
34. Chadwik, O.A. and Graham, R.C. (2000) Pedogenic Processes. In: Sumner, M., Ed., CRC Handbook of Soil Science, CRC Press, New York, E41-E73.

CHAPTER 9

Soil Respiration in Drying of an Organic Soil

Zofia Fischer[1], Pavel Blažka[2*]

[1]*Department of Biotechnology and Environment, John Paul II, Catholic University of Lublin, Lublin, Poland*

[2]*Department of Science, University of South Bohemia, České Budějovice, Czech*

ABSTRACT

Drying of soil was linearly related to time, soil volume decreased and ratio of air within the soils increased. Respiration was related with decreasing humidity, storage of CO_2 in soil water results in RQ < 0.5 in the larger soil items at least for a while. Rate of drying decreased in the second part of the process. RQ increased as the CO_2 stored was aerated when its solvent-water evaporated and access of air into the soil increased; eventually RQ = 1.0 in the last days of the experiment. Respiration of the experimental soil stopped when GWC reached 0.15. ΣRQ for the whole process is about 0.7, a bit higher in experiments with less soil suggesting less anoxia.

Keywords: RQ, Anoxia, Storage of CO_2, Changes of Soil Volume

1. INTRODUCTION

In the present contribution our original idea was to study process of drying more in detail. Soils in natural settings are changing; this may be approached not only in field work, but also in lab experiments, where the conditions may be set more accurately and less rely on unpredictable weather etc. and show how they are reflected in the process, particularly in soil respiration. During the work we realized however, that some concepts are generally not quite clear or understood and we propose some clarification, both by pointing to literature not frequently referred in soil science and/or through experimental design.

Respiration was/is considered as a general measure of soil activity (somewhat mystical statement, preferably of soil organisms), recirculation of nutrients from plant residues, while drying of soil is associated with wet and dry periods, seasons, hence more recently with global changes.

Soil respiration was studied many times; the literature on it is quite extensive [1] - [3] . It is based both on field measurements and on laboratory experimentation. The latter is mostly associated with sieving to get more reproducible results, and also with mixing and thus aeration of the soil material. We tried to overcome the sieving effect by slightly compressing the sieved soil by hand [4] . We have seen no other reference on the effect of compression in literature. We want to however first make some introductory comments to the subject respiration. The word respiration is historically (since 18th century) associated primarily with oxygen consumption by men and animals, more recently in biochemistry with oxidation of protons to water in the oxidative phosphorylation pathway as part of the general scheme of energy metabolism in most organisms, prokaryotes and eukaryotes; plants, animals and fungi—except only obligatory anaerobes [5] - [7] . Respiration is a heterotrophic process and also root respiration is not autotrophic e.g. [3] , [8] though it is run by organisms, which largely live autotrophically; all plants are however heterotrophs at night. Soil respiration, meaning respiration of the whole soil community, was introduced later (first half of 20^{th}century—Suchtelen [9] was probably among the first). Respiration is anyway a biological process; all biological principles are to be included, not discarded, but it is also a part of soil biology, a composite science as soil physics, biochemistry, geology etc.; all parts have to be observed with the same scrutiny, accuracy and respect to achieve real progress in soil science as a whole.

The sequence/wording "anaerobic respiration" (meaning reduction of e.g. nitrate, nitrite or sulfate replacing dioxygen as proton acceptor) is used quite frequently, though it is not particularly accurate, nitrate and sulfate reduction or proton acceptance would be more to the point, but may be perhaps tolerated.

Respiratory production of carbon dioxide occurs largely in the tricarboxylic acids cycle, is associated through oxidative phosphorylation with oxygen consumption, their ratio in terms of absolute values of moles or volumes is the respiratory quotient ($RQ = [\text{production } CO_2]/[\text{consumption } O_2]$).

Introduction of reliable IR non-dispersive spectrophotometers (In fact already 1850 paper based on CO_2) enabling measurements of small concentrations and changes of carbon dioxide concentrations to soil science provoked some shifts in terminology; soil biological activity—measured as production of carbon dioxide—is now quite frequently referred as soil respiration e.g. [2] [8] [10] , most clearly formulated by Gaumont-Guay et al. [11] ("The efflux of carbon dioxide (CO_2) from the soil, referred to hereafter as soil respiration…"). There are however quite a few sources of carbon dioxide in soils, besides processes involving primarily inorganic substrates, e.g. acidification of carbonates by acids [12] , there is also a number of fermentation types occurring when oxygen availability is limited. Both respiration and fermentation are the only processes providing energy to organisms (as macroergic P-bonds [~P], largely ATP). Fermentation in soils contemporary with respiration is apparently quite frequent, particularly when soils become wet and both involve also consumption of some substrate(s); it is therefore surprising, if there is no information on SOM given, particularly in dry environment where very low SOM may be expected e.g. [13] . Carbon dioxide produced in fermentation is associated with much smaller gain in useful energy (about 3 to 4-times) compared to respiration. Respiration supplemented by fermentation is likely the most frequent situation in soils with inadequate oxygen availability; though even low partial tensions of O_2 are mostly adequate for microorganisms. This results in $RQ \geq 1$ and decreasing production of ~P. On the other hand, many products of fermentations may be readily available substrates in respiration, if dioxygen becomes available again. This is also reflected in the interpretation of the RQ. Very often limits for RQ are given e.g. within 0.7 till 1.3. In fact in anaerobes (obligatory or facultative, microbial or eukaryotic) RQ may be much higher (e.g. >10), in strict anoxia it cannot be defined at all, as dividing by 0 (O_2 consumption) is a prohibited math operation. Two most important types of fermentation for soil may be lactic and ethanolic (alcoholic). While respiration of lactate and many other products of fermentation (including e.g. acetate) results in RQ = 1, like most carbohydrates; for catabolism of ethanol RQ = 0.57; in reduction of methane RQ = 0.50. Efflux of CO_2 is in fact a mix of respiration, fermentation and abiotic processes leading to absorption or release of CO_2. Field measurements of CO_2 release to the atmosphere are useful particularly for understanding of the role of soils in the global carbon cycle. Experimental conditions (e.g., participation of autotrophs, soil humidity, differences of temperature between the "dome" and the neighboring free area…), are however to be thoroughly specified and the CO_2 efflux is not considered as soil respiration.

In soil literature retention of CO_2 in soil water due its high solubility (distribution coefficient air/water = approx 1 compared to dioxygen 0.03) is mentioned only exceptionally. In soils with pH > 4 dissolved bicarbonate becomes the dominating form of the carbonate system [14] . What we measure by any instrument is the apparent production of CO_2 and RQ modified by diffusion and solubility interference. RQ is a ratio, relationship between the differences of the two components and their ratio are thus non-linear, rather exponential. Jenkins and Adams [15] are thus not correct in saying "RQ is free of empirical assumptions...", Araujo et al. [16] in accepting without comment measured RQ in the range RQ = 0.2 till 1.5 depending on the soil degradation level. On the low side remediation of hydrocarbon degradation of soils results in low RQ [17] , again below the theoretical limit 0.65 for hexadecane; methane is the most reduced hydrocarbon, its oxidation (assuming only one process) results in RQ = 0.5; this should be considered as the low limit of acceptable RQ data. On the other side of possible range synthesis of lipids from carbohydrates are sometimes considered with RQ ≈ 1.3 as possible maximum. It is probably correct for respiration; we must however expect hypoxic or anaerobic conditions particularly in wet soils. Soil organisms are forced to get additional energy to what they get from decreased or even stopped respiration through some fermentation; with prevailing fermentation RQ may be well over 10. In deep anoxia RQ cannot be calculated at all, as CO_2 is produced, but there is no oxygen consumption, division by 0 is a prohibited math operation. Let us just remind summed equation of ethanolic fermentation:

$$C_6H_{12}O_6 = 2\,CH_3 \cdot CH_2OH + 2\,CO_2$$

Being aware of interferences in measurements is particularly important in studies of processes, like various kinds of acclimation to changing temperature and humidity or in the budget of a process.

We want to analyze particular parts of the drying process, not only desiccation, but also changes of soil volume, access of air/oxygen to soil aggregates, storage capacity for CO_2 within the soil etc. and their impact on measured respiration data. This applies both to laboratory and field setups and involves size of soil items in both. Our soil is to some extent exceptional in high SOM content (around 50% of dry weight), many relationships studied in less organic soils need not to be applicable, also some of our conclusions are necessarily not of general use.

Drying of soils is a natural part of regular seasonal events, in observations and/or experiments many papers tend to limit themselves to just momentarily flashes or steady states; we tried however to estimate the budget of the drying process to go through different situations, states of the soil, get rid of many obstacles in getting real metabolic data. Incidentally, we came across low rates, their statistical treatment and time scales.

2. MATERIAL AND METHODS

2.1. The Site and Soil Sampling

The location is on the northern outskirts of the city of Warsaw (Poland). Geographic coordinates are: 52°20'212 N, 020°51'260 E, elevation 81 m. Refer to Blazka and Fischer [18] . The aim of the study is understands of mechanisms of drying through laboratory experimentation and not description of a particular locality. We used the soil for experiments as a laboratory model. The soil is highly organic; nearly 50% LOI, most of the mineral fraction is sand. It comes from a wet site in a dense forest, from the very active A1 zone, depth about 10 cm below surface, it was 5 mm sieved. We got our samples respecting suggestions by Carter and Gregorich [19] .

2.2. Laboratory Analyses

2.2.1. Humidity

(GWC—gravimetric water content, WHC—water holding capacity, dm—dry mass).

Our procedure was based essentially on Grace et al., Priha and Smolander [20] [21] with minor modifications [18] . GWC and other contents are given as ratios, but not %. Dry mass was determined by drying at 110°C to constant weight.

2.2.2. Soil Organic Matter (SOM and Density)

SOM was estimated as loss of weight in ignition to constant weight (LOI) at 500°C on selected samples. Mean densities of SOM (ρ = 1.3) and of the mineral fraction of dry weight (ρ = 2.65), are from Ilsted et al. [4] and many others.

Organic matter = LOI was between 0.45 and 0.50.

ρ(soil) = (LOI * 1.30 + (1 – LOI) * 2.65 = 1.96.

2.2.3. Respiration

(OCD—oxygen consumption, PCO_2—production of CO_2, RQ—respiratory quotient).

We used the volumetric method with the Scholander type respirometer to get both the oxygen consumption and carbon dioxide production. The capacity of its cylindrical flasks is 50 ml (inside radius r = 2 cm, V = 4 cm^2) and we refer data as $\mu l \cdot g^{-1}$(dry mass)$\cdot h^{-1}$. Each experimental series consisted of two parts. In the first, without hydroxide in the flasks, changes in the volumes were recorded mostly for 5 hours, this gave the sum of CO_2 changes and O_2 consumption, in the second part after placing filter paper soaked with NaOH solution into the respirometric chamber, we recorded the oxygen consumption (OCD) for another 5 hours interval, readings each 30 minutes. This involves assumption of slow rates of metabolic changes. The bath temperature was 17.5°C. Sensitivity of the volumetric respirometer is effectively not much worse (<0.1 µl gas$\cdot g^{-1} \cdot hour^{-1}$) than of the IR meters, though rates of providing data are not comparable.

While data of the second part were the OCD, PCO_2 was calculated as first reading (without hydroxide) + second reading (with hydroxide), respecting the signs in readings and calculations. We then plotted the results against time of readings, the slope coefficient of the regression (β_1) was the rate (O_2 consumption-OCD or PCO_2); the respective regression coefficients were all $r > 0.95$, they were later corrected for pressure and temperature, multiplied by the calibration factor and related per gram dry soil and hour. Weights of respiration vials with soil were taken regularly; tare was subtracted to get rates of drying.

Later in Results and discussion we will consider to what extent the data measured might be biased by diffusion or other transport problems within the soil items, the measured data may then be classified as apparent data (AOCD, $APCO_2$ and ARQ).

2.2.4. Soil (Vs), Air (Va), Pore Volume (Vpo)

Wet soil was gently pressed before start, there were no fissures between soil and wall of respirometric flask, they appeared however later during drying. We measured them by a series of blades (used in fine mechanics to measure narrow crevices, range 0.05 till 2 mm) on four sides of the vial, their mean was subtracted from r = 2 cm to get the radius of the base. The height (he) of the cylinders was estimated using a plate put on the top of the open vial and fixed to center by 4 screws projecting down on the outside of the flask, there were 5 evenly distributed holes on the plate, through which the distance from the top of the plate to soil top was measured by a rod, read by caliper and subtracted from the distance bottom of flask-top of plate. Mean of the 5 readings was height of the soil (he). Final calculation as cylinder volume:

$$\text{Soil volume } (Vs): Vs = \pi * r^2 * he \tag{1}$$

$$\text{Pore volume(porosity) Vpo} = \text{Vs} - ((\text{Vs} - \text{Ws})/\rho) = \text{Vs} - ((\text{Vs} - \text{H}_2\text{O})/\rho) \quad (2)$$

Ws is mass of wet soil.

$$\text{Volume of air Va} = \text{Vpo} - \text{H}_2\text{O} \quad (3)$$

2.3. Experimental

Changes of GWC in experiment are included as background for other parameters of the soil studied. We refer four principal experiments, which were preceded by several preliminary checks or pilot experiments, all done on one and the same kind of soil collected from one spot. All were aimed towards analysis of changes in soil respiration (OCD, PCO_2 and RQ) in drying of the soil. The main difference between the experiments was in the amount of soil in flasks and therefore were coded as LS (less soil) and MS (more soil) respectively. Outline of the experiments is in Table 1. Drying proceeded in MS1 and MS2 in two rates, up to about half of the experiments the rate (a) was somewhat faster, then later (rate b). Drying was terminated when oxygen uptake and CO_2production dropped below the sensitivity of the instrument, except LS1 terminated earlier. There were 12 (LS1, MS1) or later 6 (LS2, MS2) parallel respirometers in each series. The basic data differ by an order of magnitude and change during the experiment; therefore we compared the coefficients of variance-cv's. It turned out that the cv's of the means based on six replicates are smaller than those of 12 replicates (Table 2). As standard deviations (second roots of variance) are involved in calculations of the t-test, we consider the data as comparable (Table 2). Weights of vials for GWC estimates were taken not necessarily on the day of respiration measurement. Therefore GWC data for comparison with respiration were interpolated from the respective regression of GWC to time (all $r > 0.99$).

After each 10 days WHC was estimated in 10 of the vials in the first series (MS1, LS1). In experiment MS2 and LS2 we estimated WHC at the beginning and at termination of the experiment.

Additionally we made a 14-days drying experiment (MP—mud pies) starting at GWC = 1.0 and finished at GWC = 0.13 at the respiration stop. There were two modes of soil drying, each in six respirometers: a) soil (350 g dry weight) dried freely distributed on a plate, aliquots were taken for each respirometric measurements (per 6 * 5 g dry weight), returned to the plate and mixed; b) In the second mode wet soil was pressed by hand into the respirometric vials (per 5 g to each vial), it was left to dry freely between measurements each second day.

Table 1. Outline of the experiments.

	LS1	LS2	MS1	MS2
dry mass [g] mean; se	3.8; 0.0294	3.26; 0.025	15.22; 0.0751	14.61; 0.147
days of experiment	24	13	52	42
starts of experiment	Aug 25, 14	Feb 23, 15	Nov 11, 14	Feb 23, 15
1st linear section [d]	24	13	32	29
°C air	17.2 - 19.0	18.8 - 19.7	17.0 - 18.8	18.4 - 19.7
β_1 GWC rate a [m]	−1.845	−4.43	−1.632	−1.257
β_1 GWC rate b [m]	x	x	−0.721	−0.412
mud-pies formation [d]	8	6	16	20
humidity air %	50 - 65	40	45 - 62	40 - 45
GWC start	2.4	1.54	2.34	1.58
GWC end	0.644	0.12	0.15	0.16
GWC estimates N	7	5	15	15
GWC replications	12	6	12	6

GWC rates are β_1-coefficients in linear regressions per month, a stands for the first, b for the second linear section. In the LS was one section. LS1 was finished prematurely.

Table 2. Cv's of means of cv's in MS1 and MS2 experiments for OCD, PCO_2 and GWC to compare variation in data with 12 or 6 replicates.

	OCD	PCO_2	GWC	replicates	determinations
MS1	0.2919	0.3581	0.1275	12	14
MS2	0.1373	0.1401	0.0511	6	13

We also checked respiration in strict anoxia (SA) comparing respiration of a fresh collected soil with respiration after flushing the respirometers by pure dinitrogen (Air products.com, purity 5.2).

2.4. Statistics

Was rather basic, it involved averages = means (m), standard deviations (s), coefficients of variation (cv) and standard errors of the mean (se) calculated for all series of data obtained, referred here or not. Two sided t-test was applied for checks of differences of starting and closing data of series and for checks of differences in neighboring data within series. In final evaluation principal (Table 2) rates were related to month: (30days per month).

This came out from the discrepancy in evaluation of rates in changes in GWC, Vs, Vpo related to day which was our working time unit. We were unable to reject the 0-hypothesis (slope-rate of change of GWC is not different from zero—(from horizontal plane line), whereas paired T-test showed significant differences not only between start and end of experiment ($p < 0.001$), but also between most of neighboring data in the series (Table 3). In general the choice of reference units data is frequently intuitional, comparison of two or more statistical treatments might be a more general tool.

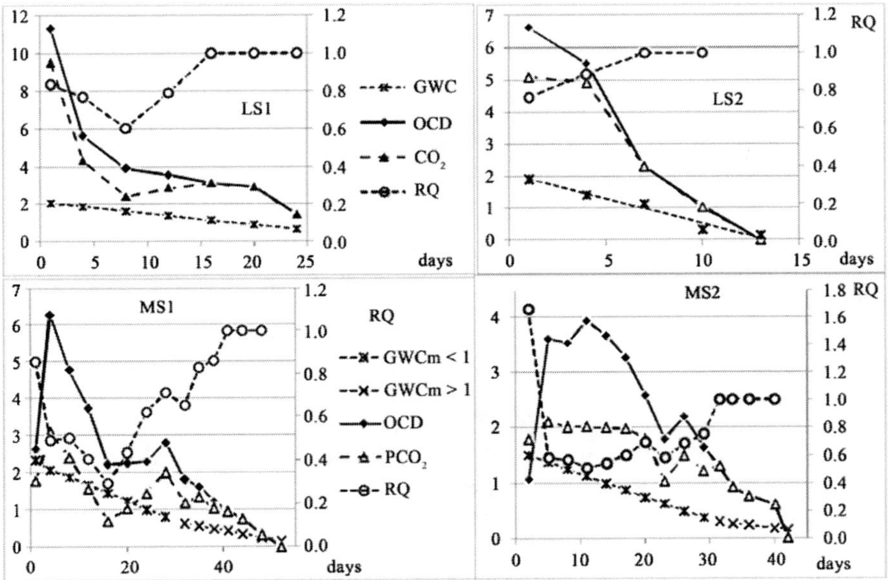

Figure 1. Changes of the gravimetric water content-GWC (as ratios), OCD (oxygen consumption), PCO_2 (production of carbon dioxide)—both as μl·g [dm]$^{-1}$·h^{-1}, and RQ (alternative scales) in the four experiments. Rates of changes in GWC are in Table 1 and Table 3, rates of OCD and PCO_2changes as related to GWC are in Table 5, overall budget of the experiments in Table 6. The LS1 experiment was finished before respiration dropped to zero. All standard errors are within the point icons. Specification of symbols for LS2 and MS2 experiments are in parts for LS1 and MS1 respectively.

3. RESULTS AND DISCUSSION

Results are largely based on the four drying experiments specified in Methods, Experimental, Table 1 and in Figure 1. Additionally we performed two checks to get respiration data of the hard mud pies and the small sandy like aggregates and demonstrate CO_2 production in strict anoxia. Also we assured us that the hard dry mud pies are not just gypsum or carbonate.

Table 3. Rates (slopes) GWC, Vs, OCD per week and month for MS2 and LS2 compared to 2-sided t-test for start-end and neighboring days.

MS2	β_1 (slope)/day	N = 14 all p	slopes/month	p	T-test of pairs start-end	T-test of neighboring days			
						$p > 0.05$	$p < 0.05$	$p < 0.01$	$p < 0.001$
GWC	−0.0361	>0.05	−1.083	<0.005	$p < 0.001$	0		0	14
Vs	−0.4255	>0.05	−12.76	<0.001	$p < 0.001$	6	5	1	2
OCD	−0.0733	>0.05	−2.319	<0.001	$p < 0.001$	3		1	10
LS2		N = 5 for GWC and Vs; N = 4 for OCD							
GWC	−0.1138	>0.05	−3.41	<0.005	$p < 0.001$	0	0	0	5
Vs	−0.3963	>0.05	−11.89	<0.001	$p < 0.001$	0	0	2	3
OCD	−0.5904	>0.05	−17.71	<0.001	$p < 0.001$	0	1	0	3

Significance of slopes related to day and month, compared to 2-sided T-test for data pairs start-end of experiment and neighboring pairs during the experiment.

3.1. Humidity

Relationships of humidity (GWC) are therefore the first to be examined more in detail. Its changes related to time in terms of days[d] proved to be controversial; this resulted in use of month [m] as time reference unit in some selected relationships (Table 3).

GWC decreased during the experiments linearly with time for approximately 2/3 of the course, towards the end in the MS experiments, there was a significant drop in rate (Table 1). The course was a single linear function with respect to time in the shorter LS experiments. A slowing of the function is indicated in LS2, there are, however, too few data to get a significant picture (Figure 1).

3.2. Soil Volume

So far cracks in dry soils were often referred from the field in the literature, but there was/is much less interest in experiments on changes of soil volume in lab. There are more recent works on the impact of repeated drying and rewetting cycles on characteristic of soils resulting so far in non-complete understanding of these processes. We registered shrinking of soil during LS1 and MS1 experiments and estimated it with assumptions of linearity; nevertheless these estimates for MS1 and LS1 are in good agreement with the data from the second series of experiments. All four experiments have shown significant shrinking and changes of pores (porosity). The LS2 experiment is less successful due to high rate of drying and GWC decrease, therefore it was short and only few data were obtained (Table 2 and Table 3).

Soil humidity (GWC) appears to be the governing parameter in shrinking of soil and respiration of soil biota. Shrinking [soil volume Vs, cm^3] and volume of calculated pores, both Vpo and Va, are linearly related with GWC, only Va in LS2 is not significant (Table 4). Regressions of Vs to time [d] were significant.

3.3. Changes of Soil Consistency

Formation of "mud-pies", hard stone—like concretions of dried soil, occurred in all four experiments. In MS1 and MS2 it begun approximately in 1/4 of the drying time, resulted from shrinking of soil, fissures appeared between soil and the walls of respirometric flasks, thus more of the soil surface was opened to the gas phase in the flasks. The rate of drying was also related to the amount of soil in the flasks. In the MS1 and MS2 series there is a higher soil layer, longer diffusion path and supposed humidity zonation. In the LS series drying and course of the experiments were faster (Table 1); the "mud pies" eventually formed just a narrow plate of dried soil.

Table 4. Soil volume (Vs), porosity (Vpo) and air filled porosity (Va), their % share in Vs, rates of change (β_1, r), probability of no change (p).

	LS1	LS2	MS1	MS2
Vs; start	9.79 (d8)	8.042	37.7 (d16)	32.735
N; se	9;	9; 0.33	13;	24; 0.509
Vs end	8.17 (d20)	3.035	25.4	15.987
se	x	0.222	x	0.245
Vs-GWC; β_1	1.57	3.74	7.45	11.73
−r	0.944	0.977	0.941	0.990
$p<$	0.025	0.001	0.001	0.001
%Vpo in Vs start	84.2	83.2	87.5	86.38
%Vpo in Vs end	62.3	56.5	50.1	58.32
Vpo-GWC; β_1	2.1775	3.15	10.696	13.252
−r	0.955	0.9947	0.995	0.8767
$p<$	0.001	0.001	0.001	0.001
%Va in Vs start	15.18	16.13	10.3	15.13
%Va in Vs end	36.20	41.79	45.01	68.75
Va-GWC; β_1	−1.13	−0.139	−3.572	−1.468
−r	0.983	0.347	0.964	0.877
p	<0.001	>0.05	<0.001	<0.001

Va did not change significantly in drying, ratio to Vs changed just due to decrease of Vs.

All these processes result in increasing access of air into the soil preparation, in a much better ventilation of the dried soils than is in the wet soils and contributed to formation of the "mud pies" during the drying experiments.

It's a common knowledge, that wet soils are soft, pastry; easily adopt form of container under only very slight pressure. When dried they become hard, stone-like, if slightly pressed together before drying; they resist pressures approximately like wall nuts, larger "mud pies" are more resistant than small ones. Sieved soil loosely distributed on a plate acquires in drying a sandy character; its total surface is much greater than that of the "mud pie".

To check the character of the hardiness of the mud pies we chopped off 5.0 g material from the top of the 25 g "mud pie" and took an aliquot of the same weight from the sandy like dried soil. Both were suspended in 50 ml of pure water and extracted for 2 hours and 30 minutes. Both soils looked hydrophobic at the beginning, after approx. 45 minutes they turned more polar. Some air bubbles were attached to dark, presumably light organic soil particles. The extracts were separated from the rests of soils by filtration through Macherey-Nagel filter paper MN 640 m. Both filtrates were yellowish brown, the "mud pie" extract was distinctly darker, but not brown, both very slightly colloidal. Most of the residues remained in the Erlenmeyer's with the heavy sand fraction, much less on the filters. We measured pH, conductivity and redox potential in both extracts: pH = 4.48 and 4.53, conductivity = 54 and 116 $\mu S \cdot cm^{-1}$ in "sand" and "mud pie" respectively, redox was E = 320 mV in both. Addition of 1 ml 0.2 molar solution of $BaCl_2$ to each of the extracts did not result in formation of any precipitate. We conclude, that the top of the "mud pie" probably contained more organic matter than "the sandy" fraction (more intense humic coloration of the extract), there was certainly no carbonate nor gypsum formed in the drying soils, could not been responsible for formation of the "mud pies". The two major components of our soil are the postglacial sand, which however remained unchanged in all the procedures as a separate and non-reactive fraction. The second major component is SOM.

We want suggest replacement of H-bonds (water bridges) in wet soil by some other weak interactions (non- covalent bonds) during drying, possibly by hydrophobic weak bonds. A slight pressure and greater mass of wet soil brings soil aggregates closer together and enables formation of other particle interactions when O-(water-) bridges are abolished in drying.

For considerations of bonding within soil aggregates a change in scale of the aggregates might be useful. Remind that there are 6E23 single molecules per each mole (Avogadro number), thus in 1 µg of SOM 6E14 molecules of organic compounds of mean molecular mass 1000 may be accommodated. Soil science does not seem to be generally ready for thinking in such scales (unfortunate Table 2.1 in Voroney [22] , containing organic molecules [10^{-9} m] larger than viruses, $\Psi > -1500$ kPa not only in Chowdhury [13] etc.).

3.4. Respiration

3.4.1. Relationships to GWC

Relationships of gas exchange (OCD, PCO_2) with time and GWC are more complex, the respective regression coefficients are lower compared to regressions of GWC and time; the overall tendency of respiration (OCD and PCO_2) is a decrease with progressing drying as water is a primary essential-even substrate-of life (Table 5, Figure 1). Data on respiration, both OCD and PCO_2, follow common trends: high values immediately at the begin ning of experiments in both LS, or after three or five days in the MS series. There is a more or less stable respiration in the central part of the course in MS1 and MS2 (Table 2 and Table 4). Basic statistics of respiration data show small variance, se's in all presented and vast majority of data not shown are within the point icons. The data in figures are of similar magnitude with those published by selected authors for forest soils considering problems in recalculation of area related data to mass related. Butler et al. [8] consider "rainfall as main driver of soil CO_2 efflux and its components".

Table 5. Regression coefficients (β_1, r) for respiration to GWC.

r/GWC	LS1	LS2	MS1 all	MS1 after d2	MS2 all	MS2 after d3
OCD β_1	−5.104	−3.612	−1.970	−2.560	−2.000	−2.950
r	0.833	0.956	0.837	0.936	0.712	0.949
PCO_2 β_1	−3.704	−0.877	−0.816	−0.978	−1.149	−1.351
r	0.735	0.953	0.732	0.765	0.842	0.881

For MS experiments coefficients are for all data and also for dates after d2 and d3 due to the delayed Birch effect.

3.4.2. CO_2 Storage

Distribution coefficient between gas and water for CO_2 is very close to one at the temperature of the experiments, storage capacity for CO_2 in soil water is therefore equal to GWC * dry mass of soil. It is the volume of water in respirometer (microcosm), about 20 ml in the first days of the MS experiments or 5 ml in the both LS; apparent production in two or three days.

Another estimate of CO_2 stored may result from the difference between AOCD and $APCO_2$ during the flush of respiration when soil is still wet, in

MS1 days 3 to12. Assuming real RQ = 1.0 it is 2.81 ml CO_2 stored in 28.8 ml H_2O, the storage capacity is thus used in less than 10%.

Similar data for the MS2 experiment is for days 5 to 12 and show 2.79 ml CO_2 stored in 12.7 ml water, storage capacity used in approximately 22%. Assuming real RQ = 0.7 result in 1.97 ml CO_2stored in MS1 and 1.95 ml CO_2 in MS2 respectively. Details of calculations are not shown.

The storage capacity for CO_2 was falling in experiments with progressing drying. Angert et al. [23] compared measured data of RQ (ARQ) with soil air analysis showing relatively low pCO_2. Their assumption on equilibrium between headspace (soil air) and wet acidic soil might not be fulfilled; there might be physical separation between the sections of the soil gas and the watered section, where CO_2 is primarily produced. Also a careful acidification and mixing of a tightly closed dilute solution of bicarbonate without gas phase need not result in production of gas bubbles. This may apply to a wet soil, where CO_2 is produced by microorganisms. ARQ data of Angert et al. [23] agree with our RQ in the respiration budget, there is also agreement on denial of carbonate chemistry as cause in the background of RQ < 1, we propose oxidation of methane and particularly ethanol produced in fermentation as hypothetic substrates in respiration with RQ < 1. We also seem to agree in accent on processes in studies. Sierra and Renault [14] conclude, that production is negligible compared to dissolved CO_2, this does not apply to acidic soils (pH < 4), where bicarbonate is practically missing. Therefore we did not consider bicarbonate storage, as estimates in our soil were always pH < 4.0. In acid soils however even a small volume stored may result in serious shifts in RQ.

3.4.3. Anoxia in Soil

To show gas exchange in strictly anoxic soil we placed 21 g fresh collected sieved wet soil (7.3 g dm) into each of four respirometers. Two of the respirometers were flushed with pure dinitrogen 50 $ml \cdot min^{-1}$, for 10 minutes; we registered O_2 consumption and CO_2 production [with and without NaOH trap inside]. The other pair of respirometers was a control to see regular soil respiration. There were 3.83 and 3.89 $\mu l \cdot g^{-1}(dm) \cdot h^{-1}$ respectively gas produced in the N_2 flushed respirometers without the hydroxide trap. There was however no gas production, nor consumption when hydroxide trap was placed in both respirometers; this also shows that the gas produced without the trap is absorbed by hydroxide, it is most probably CO_2 and not CH_4. RQ cannot be determined as shown in Introduction. The regular respiratory control in the second pair of respirometers provided 8.92 and 8.56 $\mu l\ O_2 \cdot g^{-1}(dm) \cdot h^{-1}$ respectively and 4.51 $\mu l\ CO_2 \cdot g^{-1}(dm) \cdot h^{-1}$ in both respirometers, this is consistent with data in Figure 1.

3.4.4. Respiration and Size of "Mud Pies"

Another pilot experiment aimed to differences of respiration in highly dried soils as "mud pies" or small sandy like aggregates. In the "mud pies" both OCD and PCO_2 are some 20% lower than in the sandy sieved small aggregates. Fast drops of simultaneous decrease of OCD and PCO_2 by approximately 0.6 - 0.8 $\mu l \cdot g^{-1} \cdot h^{-1}$ occurred at 0.63 GWC in "mud pies", at 0.43 in "sand". Drying in "mud pies" and associated decrease of respiration is slower in "mud pies". There is no contradiction to Figure 1.

3.4.5. "Birch Effect"

High respiration data occurring after changed conditions (sieving, watering...) are referred as the Birch effect [24] . In our experiments it occurred in LS1 on d1, in MS1 on d 5 - 14 and also at lower GWC = 1.41 starting value in LS2 and MS2 thus at a somewhat lower moisture throughout the drying experiment (Figure 1). The delayed respiration flush in both MS experiments resulted in distorsions of the relationships of respiration to GWC, this was overcome by skipping the d 1 or d 2 data from the relationships in MS2 and MS1 (Table 6). The delay of flush in both MS experiments was probably caused by pressing the greater amount of soil at start resulting in formation of solid "mud pies", while the small amounts of soil in both LS resulted in a thin layer of soil with minor limitation in gas traffic in the first days. The slightly smaller humidity in MS2 compared to MS1 was of minor importance in this range. We however feel that the impression of extraordinarily of high starting RQ's in MS2 is partly misleading, it comes rather from general characteristics of ratios; OCD and PCO_2 data are of similar magnitude (1.78 and 1.07 respectively), not strikingly different indeed.

High flushes of respiration following change of conditions are debated for last 50 years [24] . So far they were attributed to increased availability of respiration substrates from organisms which died in preceding dries, even if the drop in water content was small [25] or to repeated drying and rewetting (e.g. [26]) or simply to watering of soil (e.g. [13]). Even the detailed study by Unger [27] which includes interaction of soil and the vegetation using highly advanced methodology does not seem to be the last word in the debate. Flush took place in all four experiments in fresh and sieved samples, but not watered, of course sampling was at the very beginning (Figure 1, Table 5, first section of the course). Respiration decreased dramatically in the next readings. Therefore we want to propose a hypothesis not relying primarily on the experimental conditions, but rather on the organisms themselves: Before the collection of soil organisms were adapted to low respiration rates in their natural settings, supplemented by fermentation. Sieving (or in other papers published-water added) allows more air to enter the soil; previously accumulated CO_2 is released to air or absorbed in water. The overall tendency during the experiments is decrease of AOCD and $APCO_2$ with decreasing GWC, the data of the first part of experiments are however influenced by CO_2 produced and stored in soil water and Va available, at this stage it was not included in the changes of gas contents measured. Some loss of humidity (GWC) in sieving is highly probable. Soil biota may require time to adapt to available dioxygen, presumably by activation/reconstruction of respiratory mechanisms (e.g. enzyme complexes of oxidative phosphorylation pathway) which appear to be of small use in more enzyme copies in wet soils in natural settings with limited oxygen availability.

3.4.6. Summing up OCD and PCO_2 in the Process

Summed OCD and PCO_2 for the whole drying experiment showed in all four experiments RQ around 0.7 (Table 6). ΣRQ is somewhat higher in the faster drying experiments (LS), suggesting less hypoxia. This may support the hypothesis presented in introduction on a mixed status of organisms in wet soils, respiration contemporarily with fermentation and use of the fermentation products (e.g. ethanol) as substrates in respiration when O_2 becomes available.

Studies of some process which take into account all data, characterizing the variable soil condition during a process may provide more information, than short measurements or flashes of soil metabolism in ± constant conditions. Type of instrumentation may be of secondary importance only?modern methods of O_2 estimation are used only exceptionally [27] .

Nevertheless if there is O_2 consumption, anoxia is not in the total mass of soil, possibly in a limited area in center of the soil bloc.

3.4.7. Respiration Ends at Limiting GWC

There is a small, but significant peak of respiration in at the beginning of the decreasing phase at d 26 in MS1; in MS2 at d 26. A similar peak was registered by Thomson et al. [10] on the 5-th day of their 7-days experiment, which did not continue to a respiration stop. We propose to interpret as increased respiration due to increasing access of air to the drying soil, in our experiments shortly overrun but proceeding desiccation of the soil and confirmed by fast drops afterwards.

Table 6. Sums of OCD, COP, and ΣRQ in four experiments [ml].

	LS1	LS2	MS1	MS2
ΣOCD	7.67	15.14	42.53	33.13
ΣCOP	6.31	13.14	23.97	22.41
ΣRQ	0.82	0.87	0.56	0.68

ΣRQ is the ratio of the two sums. Duration of experiments in Table 3.

All four series (including the prematurely finished LS1) end up with RQ = 1 when GWC is approximately = 1 in LS experiments, but GWC ≈ 0.6 in MS (Figure 1).

CO_2 retained in the first section of experiments in water is completely aerated from the soil towards the end when soil is very dry; when respiration seizes at GWC ca 0.15, there is no measurable PCO_2, as is no OCD.

All drying experiments were terminated when GWC dropped to 0.15 and respiration stopped. The experiment LS1 was in fact finished prematurely, nevertheless its final RQ = 1. Also in a preliminary experiment not reported here (summer 2014) respiration (both OCD and PCO_2) seized at GWC = 0.15 and RQ = 1. The respiration limit at GWC = 0.15 need not be a universal value for many soils, Chowdhury et al. [13] reported respiration in Australian soils at recalculated GWC = 0.025.

4. CONCLUSIONS

There is a distinct peak of respiration rate at the beginning of the experiment (Birch effect); it occurred with some delay in the larger soil blocks. The effect was not associated with increased humidity, rather with increased availability of oxygen within the soil; therefore we suggest interpreting as reaction of the enzymatic systems of the soil organisms. The respiration rate decreased linearly with decreasing humidity of the soil; RQ dropped below 0.5 in both larger soil blocks for some time; reason is evidently in storage of CO_2 produced in soil water. Rate of respiration registered is also related to consistency of soil; hard "mud pies" respire some 20% less than the small aggregates of sandy character.

Respiration ceased when GWC dropped down to 0.15. RQ was increasing in the second part of the experiments; a few days before the end it reached 1.00 and remained so till the last moment. We think it comes from CO_2 stored in the earlier parts of the drying process which is aerated when amount of water within the soil decreased and air in soil got higher. ΣRQ for the whole budget of the experiments is about 0.7; it is somewhat higher for the smaller soil items (LS) suggesting less hypoxia and may be consistent with partial use of fermentation products like ethanol (RQ = 0.57).

CITE THIS PAPER

Zofia Fischer, Pavel Blažka (2015) Soil Respiration in Drying of an Organic Soil. *Open Journal of Soil Science*,**05**,181-192.

REFERENCES

1. Davidson, E.A., Savage, K., Verchot, L.V. and Navarro, R. (2002) Minimizing Artifacts and Biases in Chamber-Based Measurements of Soil Respiration. Agricultural and Forest Meteorology, 113, 21-37.
2. Luo, Y. and Zhou, X. (2010) Soil Respiration and the Environment. Academic Press, Burlington.
3. Luan, J., Liu, S., Zhu, X., Wang, J. and Liu, K. (2012). Roles of Biotic and Abiotic Variables in Determining Spatial Variation of Soil Respiration in Secondary Oak and Planted Pine Forests. Soil Biology and Biochemistry, 44, 143-150.
4. Ilstedt, U., Nordgren, A. and Malmer, A. (2000) Optimum Soil Water for Soil Respiration before and after Amendment with Glucose in Humid Tropical Acrisols and a Boreal Mor Layer. Soil Biology & Biochemistry, 32, 1591-1599.
5. Campbell, N.A. (1993) Biology. 3rd Edition, Benjamin/Cummings, Francis Group, Redwood City.
6. Alberts, B. and Johnson, A. (2002) The Molecular Biology of the Cell. Garland, Taylor and Francis Group, New York.
7. McGill, W.B. (2007) The Physiology and Biochemistry of Soil Organisms. In: Paul, E.A., Ed., Soil Microbiology Ecology, and Biochemistry, 3rd Edition, Academic Press, Burlington.
8. Butler, A., Meir, P., Saiz, G., Maracahipes, L., Marimon, B.S. and Grace, J. (2012) Annual Variation in Soil Respiration and Its Component Parts in Two Structurally Contrasting Woody Savannas in Central Brazil. Plant Soil, 352, 129-142.
9. van Suchtelen, H. (1910) Uber die Messung der Lebenstatigkeit der aerobiotischen Bakterienim Boden durch die Kohlensaureproduktion. Centrablattfur Bakteriologie, 28, 46-89. Cited from: Lundegardh, H. (1926) Carbon Dioxide Evolution of Soil and Crop Growth. Soil Science, 23, 417-453.
10. Thomson, B.C., Ostle, N.J., McNamara, N.P., Whiteley, A.S. and Griffith, R.I. (2010) Effects of Sieving, Drying and Rewetting upon Soil Bacterial Community Structure and Respiration Rates. Journal of Microbiological Methods, 83, 69-73.
11. Gaumont-Guay, D.T., Black, T.A., Tim, J., Griffis, T.J., Alan, G., Barr, A.G., Jassal, R.S. and Nesic, Z. (2006) Interpreting the Dependence of Soil Respiration on Soil Temperature and Water Content in a Boreal Aspen Stand. Agricultural and Forest Meteorology, 140, 220-235.
12. Walker, T.S., Harsh, P.B., Grotewold, E. and Vivanco, J.M. (2003) Update on Root Exudation and Rhizosphere Biology. Plant Physiology, 132, 44-51.
13. Chowdhury, N., Burns, R.G. and Marschner, P. (2011) Recovery of Soil Respiration after Drying. Plant Soil, 348, 269-279.

14. Sierra, J. and Renault, P. (1995) Oxygen Consumption by Soil Microorganisms as Affected by Oxygen and Carbon Dioxide Levels. Applied Soil Ecology, 2, 175-184.
15. Jenkins, M.E. and Adams, M.A. (2011) Respiratory Quotients and Q10 of Soil Respiration in Sub-Alpine Australia Reflect Influences of Vegetation Types. Soil Biology & Biochemistry, 43, 1266-1274.
16. Araujo, A.S.F., Cesarz, S., Leite, L.F.C., Borges, C.D., Tsai, S.M. and Eisenhauer, N. (2013) Soil Microbial Properties and Temporal Stability in Degraded and Restored Lands of Northeast Brazil. Soil Biology & Biochemistry, 55, 175-181.
17. Lamy, E., Tran, T.C., Mottelet, S., Pauss, A. and Schoefs, O. (2013) Relationships of Respiratory Quotient to Microbial Biomass and Hydrocarbon Contaminant Degradation during Soil Bioremediation. International Biodeterioration & Biodegradation, 83, 85-91.
18. Blazka, P. and Fischer, Z. (2014) Moisture, Water Holding, Drying and Wetting in Forest Soils. Open Journal of Soil Science, 4, 174-184.
19. Carter, M.R. and Gregorich, E.G., Eds. (2008) Soil Sampling and Methods of Analysis. CRC Press, Taylor & Francis Group, Boca Raton.
20. Grace, C., Hart, M. and Brookes P.C. (2006) Laboratory Manual of the Soil Microbial Biomass Group. Rothamsted Research.
21. Priha, O. and Smolander, A. (2003) Short-Term Uptake of 15NH4 into Soil Microbes and Seedlings of Pine, Spruce and Birch in Potted Soils. Biology and Fertility of Soils, 37, 324-327.
22. Voroney, R.P. (2007) The Soil Habitat. In: Paul, E.A., Ed., Soil Microbiology Ecology and Biochemistry, 3rd Edition, Elsevier, Academic Press, London, 25-49.
23. Angert, A., Yakir, D., Rodeghiero, M., Preisler, Y., Davidson, E.A. and Weiner, T. (2014) Using O2 to Study the Relationships between Soil CO2 Efflux and Soil Respiration. Biogeosciences Discussions, 11, 12039-12068.
24. Birch, H.F. (1964) Mineralisation of Plant Nitrogen Following Alternate Wet and Dry Conditions. Plant Soil, 20, 43-49.
25. Fischer, T. (2009) Substantial Rewetting Phenomena on Soil Respiration Can Be Observed at Low Water Availability. Soil Biology & Biochemistry, 41, 1577-1579.
26. Shi, A. and Marschner, P. (2014) Drying and Rewetting Frequency Influences Cumulative Respiration and Its Distribution over Time in Two Soils with Contrasting Management. Soil Biology & Biochemistry, 72, 172-179
27. Unger, S., Maguas, C., Pereira, J.S., David, T.S. and Werner, C. (2012) Interpreting Post-Drought Rewetting Effects on Soil and Ecosystem Carbon Dynamics in a Mediterranean Oak Savannah. Agricultural and Forest Meteorology, 154-155, 9-18.

CHAPTER 10

Role of Inorganic and Organic Fractions in Animal Manure Compost in Lead Immobilization and Microbial Activity in Soil

Masahiko Katoh,[1,2] Wataru Kitahara,[3,4] and Takeshi Sato[2]

[1]*Department of Agricultural Chemistry, School of Agriculture, Meiji University, 1-1-1 Higashi-Mita, Tama-ku, Kanagawa 214-8571, Japan*
[2]*Department of Civil Engineering, Faculty of Engineering, Gifu University, 1-1 Yanagido, Gifu 501-1193, Japan*
[3]*Department of Civil Engineering, Graduate School of Engineering, Gifu University, 1-1 Yanagido, Gifu 501-1193, Japan*
[4]*In Situ Solutions Co., Ltd., 2-5-2 Kandasudamachi, Chiyoda-ku, Tokyo 101-0041, Japan*

ABSTRACT

This study aimed to identify how the ratio of inorganic-to-organic components in animal manure compost (AMC) affected both lead immobilization and microbial activity in lead-contaminated soil. When AMC containing 50% or more inorganic fraction with high phosphorous content was applied to contaminated soil, the amounts of water-soluble lead in it were suppressed by over 88% from the values in the soil without compost. The residual fraction under sequential extraction increased with the inorganic fraction in the AMC; however, in those AMCs, the levels of microbial enzyme activity were the same or less than those in the control soil. The application of AMC containing 25% inorganic fraction could alter the lead phases to be more insoluble while improving microbial enzyme activities; however, no suppression of the level of water-soluble lead existed during the first 30 days. These results indicate that compost containing an inorganic component of 50% or more with high phosphorus content is suitable for immobilizing lead; however, in the case where low precipitation is expected for a month, AMC containing 25% inorganic component could be used to both immobilize lead and restore microbial activity.

1. INTRODUCTION

Lead is one of the most common and harmful heavy-metal soil contaminants worldwide, particularly near mines and shooting ranges. Lead contamination in the soil of such sites poses a risk to human and animal health as well as plant growth. Thus, its mobility and bioavailability should be reduced by appropriate treatments. In addition, the rehabilitation of soil ecosystems that have undergone destruction by lead contamination should be accomplished through the remediation of such soil. Since contamination in these sites is extensive and asset values are extremely low, chemical immobilization capable of transforming lead into less soluble phases is a cost-effective remediational approach. Thus, various immobilization materials have been studied and developed [1–7].

Animal manure compost (AMC), which is the most abundantly found organic waste material in Japan, can immobilize heavy metals and improve plant growth and microbial activity [8–16] by supplying nutrients to plant and soil biota. Therefore, in addition to lead immobilization, the application of AMC to lead-contaminated soil can help rehabilitate soil ecosystems, which inorganic immobilization materials cannot [17]. Numerous studies have investigated the effectiveness of AMC for lead immobilization [18, 19]; however, consistent conclusions on the mechanisms behind this process have not been obtained as such immobilization that greatly depends upon the AMC type, that is, AMC's physicochemical properties. In particular, the presence of both inorganic and organic components in AMC makes it complicated to understand lead immobilization.

Lead immobilization by AMC can be separated into indirect and direct mechanisms. The typical indirect mechanisms of lead immobilization are explained by the pH increase owing to the alkalinity of the inorganic component in AMC; this pH increase can promote the precipitation of lead carbonate and hydroxide minerals, resulting in a reduction of lead mobility and bioavailability [1, 20]. The direct mechanisms of lead immobilization by AMC have been considered to be the reaction of the lead with the inorganic and organic components in AMC. Phosphorus, sulfate, and iron in the inorganic component and humic substances in the organic component in AMC seem to be the sources responsible for the reaction with lead [10, 18, 21–23]. Katoh et al. [24] applied a specific method for the fractionation of the inorganic and acid-insoluble organic components from AMC to elucidate their separate contributions to lead immobilization. They indicated that the inorganic component in AMC could immobilize lead more effectively than its organic components [24]. These results imply that the inorganic component in AMC has a crucial role in lead immobilization, and AMC with a higher inorganic content is more suitable. However, to rehabilitate microbial activity, the organic component in AMC is also required since it supplies sufficient nutrients to soil microorganisms [17]. This component, however, may negatively affect lead mobility; water-soluble organic matter in AMC reacts and forms complexes with lead ions, resulting in enhancements in lead mobility. Therefore, a suitable ratio of inorganic and organic components in AMC should be clarified to reduce the mobility and bioavailability of lead and enhance the microbial activities in the soil. AMC has a wide range of the inorganic-to-organic component ratios [25]; however, to our knowledge, the optimal ratio of inorganic-to-organic components in AMC for lead immobilization and the rehabilitation of microbial activity have not been studied.

We investigated the mobility and bioavailability of the lead, lead phases, and microbial activities in soil amended with the AMCs of various inorganic-to-organic component ratios. The inorganic fraction was derived from swine manure compost and the acid-insoluble organic fraction was derived from cattle manure compost. These fractions can immobilize lead more effectively than other AMCs owing to the high content of phosphorus and mature organic matter in the inorganic and acid-insoluble organic fractions, respectively [24]. We aimed to identify how the ratio of inorganic-to-organic components in AMC affected the mobility and bioavailability of lead, lead phase, and microbial activity. On the basis of the obtained results, we discuss the optimal ratio of the inorganic and organic components in AMC to immobilize lead and rehabilitate microbial activity in soil.

2. MATERIALS AND METHODS

2.1. Preparation of Soil

The lead-contaminated soil used herein was collected from depths of 5–15 cm at a shooting range located at 35°28′6″N and 137°29′2″E in Gifu, Japan. The soil sample was air-dried, passed through a 2 mm sieve, and was used for chemical analysis and incubation tests. The selected physicochemical properties of the soil used are shown in Table 1. The soil had a sandy loam texture. The total lead content of the soil was 4.40 g kg^{-1} and the soil pH was 7.3.

Table 1: Physicochemical properties of contaminated soil used in this study (on the basis of air-dried weight).

Sand (%)	Silt (%)	Clay (%)	pH	TC (g kg^{-1})	TN (g kg^{-1})	WSOC[a] (mg kg^{-1})	WS-Pb[b] (mg kg^{-1})	Total Pb (g kg^{-1})	Total P (g kg^{-1})	Total Al (g kg^{-1})	Total Fe (g kg^{-1})	Amorphous Fe (g kg^{-1})
81.5	9.5	9.0	7.3	6.0	0.0	139	35	4.40	0.40	13.1	33.8	1.3

[a] Water-soluble organic carbon.
[b] Water-soluble lead.

2.2. Preparation of Animal Manure Compost

Commercial swine and cattle manure composts were used herein to obtain the inorganic and acid-insoluble organic fractions, respectively. We selected these composts since each fraction of the compost can immobilize lead more effectively than other composts [24]. The fractionation followed the method described by Katoh et al. [24]. In brief, the swine manure compost was combusted at 600°C for 2 h; the residue after combustion was used as the inorganic fraction. The cattle manure compost was subjected to extraction with 1 M HCl for 1 h to remove almost all inorganic components, and the residue after extraction was used as the acid-insoluble organic fraction. The pH of each fraction was adjusted to 7 to match that of the soil. Table 2 shows the selected chemical properties of each fraction [24]. The calcium, magnesium, potassium, iron, and phosphorus contents in the inorganic fraction were 152, 44, 53, 7, and 131 mg g^{-1}, respectively; moreover, the total, acid-detergent fiber, humic acid, and fulvic acid carbon contents in the acid-insoluble organic fraction were 452, 344, 25, and 52 mg g^{-1}, respectively [24]. After pH adjustment, the fractions were well mixed at the inorganic-to-acid-insoluble organic fraction ratios of 100/0, 75/25, 50/50, 25/75, and 0/100; the mixed samples were used as the AMCs herein (hereafter referred to AMC with inorganic-to-organic ratios of 100/0, 75/25, 50/50, 25/75, and 0/100).

2.3. Soil-Incubation Experiment

The contaminated soil was mixed with each AMC sample at 10 wt% and incubated at room temperature (25°C) for 184 days. Soil without AMC was also prepared as a control. Three replicates were prepared for each treatment. The water content in the soil was maintained at 60% of its maximum water-holding capacity during the incubation period. Soil samples were collected at 0, 7, 30, 90, and 184 days; the samples were freeze-dried and analyzed to determine the levels of water-soluble lead and water-soluble organic carbon. The soil sampled on day 184 was analyzed to assess lead phases by sequential extraction and microbial enzyme activities. Herein, the level of water-soluble lead was employed as an indicator of lead mobility and bioavailability in the soil, because its water-soluble form is easily mobile and utilized by plant and soil biota.

Table 2: Chemical properties of inorganic and acid-insoluble organic fractions used in this study (on the basis of air-dried weight, mg g^{-1}) [24].

Characteristic	Inorganic fraction	Acid-insoluble organic fraction
Total calcium	152	ND*1
Total magnesium	44	ND
Total potassium	53	ND
Total iron	7	ND
Total phosphorus	131	ND
Total carbon	ND	452
ADF-C*2	ND	344
Humic acid carbon	ND	25
Fulvic acid carbon	ND	52

*1 Not determined.
*2 Acid detergent fiber carbon.

Another soil-incubation experiment was conducted to evaluate CO_2 emission from the soil with the AMCs with inorganic-to-organic ratios of 100/0, 50/50, and 0/100. A 100 mL polypropylene bottle including the soil with and without the AMC, and two glass beakers—one having 0.5 M NaOH and the other having 0.5 M H_2SO_4—were placed into a 5 L glass bottle and incubated at room temperature (25°C). A 5 L bottle without soil was also prepared. Three replicates were prepared for each treatment. At days 7, 14, 30, 60, 120, and 194, the glass beaker containing 0.5 M NaOH was replaced with a fresh 0.5 M NaOH-containing beaker. The collected NaOH was titrated with 0.2 M HCl and the CO_2 emission was calculated from the following equation:

$$C_s = \frac{V \times F - C_b}{S}, \tag{1}$$

where is the amount of CO_2 emission (mol g-soil^{-1}), is the amount of 0.2 M HCl titrated (mL), is the factor of 0.2 M HCl (mol-CO_2 mL^{-1}), is the amount of CO_2 emission in the blank (mol), and is the amount of soil (10.0 g). The CO_2 emission derived from AMC was calculated by subtracting the emission by the AMC-treated soil samples from that by the control soil sample.

2.4. Analytical Methods

The soil's pH value was measured in ultra-pure water using a pH meter (MM-60M, DKK-TOA Co., Japan). Soil texture was determined using the hydrometer method [26]. The total carbon and nitrogen contents in the soil were determined using a carbon, hydrogen, and nitrogen elemental analyzer (MT-6; Yanaco New Science Inc., Japan).

The water-soluble lead and organic carbon in soil were extracted using ultra-pure water (1 : 10 soil : solution ratio) and analyzed using an inductively coupled plasma-atomic emission spectrometer (ICP-AES; ULTIMA2; HORIBA Ltd., Japan) and a total organic carbon analyzer (TOC-; Shimadzu Co., Japan). The amorphous iron was extracted using Shuman's method [27]. The total lead, phosphorus, aluminum, and iron contents in the soil were determined by acid digestion with HNO_3 and HCl using a microwave. A sequential extraction procedure was performed on the soil samples in accordance with the procedure described by Tessier et al. [28]. In brief, each fraction was extracted with a 1 M $MgCl_2$ solution (exchangeable fraction), 1 M sodium acetate solution at pH 5 (carbonate fraction), 0.04 M NH_2OH-HCl in 25% (v/v) HOAc solution in a water bath at 95°C with occasional agitation (Fe/Mn oxide fraction), 0.02 M HNO_3 solution, 5 mL 30% H_2O_2 solution in a water bath at 85°C with occasional agitation (organic fraction), and 5 mL of HNO_3 and 2 mL of HCl using a microwave oven (residual fraction). All of the extracted or digested solutions were passed through a 0.45 μm filter and analyzed to determine the elemental concentration using ICP-AES.

Three different microbial enzyme activities—dehydrogenase, urease, and saccharase—were measured in the soil sample. The dehydrogenase activity was determined in accordance with the method described by Tabatabai [29] using triphenyltetrazolium chloride. The triphenylformazan produced was measured using spectrophotometry at 485 nm. The urease activity was monitored using the method described by Kandeler and Gerber [30] by measuring the produced after the incubation of soil with urea. The saccharase activity was determined in accordance with the method described by Frankenberger and Johanson [31].

2.5. Statistical Analysis

Statistical analyses were performed using JMP Ver. 8.0.2 (SAS Institute Inc.). An analysis of variance was used to measure microbial enzyme activities. The differences between the mean values were determined using Tukey's honestly significant difference test at a 95% confidence level.

3. RESULTS

3.1. Soil pH, Water-Soluble Lead, and Organic Carbon in Soil with AMC

The pH in the control soil was not significantly changed and ranged from 7.5 to 7.7 throughout the incubation period (Table 3). Moreover, the soil pH levels in all the AMCs were not significantly different from that in the control, indicating that the soil pH did not affect the enhancement or reduction in the lead mobility and bioavailability in the soil with AMC.

Table 3: pH in soils amended with simulated compost during incubation period ().

Compost	Days of incubation				
(Inorganic/organic)	0	7	30	90	184
100/0	7.7 ± 0.0	7.5 ± 0.0	7.6 ± 0.1	7.8 ± 0.0	7.8 ± 0.0
75/25	7.7 ± 0.0	7.5 ± 0.0	7.7 ± 0.0	7.8 ± 0.0	7.8 ± 0.1
50/50	7.6 ± 0.0	7.6 ± 0.0	7.7 ± 0.1	7.8 ± 0.0	7.6 ± 0.0
25/75	7.4 ± 0.0	7.4 ± 0.0	7.8 ± 0.0	7.8 ± 0.0	7.6 ± 0.0
0/100	7.2 ± 0.0	7.3 ± 0.0	7.6 ± 0.1	7.8 ± 0.0	7.5 ± 0.2
Control	7.5 ± 0.1	7.6 ± 0.0	7.7 ± 0.0	7.5 ± 0.1	7.5 ± 0.2

Figure 1(a) shows the amount of water-soluble lead in the soil with AMC. The amount of water-soluble lead in the control soil ranged from 35 to 44 mg kg^{-1} throughout the incubation period. The amount of water-soluble lead in the soil treated with AMC with an inorganic-to-organic ratio of 0/100 was higher than that in the control soil throughout the incubation period. The amount of water-soluble lead in the soil treated by AMC with an inorganic-to-organic ratio of 25/75 was equal to or at a higher level than that in the control soil before day 30, but it decreased to 36% of that in the control soil on day 184. The amounts of water-soluble lead in the soil samples treated by AMC with inorganic-to-organic ratios of 50/50, 75/25, and 100/0 were 3.4%–12%, 0.1%–0.8%, and 0.1%–0.2%, respectively, of the value in the control soil. The amount of water-soluble organic carbon in the soil is shown in Figure 1(b). At the beginning of the incubation period, the amounts of water-soluble organic carbon in the soils treated with different AMC compositions were arranged in the following order: inorganic/organic = 0/100 > inorganic/organic = 25/75 > inorganic/organic = 50/50 > inorganic/organic = 75/25 > control > inorganic/organic = 100/0. The amounts of water-soluble organic carbon in the soil treated by AMC with inorganic-to-organic ratios of 50/50, 75/25, and 100/0 remained at their initial values during the incubation period. The amounts of water-soluble organic carbon in the soil treated by AMC with inorganic-to-organic ratios of 0/100 and 25/75 had rapidly decreased between days 0 and 7, after which gradual decreases were observed. The decreases in water-soluble organic carbon between days 0 and 184 were 1,071 and 615 mg kg^{-1} under inorganic-to-organic ratios of 0/100 and 25/75, respectively.

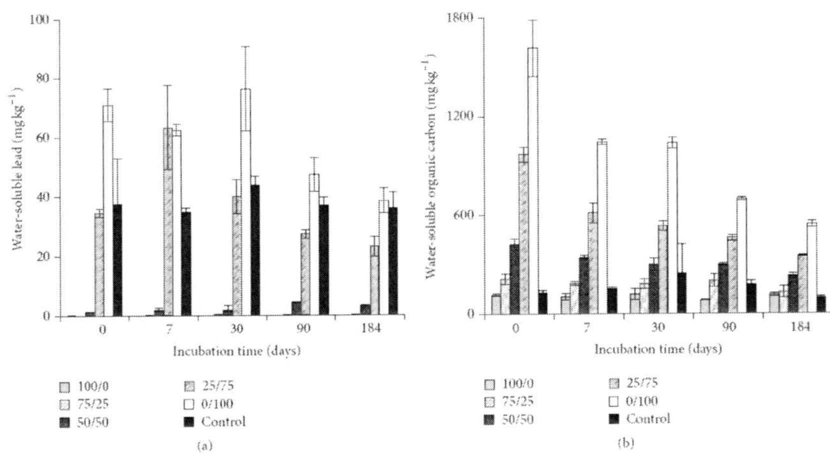

Figure 1: Water-soluble lead (a) and organic carbon (b) in soil amended with AMC of inorganic/organic = 100/0, 75/25, 50/50, 25/75, and 0/100 and without AMC (control). Vertical bars indicate the standard error.

3.2. CO_2 Emission from Soil Amended with AMC

The cumulated CO_2 emission derived from AMC is shown in Figure 2. The level of CO_2 emission from the AMC with an inorganic-to-organic ratio of 0/100 increased during the early stage of incubation and then remained at approximately 500 mg-C kg^{-1}. The CO_2 emission trends from the AMC with inorganic-to-organic ratios of 50/50 and 100/0 were similar; they slightly increased at the beginning of incubation and then remained at approximately 100 mg-C kg^{-1}.

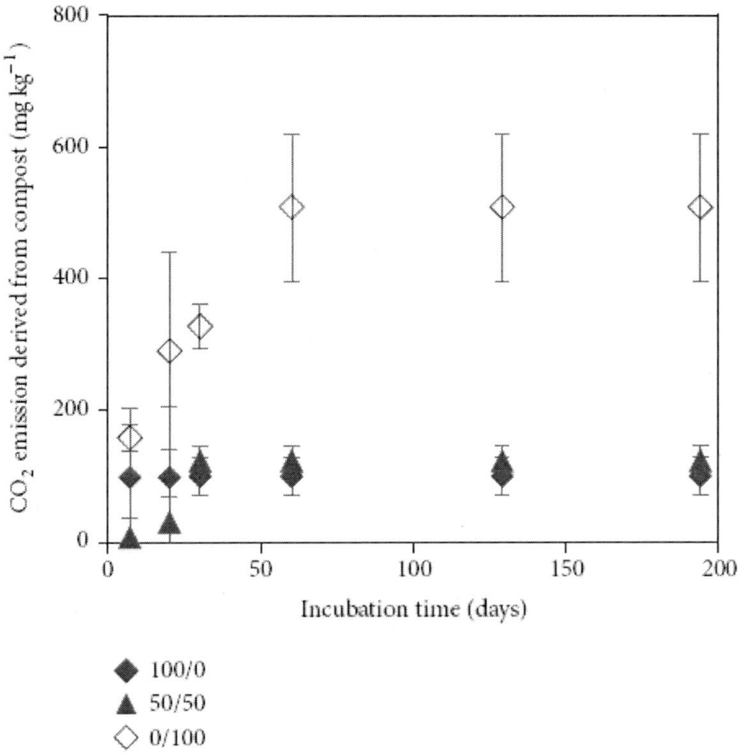

Figure 2: Cumulative CO_2 emission derived from AMC with inorganic/organic = 100/0, 50/50, and 0/100. Vertical bars indicate the standard error.

3.3. Lead Phases by Sequential Extraction in Lead-Contaminated Soil

Figure 3 shows the lead fraction results under sequential extraction. The average recovery, which was defined herein as the ratio of the sum total level of each fraction (Figure 3) to the total level of lead in the soil (Table 1), was %. The composition of the control soil was as follows: 12.8% exchangeable fraction, 51.2% carbonate fraction, 20.7% Fe/Mn oxide fraction, 9.3% organic fraction, and 6.0% residual fraction. The percentage of organic fraction by sequential extraction in the soil treated by AMC with an inorganic-to-organic ratio of 0/100 increased compared with that in the control soil, whereas that of carbonate decreased. The addition of AMC with a higher ratio of inorganic fraction resulted in a greater enhancement in the residual percentage. Furthermore, in soils with the AMCs containing an inorganic fraction of 25% or more, the percentage of organic fraction was at the same level as that in the control soil.

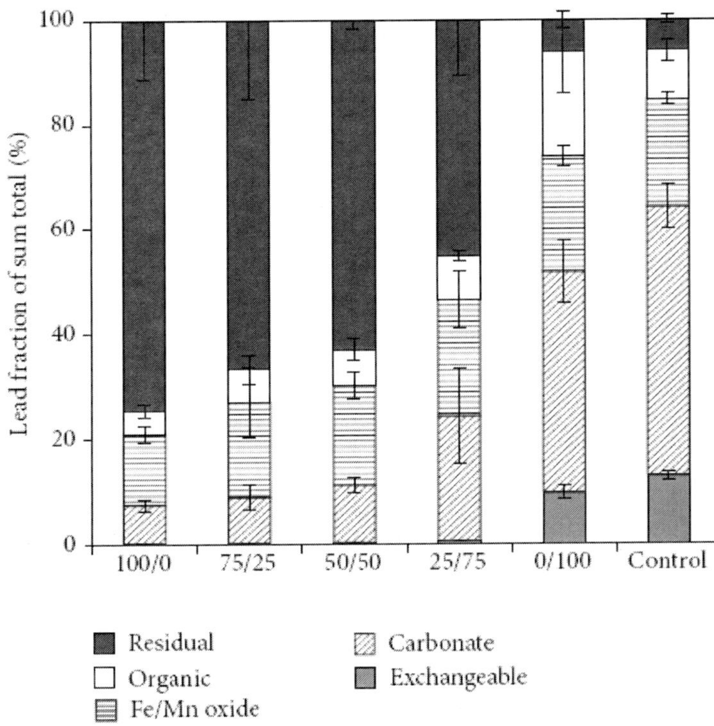

Figure 3: Sequential extraction of lead from soil amended with AMC of inorganic/organic = 100/0, 75/25, 50/50, 25/75, and 0/100 and without AMC (control) after 184 days of incubation. Vertical bars indicate the standard error.

3.4. Microbial Enzyme Activities in Lead-Contaminated Soil

The dehydrogenase and urease activities in the soil treated by AMC with inorganic-to-organic ratios of 0/100 and 25/75, respectively, were significantly higher than that in the control soil; however, those in the soil treated by AMC with inorganic-to-organic ratios of 75/25 and 100/0 were significantly lower or on the same level (Figure 4).

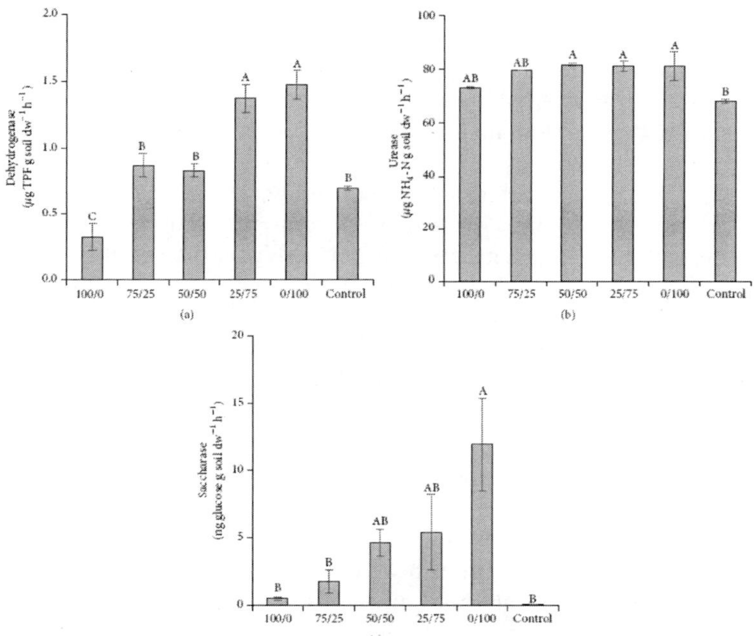

Figure 4: Enzyme activities of dehydrogenase (a), urease (b), and saccharase (c) in soil amended with AMC of inorganic/organic = 100/0, 75/25, 50/50, 25/75, and 0/100 and without AMC (control) after 184 days of incubation. Vertical bars indicate the standard error. Different letters indicate significant difference at .

4. DISCUSSION

4.1. Role of Organic Component in Animal Manure Compost Lead Immobilization and Rehabilitation of Microbial Activity

The water-soluble lead level in the soil treated by AMC with an inorganic-to-organic ratio of 0/100 did not become lower than that in the control soil during the incubation period (Figure 1(a)), whereas the lead phases in the soil by sequential extraction were altered to be more insoluble; the organic fraction was increased (Figure 3). Moreover, at the early stage of incubation, the water-soluble lead level remained at a higher level in the soil treated by AMC with an inorganic-to-organic ratio of 25/75. However, the water-soluble lead level decreased as the incubation time increased in soil with both composts. These observations would be explained by the higher level of water-soluble organic matter and its sorption and decomposition in the soil thereby increasing the incubation time. The water-soluble organic matter can easily form complexes with lead ions, and its complexes could enhance the lead mobility in the soil [17, 32, 33]. Therefore, at the early stage of incubation, the high level of water-soluble organic carbon would result in an enhancement of the lead mobility in the soil treated by AMC with inorganic-to-organic ratios of 0/100 and 25/75. The water-soluble organic matter was derived from the compost containing humic and fulvic acids with relatively low molecular weight [34, 35]. These organic compounds are also readily decomposable and sorbed on the soil surface [36, 37]. According to the result for CO_2 emission (Figure 2), the amount of CO_2 emission was lower than that of water-soluble organic carbon-decrease from day 0 to day 184, suggesting that some of the water-soluble organic matter derived from the compost was decomposed and some of it was sorbed in the soil with the increase in the incubation period. Thus, the level of water-soluble lead would decrease with that water-soluble organic carbon in the soil treated by AMC with inorganic-to-organic ratios of 0/100 and 25/75. It has been known that the compost amendment induced the very short-term leaching pulses of lead after the application [38], and lead was redistributed to the soil component by the decomposition of dissolved organic matter [39]. These results suggest that the organic component in the AMC does not considerably contribute to the immobilization of lead and suppression of lead mobility and bioavailability. This was comparable with the results of Schwab et al. [19] and Levonmäki et al. [40], who reported that water-soluble organic matter enhanced the lead mobility by the formation of complexation. However, all of the microbial enzyme activities measured herein increased in the soil treated by AMC with an inorganic-to-organic ratio of 0/100, despite the fact that the level of water-soluble lead was higher. The dehydrogenase activity, which is an indicator of the overall microbial activity [8, 9, 41–43], in the soil treated by AMC containing 50% or more inorganic fraction was the same level as that in the control soil, suggesting that the inorganic component in the AMC did not significantly contribute to the enhancement in the microbial activity, despite the fact that the lead mobility and bioavailability are greatly decreased by its addition. Farrell et al. [44] also showed that the microbial enzyme activities in the soil amended with the compost became higher than that in the inorganic material. The water-soluble organic matter derived from the compost is utilized as a nutrient source by microorganisms; thus, the addition of AMC with high organic matter content would induce the rehabilitation of microbial activity owing to the large amount of readily decomposable organic matter.

4.2. Role of the Inorganic Component in Animal Manure Compost on Lead Immobilization

The addition of AMC containing 50% or more inorganic fraction reduced the water-soluble lead by over 88% during the incubation period (Figure 1(a)). Moreover, the addition of AMC containing 25% or more inorganic fraction could alter lead phases to be more insoluble. These results indicate that AMC containing 50% or more inorganic fraction could immobilize lead and reduce the lead bioavailability in the soil. The inorganic fraction used herein contained a considerable amount of phosphorus, which results in the precipitation of lead phosphate minerals such as pyromorphite [24]. Pyromorphite is thermodynamically stable with a solubility product of [45]. Therefore, the phosphorus in the AMC's inorganic component was responsible for the increase in the percentage of the residual fraction and decrease in the water-soluble lead level. This is supported by the results of Walker et al. [10], Liu et al. [11], and Clemente et al. [18], who suggested that the immobilization of lead and cadmium would be due to the precipitation of insoluble phosphate salts. The percentage of Fe/Mn oxide and organic fractions by sequential extraction in the soil with AMC containing 25% or more inorganic fraction was lesser than or approximately equal to that in the control soil, although the residual fraction increased with increase in the inorganic fraction ratio in AMC. Moreover, in the soil treated by AMC containing 50% or less inorganic fraction, the level of water-soluble organic carbon was higher than that in the control soil. These results suggest that the phosphorus in the inorganic component of AMC could immobilize lead-precipitating lead phosphate minerals even if the AMC contained more organic than inorganic components. This may be because the magnitude of lead immobilization by the inorganic component of AMC suppressed the magnitude of the facilitation of lead mobility by the organic component in AMC.

The level of water-soluble lead in the soil treated by AMC with an inorganic-to-organic ratio of 25/75 was higher than that in the control soil during the first 30 days of incubation, whereas the lead phases were altered to be more insoluble. This was attributed to incomplete immobilization. Scheckel et al. [46] demonstrated that not all lead in soil could be immobilized by phosphorus material, according to extended X-ray-absorption fine-structure analysis. The percentages of exchangeable and carbonate fractions by sequential extraction in the soil treated by AMC with an inorganic-to-organic ratio of 25/75 were higher than that in the soil treated by AMC that was 50% or more inorganic fraction. The levels of water-soluble organic carbon in the soil treated by AMC with an inorganic-to-organic ratio 25/75 were also higher, particularly during the first 30 days of incubation. The higher amount of water-soluble organic matter formed complexes with lead dissolved from the exchangeable and carbonate fractions, resulting in an enhancement in the level of water-soluble lead in the early stage of incubation. With the increasing incubation time, water-soluble organic matter was decomposed and the level of water-soluble lead decreased below that in the control soil.

4.3. Optimal Ratio of Inorganic and Organic Components in Animal Manure Compost for Lead Immobilization and the Rehabilitation of Microbial Activity

Various inorganic percentages are used in the AMC, ranging from 7.3 to 82.8% [25]. The ranges of the inorganic fraction of AMC in this study fall within this range. On the basis of this study's findings, an inorganic component with a high phosphorus content of 50% or more is required to alter lead phases to be more insoluble and reduce the water-soluble lead level to less than that in the soil without compost, although the microbial activities are not enhanced. The importance of an inorganic component for the immobilization of heavy metals including lead in soil is pointed out by other researchers [21, 47, 48]. In contrast, 25% inorganic AMC can alter lead phases to be more insoluble and simultaneously enhance microbial activities, whereas the water-soluble lead level is higher than that in the soil without compost during the first 30 days after the application. This is consistent with the result of Katoh et al. [17], who indicated that the application of 25% inorganic swine manure compost could alter lead phases to be more insoluble and improve plant growth and microbial enzyme activities; however, the level of water-soluble lead was higher than that in the cattle manure compost during the early stage of incubation (90 days after application). These findings suggest that AMCs cannot immobilize both lead and restore microbial activity. Therefore, to immobilize lead and suppress the lead mobility and bioavailability using AMC, AMC that is 50% inorganic or more and contains a large phosphorus component should be applied to contaminated sites. However, in the case of low precipitation over a month, for example, during a dry season, 25% inorganic AMC should be used to both immobilize lead and restore microbial activity.

5. CONCLUSION

The amount of water-soluble lead in the soil treated by 100% organic AMC remained higher than that in the control soil throughout the 184-day incubation period, although it tended to decrease with increasing incubation time. The amount of water-soluble lead in the soil treated by AMC with an inorganic-to-organic ratio of 25/75 was higher than that in the soil without compost during the early stage of incubation, but it reached a lower level than that in the soil without compost after 90 days. The amounts of water-soluble lead in the soil treated by AMC containing 50% or more inorganic fraction were suppressed by over 88% from the value in the soil without compost and remained low throughout the incubation period. The reduction in the level of water-soluble lead in the soil treated by AMC with inorganic-to-organic ratios of 0/100 and 25/75 would be explained by the decomposition of water-soluble organic matter. The ratio of the residual fraction after sequential extraction was enhanced and the readily soluble lead fractions (exchangeable and carbonate fractions) were reduced in soils treated by AMC containing 25% or more inorganic fraction. In soil treated with fully organic AMC, the readily soluble lead fraction in sequential extraction did not change in comparison with that in the control soil, whereas the organic fraction in sequential extraction was enhanced. The compost containing 25% or less inorganic fraction could improve the microbial activity, but the compost containing 50% or more inorganic fraction could not. These results indicate that the compost containing 50% or more inorganic fraction with a high phosphorus content is suitable for immobilizing lead and reducing lead mobility and bioavailability in the soil. However, to immobilize lead and improve the microbial activity at the same time, the AMC with 25% inorganic and 75% organic components should be used as a lead immobilization material, but note that higher lead mobility at the initial stage after the application should be considered.

CONFLICT OF INTERESTS

The authors declare that there is no conflict of interests regarding the publication of this paper.

ACKNOWLEDGMENTS

The ICP-AES and CHN elemental analyzer instruments used for the chemical analysis in this study were made available by the Division of Instrumental

Analysis at Gifu University. The authors are grateful to Professor F. Li and Professor T. Yamada (Gifu University) for allowing the use of the TOC analyzer. This study was supported by the Japan Society for the Promotion of Science (JSPS) KAKENHI [Grant no. 23710089].

REFERENCES

1. D. J. Walker, R. Clemente, and M. P. Bernal, "Contrasting effects of manure and compost on soil pH, heavy metal availability and growth of Chenopodium album L. in a soil contaminated by pyritic mine waste," Chemosphere, vol. 57, no. 3, pp. 215–224, 2004.
2. J. H. Park, N. Bolan, M. Megharaj, and R. Naidu, "Comparative value of phosphate sources on the immobilization of lead, and leaching of lead and phosphorus in lead contaminated soils," Science of the Total Environment, vol. 409, no. 4, pp. 853–860, 2011.
3. X. Cao, A. Wahbi, L. Ma, B. Li, and Y. Yang, "Immobilization of Zn, Cu, and Pb in contaminated soils using phosphate rock and phosphoric acid," Journal of Hazardous Materials, vol. 164, no. 2-3, pp. 555–564, 2009.
4. X. Cao, D. Dermatas, X. Xu, and G. Shen, "Immobilization of lead in shooting range soils by means of cement, quicklime, and phosphate amendments," Environmental Science and Pollution Research, vol. 15, no. 2, pp. 120–127, 2008.
5. G. M. Hettiarachchi, G. M. Pierzynski, and M. D. Ransom, "In situ stabilization of soil lead using phosphorus and manganese oxide," Environmental Science and Technology, vol. 34, no. 21, pp. 4614–4619, 2000.
6. A. Davis, L. E. Eary, and S. Helgen, "Assessing the efficacy of lime amendment to geochemically stabilize mine tailings," Environmental Science and Technology, vol. 33, no. 15, pp. 2626–2632, 1999.
7. E. Lombi, F.-J. Zhao, G. Zhang et al., "In situ fixation of metals in soils using bauxite residue: chemical assessment," Environmental Pollution, vol. 118, no. 3, pp. 435–443, 2002.
8. P. Alvarenga, A. P. Gonçalves, R. M. Fernandes et al., "Evaluation of composts and liming materials in the phytostabilization of a mine soil using perennial ryegrass," Science of the Total Environment, vol. 406, no. 1-2, pp. 43–56, 2008.
9. P. Alvarenga, A. P. Gonçalves, R. M. Fernandes et al., "Organic residue as immobilizing agents in aided phytostabilization: (I) Effects on soil chemical characteristics," Chemosphere, vol. 74, no. 10, pp. 1292–1300, 2009.
10. D. J. Walker, R. Clemente, A. Roig, and M. P. Bernal, "The effects of soil amendments on heavy metal bioavailability in two contaminated Mediterranean soils," Environmental Pollution, vol. 122, no. 2, pp. 303–312, 2003.
11. L. Liu, H. Chen, P. Cai, W. Liang, and Q. Huang, "Immobilization and phytotoxicity of Cd in contaminated soil amended with chicken manure compost," Journal of Hazardous Materials, vol. 163, no. 2-3, pp. 563–567, 2009.
12. A. Sato, H. Takeda, W. Oyanagi, E. Nishihara, and M. Murakami, "Reduction of cadmium uptake in spinach (Spinacia oleracea L.) by soil amendment with animal waste compost," Journal of Hazardous Materials, vol. 181, no. 1–3, pp. 298–304, 2010.
13. H.-S. Chen, Q.-Y. Huang, L.-N. Liu, P. Cai, W. Liang, and M. Li, "Poultry manure compost alleviates the phytotoxicity of soil cadmium: influence on growth of pakchoi (Brassica chinensis L.)," Pedosphere, vol. 20, no. 1, pp. 63–70, 2010.
14. R. Clemente, C. Almela, and M. P. Bernal, "A remediation strategy based on active phytoremediation followed by natural attenuation in a soil contaminated by pyrite waste," Environmental Pollution, vol. 143, no. 3, pp. 397–406, 2006.

15. E. Doelsch, A. Masion, G. Moussard, C. Chevassus-Rosset, and O. Wojciechowicz, "Impact of pig slurry and green waste compost application on heavy metal exchangeable fractions in tropical soils," Geoderma, vol. 155, no. 3-4, pp. 390–400, 2010.
16. R. P. Narwal and B. R. Singh, "Effect of organic materials on partitioning, extractability and plant uptake of metals in an alum shale soil," Water, Air, and Soil Pollution, vol. 103, no. 1–4, pp. 405–421, 1998.
17. M. Katoh, W. Kitahara, R. Yagi, and T. Sato, "Suitable chemical properties of animal manure compost to facilitate Pb immobilization in soil," Soil and Sediment Contamination, vol. 23, no. 5, pp. 523–539, 2014.
18. R. Clemente, Á. Escolar, and M. P. Bernal, "Heavy metals fractionation and organic matter mineralisation in contaminated calcareous soil amended with organic materials," Bioresource Technology, vol. 97, no. 15, pp. 1894–1901, 2006.
19. P. Schwab, D. Zhu, and M. K. Banks, "Heavy metal leaching from mine tailings as affected by organic amendments," Bioresource Technology, vol. 98, no. 15, pp. 2935–2941, 2007.
20. R. Clemente and M. P. Bernal, "Fractionation of heavy metals and distribution of organic carbon in two contaminated soils amended with humic acids," Chemosphere, vol. 64, no. 8, pp. 1264–1273, 2006.
21. A. Baghaie, A. H. Khoshgoftarmanesh, M. Afyuni, and R. Schulin, "The role of organic and inorganic fractions of cow manure and biosolids on lead sorption," Soil Science and Plant Nutrition, vol. 57, no. 1, pp. 11–18, 2011.
22. S. Deiana, C. Gressa, B. Manunza, R. Rausa, and R. Seever, "Analytical and spectroscopic characterization of humic acids extracted from sewage sludge, manure, and worm compost," Soil Science, vol. 150, no. 1, pp. 419–424, 1990.
23. M. A. A. Zaini, R. Okayama, and M. Machida, "Adsorption of aqueous metal ions on cattle-manure-compost based activated carbons," Journal of Hazardous Materials, vol. 170, no. 2-3, pp. 1119–1124, 2009.
24. M. Katoh, W. Kitahara, and T. Sato, "Sorption of lead in animal manure compost: contributions of inorganic and organic fractions," Water, Air, and Soil Pollution, vol. 225, no. 1, article 1828, 2014.
25. T. Yamaguchi, Y. Harada, and M. Tsuiki, "Basic data of animal waste composts," Miscellaneous Publication of the National Agriculture Research Center, vol. 41, pp. 1–178, 2000 (Japanese).
26. G. W. Gee and J. M. Bauder, "Partical-size analysis," in Methods of Soil Analysis, Part 1, A. L. Page, R. H. Miller, and D. R. Keeney, Eds., pp. 383–411, American Society of Agronomy, Madison, Wis, USA, 1986.
27. L. M. Shuman, "Fractionation method for soil microelements," Soil Science, vol. 140, no. 1, pp. 11–22, 1985.
28. A. Tessier, P. G. C. Campbell, and M. Blsson, "Sequential extraction procedure for the speciation of particulate trace metals," Analytical Chemistry, vol. 51, no. 7, pp. 844–851, 1979.
29. M. A. Tabatabai, "Soil enzymes," in Methods of Soil Analysis, Part 2, S. H. Mickelson and J. M. Bigham, Eds., pp. 77–83, American Society of Agronomy, Madison, Wis, USA, 1994.
30. E. Kandeler and H. Gerber, "Short-term assay of soil urease activity using colorimetric determination of ammonium," Biology and Fertility of Soils, vol. 6, no. 1, pp. 68–72, 1988.
31. W. T. Frankenberger and J. B. Johanson, "Method of measuring invertase activity in soils," Plant and Soil, vol. 74, no. 3, pp. 313–323, 1983.
32. S. Sauvé, M. McBride, and W. Hendershot, "Soil solution speciation of lead (II): effects of organic matter and pH," Soil Science Society of America Journal, vol. 62, no. 3, pp. 618–621, 1998.
33. H. Wang, X. Shan, T. Liu et al., "Organic acids enhance the uptake of lead by wheat roots," Planta, vol. 225, no. 6, pp. 1483–1494, 2007.

34. M. Aoyama, "Fractionation of water-soluble organic substances formed during plant residue decomposition and high performance size exclusion chromatography of the fractions," Soil Science and Plant Nutrition, vol. 42, no. 1, pp. 31–40, 1996.
35. M. Aoyama, "Properties of fine and water-soluble fractions of several composts II. Organic forms of nitrogen, neutral sugars, and muramic acid in fractions," Soil Science and Plant Nutrition, vol. 37, no. 4, pp. 629–637, 1991.
36. T. Paré, H. Dinel, M. Schnitzer, and S. Dumontet, "Transformations of carbon and nitrogen during composting of animal manure and shredded paper," Biology and Fertility of Soils, vol. 26, no. 3, pp. 173–178, 1998.
37. D. L. Jones and D. S. Brassington, "Sorption of organic acids in acid soils and its implications in the rhizosphere soil," European Journal of Soil Science, vol. 49, pp. 447–455, 1998.
38. M. Farrell, W. T. Perkins, P. J. Hobbs, G. W. Griffith, and D. L. Jones, "Migration of heavy metals in soil as influenced by compost amendments," Environmental Pollution, vol. 158, no. 1, pp. 55–64, 2010.
39. A. W. Schroth, B. C. Bostick, J. M. Kaste, and A. J. Friedland, "Lead sequestration and species redistribution during soil organic matter decomposition," Environmental Science & Technology, vol. 42, no. 10, pp. 3627–3633, 2008.
40. M. Levonmäki, H. Hartikainen, and T. Kairesalo, "Effect of organic amendment and plant roots on the solubility and mobilization of lead in soils at a shooting range," Journal of Environmental Quality, vol. 35, no. 4, pp. 1026–1031, 2006.
41. A. Pérez-de-Mora, P. Burgos, E. Madejón, F. Cabrera, P. Jaeckel, and M. Schloter, "Microbial community structure and function in a soil contaminated by heavy metals: effects of plant growth and different amendments," Soil Biology and Biochemistry, vol. 38, no. 2, pp. 327–341, 2006.
42. S. Doni, C. MacCi, E. Peruzzi, M. Arenella, B. Ceccanti, and G. Masciandaro, "In situ phytoremediation of a soil historically contaminated by metals, hydrocarbons and polychlorobiphenyls," Journal of Environmental Monitoring, vol. 14, no. 5, pp. 1383–1390, 2012.
43. J. Wyszkowska, J. Kucharski, and W. Lajszner, "The effects of copper on soil biochemical properties and its interaction with other heavy metals," Polish Journal of Environmental Studies, vol. 15, no. 6, pp. 927–934, 2006.
44. M. Farrell, G. W. Griffith, P. J. Hobbs, W. T. Perkins, and D. L. Jones, "Microbial diversity and activity are increased by compost amendment of metal-contaminated soil," FEMS Microbiology Ecology, vol. 71, no. 1, pp. 94–105, 2010.
45. P. Miretzky and A. Fernandez-Cirelli, "Phosphates for Pb immobilization in soils: a review," Environmental Chemistry Letters, vol. 6, no. 3, pp. 121–133, 2008.
46. K. G. Scheckel, J. A. Ryan, D. Allen, and N. V. Lescano, "Determining speciation of Pb in phosphate-amended soils: method limitations," Science of the Total Environment, vol. 350, no. 1-3, pp. 261–272, 2005.
47. P. Castaldi, L. Santona, and P. Melis, "Heavy metal immobilization by chemical amendments in a polluted soil and influence on white lupin growth," Chemosphere, vol. 60, no. 3, pp. 365–371, 2005.
48. A. Ruttens, M. Mench, J. V. Colpaert, J. Boisson, R. Carleer, and J. Vangronsveld, "Phytostabilization of a metal contaminated sandy soil. I: influence of compost and/or inorganic metal immobilizing soil amendments on phytotoxicity and plant availability of metals," Environmental Pollution, vol. 144, no. 2, pp. 524–532, 2006.

CHAPTER 11

Soil Contamination, Risk Assessment and Remediation

Muhammad Aqeel Ashraf[1], Mohd. Jamil Maah[2] and Ismail Yusoff[1]

[1] *Department of Geology, Faculty of Science, University of Malaya, Kuala Lumpur, Malaysia*

[2] *Department of Chemistry, Faculty of Science, University of Malaya, Kuala Lumpur, Malaysia*

1. INTRODUCTION

Environment pollution is a burning topic of the day. Air, water and soil are being polluted alike. Soil being a "universal sink" bears the greatest burden of environmental pollution. It is getting polluted in a number of ways. There is urgency in controlling the soil pollution in order to preserve the soil fertility and increase the productivity. Pollution may be defined as an undesirable change in the physical, chemical and biological characteristics of air, water and soil which affect human life, lives of other useful living plants and animals, industrial progress, living conditions and cultural assets. A pollutant is something which adversely interfere with health, comfort, property or environment of the people. Generally most pollutants are introduced in the environment by sewage, waste, accidental discharge or else they are by-products or residues from the production of something useful. Due to this our precious natural resources like air, water and soil are getting polluted.

The basis of agriculture is Soil. All crops for human food and animal feed depend upon it. We are losing this important natural resource by the accelerated erosion 10 some extent. In addition to this the enormous quantities of man-made waste products, sludge and other product" from new waste treatment plants even polluted water are also causing or leading to soil pollution. In order to preserve the fertility and the productivity of the soil, control measures are to be taken in a herculean manner, thereby improving the health of all living beings.

Assessing the ecological risk of contaminated soil, pesticide application, sewage sludge amendment, and other human activities leading to exposure of the terrestrial environment to hazardous substances is a complicated task with numerous associated problems. Not only is terrestrial ecological risk assessment a relatively new field of science that has developed rapidly only since the mid-1980s, but it is also complicated by the fact that soil, in contrast to most aquatic environments, is very often on private lands and traded as real estate. Professional and economic divergence between the interests of scientists, stakeholders, authorities, engineers, managers, lawyers, nongovernment organizations (NGOs) and regulators is therefore not unusual. Even neglecting those aspects, a number of unresolved problems exist in the way we currently assess risk and manage the impact of anthropogenic substances in the terrestrial environment.

This chapter does not intend to present a comprehensive review of all published data from ecological studies at contaminated sites. Instead, the observations from all case studies are used in the discussion and form the basis for the final conclusion. In each case, we try to answer the following questions:

1. What is soil pollution and how it occurs?
2. How to determine the ecological risk assessment of the soil?
3. To what extent do soil screening levels (over)estimate risk?
4. Do bioassays represent a more realistic risk estimate?
5. Is it possible to make sound field surveys, or do we lack suitable reference situations?
6. What are the possible soil management methods for the polluted soils?

2. SOIL POLLUTION

Soil pollution is defined as the build-up in soils of persistent toxic compounds, chemicals, salts, radioactive materials, or disease causing agents, which have adverse effects on plant growth and animal health [1].

Soil is the thin layer of organic and inorganic materials that covers the Earth's rocky surface. The organic portion, which is derived from the decayed remains of plants and animal, is concentrated in the dark uppermost topsoil. The inorganic portion made up of rock fragments, was formed over thousands of years by physical and chemical weathering of bedrock. Productive soils are necessary for agriculture to supply the world with sufficient food [2].

There are many different ways that soil can become polluted, such as:

- Seepage from a landfill
- Discharge of industrial waste into the soil
- Percolation of contaminated water into the soil
- Rupture of underground storage tanks
- Excess application of pesticides, herbicides or fertilizer
- Solid waste seepage

The most common chemicals involved in causing soil pollution are:

- Petroleum hydrocarbons
- Heavy metals
- Pesticides
- Solvents

2.1. Inorganic Toxic Compounds

Inorganic residues in industrial waste cause serious problems as regards their disposal. They contain metals which have high potential for toxicity. Industrial activity also emits large amounts of arsenic fluorides and sulphur dioxide (SO_2) [3]. Fluorides are found in the atmosphere from superphosphate, phosphoric acid, aluminium, steel and ceramic industries. Sulphur dioxide emitted by factories and thermal plants may make soils very acidic. These metals cause leaf injury and destroy vegetation.

Copper, mercury, cadmium, lead, nickel, arsenic are the elements which can accumulate in the soil, if they get entry either through sewage, industrial waste or mine washings. Some of the fungicides containing copper and mercury also add to soil pollution. Smokes from automobiles contain lead which gets adsorbed by soil particles and is toxic to plants. The toxicity can be minimized by building up soil organic matter, adding lime to soils and keeping the soil alkaline [4].

2.2. Organic Wastes

Organic wastes of various types cause pollution hazards. Domestic garbage, municipal sewage and industrial wastes when left in heaps or improperly disposed seriously affect health of human beings, plants and animals [5-7]. Organic wastes contain borates, phosphates, detergents in large amounts. If untreated they will affect the vegetative growth of plants. The main organic contaminants are phenols and coal.

Asbestos, combustible materials, gases like methane, carbon dioxide, hydrogen sulphide, carbon monoxide, sulphur dioxide, petrol are also contaminants. The radioactive materials like uranium, thorium, strontium etc. also cause dangerous soil pollution. Fallout of strontium mostly remains on the soil and is concentrated in the sediments [8]. Decontamination procedures may include continuous cropping and use of chelate amendments. Other liquids wastes like sewage, sewage sludge, etc. are also important sources of soil problems.

a. Sewage and sewage sludge

Soil pollution is often caused by the uncontrolled disposal of sewage and other liquid wastes resulting from domestic uses of water, industrial wastes containing a variety of pollutants, agricultural effluents from animal husbandry and drainage of irrigation water and urban runoff [9-10]. Irrigation with sewage water causes profound changes in the irrigated soils. Amongst various changes that are brought about in the soil as an outlet of sewage irrigation include physical changes like leaching, changes in humus content, and porosity etc., chemical changes like soil reaction, base exchange status, salinity, quantity and availability of nutrients like nitrogen, potash, phosphorus, etc. Sewage sludges pollute the soil by accumulating the metals like lead, nickel, zinc, cadmium, etc. This may lead to the phytoxicity of plants.

a. Heavy metal pollutants

Heavy metals are elements having a density greater than five in their elemental form. They mostly find specific absorption sites in the soil where they are retained very strongly either on the inorganic or organic colloids. They are widely distributed in the environment, soils, plants, animals and in their tissues. These are essential for plants and animals in trace amounts. Mainly urban and industrial aerosols, combustion of fuels, liquid and solid from animals and human beings, mining wastes, industrial and agricultural chemicals etc. are contributing heavy metal pollution. Heavy metals are present in all uncontaminated soils as the result of weathering from their parent materials. Concentration of heavy metals in soils and plants is given in Table 1.

Table 1. Heavy metal concentration in the hithosphere, soils and plants (Ug/gm dry matter)

Sl.No	Heavy metal	Hithosphere	Soil range	Plants
1	Cadmium (Cd)	0.2	0.01-0.7	0.2-0.8
2	Cobalt (Co)	40	1-40	0.05-0.5
3	Chromium (Cr)	200	5-3000	0.2-1.0
4	Copper (Cu)	70	2-100	4-15
5	Iron (Fe)	50,000	7000-5,50,000	140
6	Mercury (Hg)	0.5	0.01-0.3	0.015
7	Manganese (Mn)	1000	100-4000	15-100
8	Molybdenum (Mo)	2.3	0.2-5	1-10
9	Nickel (Ni)	100	10-1000	1
10	Lead (Pb)	16	2-200	0.1-10
11	Tin (Sn)	40	2-100	0.3
12	Zinc (Zn)	80	10-300	8-100

In agricultural soils, however, the concentration of one or more of these elements may be significantly increased in several ways, like through applications of chemicals, sewage sludge, farm slurries, etc. Increased doses of fertilizers, pesticides or agricultural chemicals, over a period, add heavy metals to soils which may contaminate them. Certain phosphatic fertilizers frequently contain trace amounts of cadmium which may accumulate in these soils. Likewise, some fertilizers when applied to soils, they add certain heavy metals which are given in Table 2.

Table 2. Heavy metal content of fertilizers (ug/gm)

Sl.No	Fertilizer	Co	Cr	Cu	Mn	Mo	Ni	Pb	Zn
1	Nitrochalk	-	-	22	24	-	2	-	15
2	Calcium	0.1	Traces	Traces	Traces	-	-	-	1
3	Nitrate	-	-	To 10	To 5	-	-	-	-
4	Ammonium sulphate	<5	<5	0.800	0.80	<0.05 to 0.22	<5	Traces to 200	0.800
5	Super phosphate	0.02-13	0-1000	Traces to 1000	Traces to 2842	Traces to 35	Traces to 32	Traces to 92	70-3000
6	Potassium chloride	001	-	0-10	Traces-8	<0.05	<1	<1	0-3
7	Potassium sulphate	<5	<5	0-300 to 80	Traces to . 33	0.09	<5	<50	<50

The range of heavy metal contents in sludges is given in Table 3.

Table 3. . Heavy metal contents in sludges (ppm)

Sl.No	Heavy metal	Range (ppm)
1	Cadmium	< 60-1500
2	Cobalt	2-260
3	Chromium	40-8800
4	Copper	200-8000
5	Iron	6000-62,000
6	Manganese	150-2500
7	Molybdenum	2-30
8	Nickel	20-5300
9	Lead	120-3000
10	Zinc	700-49,000

The fate of heavy metals in soil will be controlled by physical and biological processes acting within the soil. Metal ions enter the soil solution from these various forms of combination in different rates they may either remain in solution or pass into the drainage water or be taken up by plants growing on the soil or be retained by the soil in sparingly soluble or insoluble forms. The organic matter of these soil have great affinity to heavy metals cations which form stable complexes thereby leading to reduced nutrient content [11-12].

2.3. Organic Pesticides

Pesticides are quite frequently used to -control several types of pests now-a-days. Pesticides may exert harmful effects to micro-organisms, as a result of which plant growth may be affected. Pesticides which are not rapidly decomposed may create such problems. Accumulation is residues of pesticides in higher concentrations are toxic. Pesticides persistence in soil and movement into water streams may also lead to their entry into foods and create health hazards. Pesticides particularly aromatic organic compounds are not degraded rapidly and therefore, have a long persistence time which can be seen in Table 4.

Table 4. Persistence time for some selected pesticides

Sl.No	Pesticide	Persistence time
1	BHC	11 yrs
2	DDT	10 yrs
3	2,4-D	2-8 weeks
4	Aldrin	9 yrs
5	Diuron	16 months
6	Atrazine	18 months
7	Siwazine	17 months
8	Chlordane	12 yrs
9	2,3 6-Trichlorobenzene (TBA)	2-5 yrs

Mercury, cadmium and arsenic are common constituents of pesticides and all these heavy metals are toxic. At present DDT and a number of organochlorine compounds used as pesticides have been declared harmful and banned in U.S.A. and England [13-14]. It is due to the persistence of their residues in soils for considerable time without losing their toxicity. This has led to higher concentration of these pesticides in vegetation, in animal flesh and milk. Eventually man has been affected. In view of their demerits, organochlorines have been replaced by organophosphate pesticides which are more toxic, but do not leave any residue. They do not pollute the soil. The rodenticides too add to soil pollution. A major method of checking this pesticidal pollution is to increase the organic matter content of the sol and choose such pesticides which are non-persistent and leave no harmful residue.

2.4. Types of Soil Pollution

- Agricultural Soil Pollution

i. pollution of surface soil

ii. pollution of underground soil

- Soil pollution by industrial effluents and solid wastes

i. pollution of surface soil

ii. disturbances in soil profile

- Pollution due to urban activities

i. pollution of surface soil

ii. pollution of underground soil

2.5. SOURCES OF SOIL POLLUTION

The sources which pollute the soil are twofold: Agricultural sources and non-agricultural sources. Figure 1 shows the different sources for the soil pollution.

a. Agricultural sources

Soil pollution comes from different sources including agriculture and animal husbandry. Some of the agricultural practices lead to soil pollution. They are animal wastes, use of long lived pesticides, herbicides, fungicides, nematocides, etc. fertilizers and some agricultural practices.

a. Non-agricultural sources

Soil pollution by non-agricultural sources is usually the direct result of urban sprawl caused by rapidly increasing population and a rapidly per capita output of waste related to our modem way of life. Its materials that find their entry into the soil system have long persistence and accumulate in toxic concentration and thus become sources of pollution. Some of those most important soil pollutants are inorganic toxic compounds.

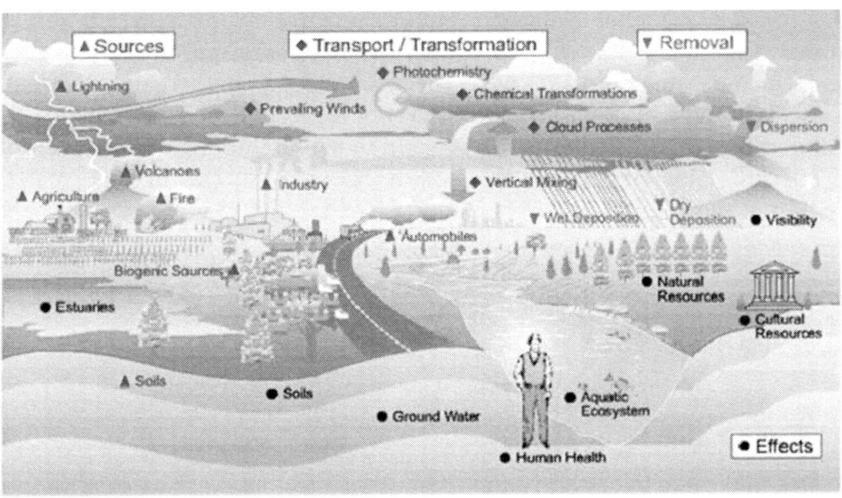

Figure 1. Sources of Soil Pollution [5]

2.6. Causes of Soil Pollution

Soil pollution is caused by the presence of man-made chemicals or other alteration in the natural soil environment. This type of contamination typically arises from the rupture of underground storage links, application of pesticides, and percolation of contaminated surface water to subsurface strata, oil and fuel dumping, leaching of wastes from landfills or direct discharge of industrial wastes to the soil. The most common chemicals involved are petroleum hydrocarbons, solvents, pesticides, lead and other heavy metals. This occurrence of this phenomenon is correlated with the degree of industrialization and intensities of chemical usage. A soil pollutant is any factor which deteriorates the quality, texture and mineral content of the soil or which disturbs the biological balance of the organisms in the soil. Pollution in soil has adverse effect on plant growth.

Pollution in soil is associated with

- Indiscriminate use of fertilizers
- Indiscriminate use of pesticides, insecticides and herbicides
- Dumping of large quantities of solid waste
- Deforestation and soil erosion

a. Indiscriminate use of fertilizers

Oxygen from air and water but other necessary nutrients like nitrogen, phosphorus, potassium, calcium, magnesium, sulfur and more must be obtained from the soil. Farmers generally use fertilizers to correct soil deficiencies. Fertilizers contaminate the soil with impurities, which come from the raw materials used for their manufacture. Mixed fertilizers often contain ammonium nitrate (NH_4NO_3), phosphorus as P_2O_5, and potassium as K_2O. For instance, As, Pb and Cd present in traces in rock phosphate mineral get transferred to super phosphate fertilizer. Since the metals are not degradable, their accumulation in the soil above their toxic levels due to excessive use of phosphate fertilizers becomes an indestructible poison for crops. The over use of NPK fertilizers reduce quantity of vegetables and crops grown on soil over the years. It also reduces the protein content of wheat, maize, grams, etc., grown on that soil. The carbohydrate quality of such crops also gets degraded [15]. Excess potassium content in soil decreases Vitamin C and carotene content in vegetables and fruits. The vegetables and fruits grown on over fertilized soil are more prone to attacks by insects and disease.

a. Indiscriminate use of pesticides, insecticides and herbicides

Plants on which we depend for food are under attack from insects, fungi, bacteria, viruses, rodents and other animals, and must compete with weeds for nutrients. To kill unwanted populations living in or on their crops, farmers use pesticides. The first widespread insecticide use began at the end of World War II and included DDT (dichlorodiphenyltrichloroethane) and gammaxene. Insects soon became resistant to DDT and as the chemical did not decompose readily, it persisted in the environment. Since it was soluble in fat rather than water, it biomagnified up the food chain and disrupted calcium metabolism in birds, causing eggshells to be thin and fragile. As a result, large birds of prey such as the brown pelican, ospreys, falcons and eagles became endangered. DDT has been now been banned in most western countries. Ironically many of them including USA, still produce DDT for export to other developing nations whose needs outweigh the problems caused by it [16].

a. Dumping of solid wastes

In general, solid waste includes garbage, domestic refuse and discarded solid materials such as those from commercial, industrial and agricultural operations. They contain increasing amounts of paper, cardboards, plastics, glass, old construction material, packaging material and toxic or otherwise hazardous substances. Since a significant amount of urban solid waste tends to be paper and food waste, the majority is recyclable or biodegradable in landfills. Similarly, most agricultural waste is recycled and mining waste is left on site. The portion of solid waste that is hazardous such as oils, battery metals, heavy metals from smelting industries and organic solvents are the ones we have to pay particular attention to. These can in the long run, get deposited to the soils of the surrounding area and pollute them by altering their chemical and biological properties [17].

a. Deforestation

Soil Erosion occurs when the weathered soil particles are dislodged and carried away by wind or water. Deforestation, agricultural development, temperature extremes, precipitation including acid rain, and human activities contribute to this erosion. Humans speed up this process by construction, mining, cutting of timber, over cropping and overgrazing. It results in floods and cause soil erosion. Forests and grasslands are an excellent binding material that keeps the soil intact and healthy. They support many habitats and ecosystems, which provide innumerable feeding pathways or food chains to all species. Their loss would threaten food chains and the survival of many species. During the past few years quite a lot of vast green land has been converted into deserts. The precious rain forest habitats of South America, tropical Asia and Africa are coming under pressure of population growth and development (especially timber, construction and agriculture). Many scientists believe that a wealth of medicinal substances including a cure for cancer and aids, lie in these forests. Deforestation is slowly destroying the most productive flora and fauna areas in the world, which also form vast tracts of a very valuable sink for CO_2 [18].

a. Pollution due to urbanization

Pollution of surface soils materials (like vegetables, animal wastes, papers, wooden pieces, carcasses, plant twigs, leaves, cloth wastes as well as sweepings) and many non-biodegradable materials (such as plastic bags, plastic bottles, plastic wastes, glass bottles, glass pieces, stone / cement pieces) [19-20]. On a rough estimate Indian cities are producing solid city wastes to the tune of 50,000 - 80,000 metric tons every day. If left uncollected and decomposed, they are a cause of several problems such as;

- Clogging of drains: Causing serious drainage problems including the burst / leakage of drainage lines leading to health problems.
- Barrier to movement of water: Solid wastes have seriously damaged the normal movement of water thus creating problem of inundation, damage to foundation of buildings as well as public health hazards.
- Foul smell: Generated by dumping the wastes at a place.
- Increased microbial activities: Microbial decomposition of organic wastes generate large quantities of methane besides many chemicals to pollute the soil and water flowing on its surface
- When such solid wastes are hospital wastes they create many health problems: As they may have dangerous pathogen within them besides dangerous medicines, injections.

a. Pollution of underground soil

Underground soil in cities is likely to be polluted by

- Chemicals released by industrial wastes and industrial wastes
- Decomposed and partially decomposed materials of sanitary wastes

Many dangerous chemicals like cadmium, chromium, lead, arsenic, selenium products are likely to be deposited in underground soil. Similarly underground soils polluted by sanitary wastes generate many harmful chemicals. These can damage the normal activities and ecological balance in the underground soil

2.7. Effects of Soil Pollution

a. Agricultural

- Reduced soil fertility
- Reduced nitrogen fixation
- Increased erodibility

- Larger loss of soil and nutrients
- Deposition of silt in tanks and reservoirs
- Reduced crop yield
- Imbalance in soil fauna and flora

a. Industrial

- Dangerous chemicals entering underground water
- Ecological imbalance
- Release of pollutant gases
- Release of radioactive rays causing health problems
- Increased salinity
- Reduced vegetation

a. Urban

- Clogging of drains
- Inundation of areas
- Public health problems
- Pollution of drinking water sources
- Foul smell and release of gases
- Waste management problems

a. Environmental

- Soil becomes unavailable to grow food
- If contaminated soil is used to grow food, the land will usually produce lower yields
- Can cause even more harm because a lack of plants on the soil will cause more erosion
- The pollutants will change the makeup of the soil and the types of microorganisms that will live in it.
- Thus it's possible for soil pollution to change whole ecosystems

2.8. Control of Soil Pollution

The following steps have been suggested to control soil pollution. To help prevent soil erosion, we can limit construction in sensitive area. In general we would need less fertilizer and fewer pesticides if we could all adopt the three R's: Reduce, Reuse, and Recycle. This would give us less solid waste.

a. Extraction and separation techniques

In solvent extraction, the contaminated & oil is mixed with an extracting agent in general (an aqueous solution but preferably an organic solvent). Potential applications include the removal of metals such as cadmium, copper, zinc, nickel, chromium, arsenic, antimony and lead using a mineral solution, zinc lead, organo-metallic compounds and some cyanides using sodium hydroxide solution. Hydrocarbons and halogenated hydrocarbons can also be removed [21]. Contamination is often preferentially present in the finer or coarser fraction of the soil or the organic components (ex. humus), contamination can therefore, be removed in some cases using a process which separates the soil into fractions on the basis of specific gravity or particle size or settling velocity.

a. Thermal methods

In thermal methods, there are two ways of heat treatment; removal of contaminants by evaporation either by direct heat transfer from heated air or an open flame or by indirect heat transfer, and destruction of the contaminants directly or indirectly at an appropriate temperature. The gas leaving the heating appliance must be treated to destroy or remove any contaminants or unwanted products of combustion. A related process is stream stripping in which steam is injected into soil to aid evaporation of relatively volatile contaminants which may be water soluble or insoluble.

a. Chemical methods

Treatment of the soil in suspension in a suitable liquid and without sludging is the two possible methods. In these, intimate, contact between soil and chemical is essential and should be frequently done so that the process of detoxification is complete.

a. Microbial treatment methods

The microbial treatment methods appear to be more promising which can deal with whole range of organic contaminants including phenol, polychlorinated hydrocarbons, oil and oil products, dioxins, etc. There are two different ways of approaching the problems.

1. A community of microbes already existing on the site is collected and cultured in the laboratory.

2. Strains of microbes are developed in the laboratory that is capable of metabolizing particular chemicals.

Excavation of the soil prior to treatment offers the greatest scope for creating optimum conditions. The excavated soil can be placed on thin layers to various depth using standard earth moving techniques and microbes and nutrients applied using standard agricultural techniques such as fertilizing, ploughing, harrowing, etc.

a. Reducing chemical fertilizer and pesticide use

Applying bio-fertilizers and manures can reduce chemical fertilizer and pesticide use. Biological methods of pest control can also reduce the use of pesticides and thereby minimize soil pollution.

a. Reusing of materials

Materials such as glass containers, plastic bags, paper, cloth etc. can be reused at domestic levels rather than being disposed, reducing solid waste pollution.

a. Recycling and recovery of materials

This is a reasonable solution for reducing soil pollution. Materials such as paper, some kinds of plastics and glass can and are being recycled. This decreases the volume of refuse and helps in the conservation of natural resources. For example, recovery of one tonne of paper can save 17 trees.

a. Reforesting

Control of land loss and soil erosion can be attempted through restoring forest and grass cover to check wastelands, soil erosion and floods. Crop rotation or mixed cropping can improve the fertility of the land.

a. Solid waste treatment

Proper methods should be adopted for management of solid waste disposal. Industrial wastes can be treated physically, chemically and biologically until they are less hazardous. Acidic and alkaline wastes should be first neutralized; the insoluble material if biodegradable should be allowed to degrade under controlled conditions before being disposed. As a last resort, new areas for storage of hazardous waste should be investigated such as deep well injection and more secure landfills. Burying the waste in locations situated away from residential areas is the simplest and most widely used technique of solid waste management.

2.9. Soil Monitoring

The main objective of soil monitoring is to prevent and mitigate contamination by substances with the potential to exert an adverse effect on the soil itself, and on air, water and organisms that may contact the soil. Soil monitoring, within the approvals program, is directed primarily to the assessment of contaminants that have been released to the soil surface. Thus, subsurface facilities are generally not the reason for soil monitoring, but may be the reason for groundwater monitoring [22]. However, where soil contamination is known or suspected to originate from subsurface sources such as underground tanks or pipes, an assessment will be required. Where the above considerations indicate soil monitoring is required as a condition of an Approval, the proponent is required to carry out the following, as specified in the Soil Monitoring Directive: > prepare a soil monitoring proposal; > execute the approved soil monitoring plan; > interpret and report the results of the soil monitoring; and > prepare and execute a soil management plan where indicated by the results of soil monitoring. This guideline provides a background for the soil monitoring program and a description of soil management program requirements.

a. Legislative background

The soil monitoring program mostly developed under the Environmental Protection in support of the following principles: > development must be sustainable, meaning that the use of resources and the environment today must not impair prospects for their use by future generations; > the environmental impact of development must be prevented or mitigated; > polluters should bear the responsibility of paying for the costs of their actions; > remediation costs should be incorporated into financial planning so that adequate funds are available for site remediation and planners can know the true costs and benefits of source reduction programs. Recognizing that under the environmental protection is a shared responsibility, it follows that both the approval holder and the Department must have a means to assess environmental performance with respect to the above principles and requirements.

a. Soil quality standards

Environmental Protection expects that approval holders will manage their operations to prevent substance releases to soil. Substance releases to soil do occur, however, and contaminants are often present above background concentrations at industrial facilities. In view of this, Environmental Protection should have soil quality standards to guide assessment and remediation of soil contamination. Facilities that are currently uncontaminated have the opportunity to maintain conditions that allow unrestricted land-use. For these facilities, the minimum standards will be determined by the Tier I criteria or equivalent objectives. Older facilities, however, were often operated under different standards and environmental management practices than are currently acceptable.

2.10. Soil Pollution Risks

There is an increasing use of risk-oriented policies to deal with the local effects of soil pollution. The risks that such policies deal with are: human health risks and can also include eco-toxicological risks. These risks are expressed in terms of negative effects and chances between 0 and 1 that such negative effects will occur. Examples of areas where risk-oriented policies are applied to soil pollution include the United States of America [23], Canada [24] and countries in the European Union [25]. Historically, these risk oriented policies have followed the abandonment of policies aimed at restoring soils to their original 'clean' state.

Risk-based criteria or standards, developed in the framework of risk oriented policies, are applied to risks estimated with deterministic methodologies, following the steps of hazard characterization, appraisal of exposure and risk characterization, while using exposure-risk relations established beforehand. Risk-based criteria have been applied to decisions about soil remediation in the form of soil clean-up standards [26], to the use of soils for specific purposes and in the United States also to sediment management [27]. The risk-oriented policies considered here [28], assume that background exposure to pollutants carries no risk and that a specified level of soil pollution carries a maximum tolerable or maximum acceptable risk for organisms living locally. The latter is the main basis for standard setting.

In part, risk-oriented soil pollution legislation includes policy goals that are qualitative [29]. For instance, the primary UK legislation on contaminated soil defines land as contaminated in need of risk management 'if significant harm is being caused or there is a significant possibility of such harm being caused' [29]. Mostly, however policies have resulted in specific quantitative values for maximum tolerable or acceptable soil pollution. The analysis of such values used in different industrialized countries has shown that there are very large differences, roughly up to a factor [30]. According to Provoost et al. [31], these differences to a large extent originate in different political choices (e.g. including or excluding ecotoxicity) and in different assumptions as to the modeling of exposure to soil pollutants, including site related factors, such as soil type and building constructions [31].

2.10.1. Risks related to one soil pollutant

In practice, there are several matters which are at variance with the proper establishment of actual risk related to one soil pollutant. These are: the absence of standards for pollutants, neglect of background exposure, and neglect of routes of exposure to soil pollution, neglect of available dose-effect studies and neglect of biological availability. These will now be discussed in more detail.

a. Absence of quality standards

When data regarding soil pollutants are available, they should be compared with quality standards reflecting maximum tolerable risk of exposure. However, such standards are not always in place. For instance, of the volatile organic carbon compounds detected in groundwater samples by the US Geological Service, were unregulated- with no standards in place [32]. Similarly Patterson et al. [33] found a variety of brominated ethenes in Australian groundwater, all lacking standards.

a. Neglect of background exposure

For a proper estimate of soil pollution related risks, exposure to specific soil pollutants should be evaluated in combination with exposure to the same substance that is not related to local soil contamination. Several countries, such as Canada, Germany, Spain and Belgium, do indeed establish soil clean-up standards while considering background dietary and inhalatory exposure but others, e.g. Sweden, Norway and the Netherlands, do not [34]. Neglecting background exposure or specific types of background exposure may have implications for risk estimates.

a. Neglect of routes of exposure to soil pollution

In evaluating exposure to soil pollutants, assumptions regarding exposure routes are important. In this respect difference between countries may be noted. Soil clean-up standards for lead of Norway and Sweden differ in part because in Sweden the dominant exposure route is assumed to be by drinking water and in Norway it is thought to be by drinking water and ingestion of soil [34].

Inhalation of household dust and soil particles is not always taken into account in governmental decision making about risks of soil pollution. For instance, in the Netherlands inhalation of soil particles has been neglected as an exposure route, but in e.g. Spain it is not [34]. Neglect of inhalation would seem at variance with existing studies. Nawrot et al. [35] have studied the effects of cadmium pollution in soil (around former thermal zinc plants) and found a significant increase in lung cancer risk correlated with cadmium exposure. They plausibly explain this in terms of exposure of lung tissue to cadmium present in inhaled soil and household dust particles.

Household dust particles have also been found to be important in the exposure of children to pesticides in agricultural settings.

a. Neglect of available dose-effect studies

Akesson et al. [37] have analyzed the effects of low environmental cadmium exposure in an epidemiological study of Swedish women in the Lund area, being 53-64 years of age, excluding women from areas with soils heavily polluted by cadmium. Akesson et al. [37] found associations between the internal dose of cadmium and tubular and glomerular kidney effects, which may represent early signs of adverse effects. Women with diabetes seemed to be at increased risk of experiencing such early signs. In view of these data it seems plausible that at a background exposure that is common in Sweden, old women in the general population may be at risk for adverse cadmium effects [34] and that even a modest increase in cadmium exposure due to polluted soil may lead to added risk. However, when establishing soil clean-up standards in Sweden this background exposure has been neglected [34]. Nawrot et al. [35] have studied the relation between mortality and cadmium body burden in Belgium. They obtained evidence that total mortality and non-cardiovascular mortality may be elevated at cadmium body burdens which can be found among the population not living on soils that are currently considered to be a health risk. Similarly there are now strong indications that the negative effects of lead on the neurophysiologic and sexual development may well be found at the level of background exposure common in Western European and US cities [37], though soil pollution policy, at least in European counties, assumes that such background exposure is safe [14].

2.10.2. Ecotoxicological risks

Maximum acceptable or maximum tolerable ecotoxicological risks are usually derived from a limited number of studies concerning single species under laboratory conditions. Laboratory conditions may be very different from actual conditions in the field, and thus findings in the field are often at variance with laboratory studies [35]. In field studies it has been found that several factors which tend to be neglected in laboratory studies may strongly impact toxic effects of soil pollutants. These include among others: density and adaptability of populations of affected organisms, the presence of other environmental stress factors and the presence or absence of specific landscape elements such as buffer strips [35].

2.10.3. Biological availability

Biologically available pollutants determine risk [3]. Biological availability may vary strongly for different types of organisms [36]. Biological availability of a compound in a specific soil is also dependent on physical, chemical and biological and spatial factors [35]. Examples of such factors are pH, the amount and nature of organic and mineral compounds also present and the presence of organisms that can mobilize soil pollutants [37-39]. In practice, biological availability may be much at variance with total concentrations [40].

2.10.4. Combination Effects

a. Limited accounting of combination effects

As to the overall risk of soil pollutants, cumulative effects of the combination of substances present in soils should be considered. However actual standard setting practice has largely focused on criteria relating to one element or compound. In some cases there are criteria for groups of compounds [34]. Such criteria limit the amount (in g/kg soil) of groups of compounds but often do not address the possibility that the risk per unit of weight may be different for different compounds. An exception to this is criteria for the presence of halogenated dioxins and benzofurans and planar biphenyls. The establishment of risk in case of exposure to these compounds uses addition on the basis of equivalent toxicity [40]. This is a major improvement, though it has been pointed out that this approach may still underestimate the risk of neurodevelopment effects [41].

a. Importance of combination effects

Combination effects may be important in two respects. Firstly, coexisting soil contaminants may impact each others' biological availability [43]. Secondly, exposure to a combination of pollutants may be associated with antagonistic, synergistic and additive interactions of these pollutants, impacting their effect on organisms [44-47]. Some risks of pollutant mixtures can be predicted on the basis of existing knowledge. For instance there is a fair chance that there will be dose additivity when effects are receptor mediated [48]. Also in case of narcotic effects, joint-mixture ecotoxicological effects may be predicted [48]. If responses are dissimilar, response addition may be used [49]. A methodology to deal with the ecotoxicity of mixtures giving rise to both dose-additive and response-additive effects has been proposed [48]. This two step model evaluates mixture toxicity for the same mode of action with concentration additivity and the toxicity for different modes of action with response additivity. For determining the severity of ecotoxicological effects in case of heavily polluted soils (in which legal maximum tolerable levels for one or more substances are exceeded), a systematic approach to combination effects based on a mixture of concentration addition and response addition has been proposed [41].

2.10.5. Remedies For Shortcomings

Remedies would seem possible which would allow for a significant improvement in risk estimates. Unregulated substances can get standards. Standards may be regularly updated on the basis of new dose-effect studies. Risk estimates can include both background exposure and all exposure routes for local soil pollution. Estimates of biological availability can be integrated in risk assessments and improved by better testing of bioavailability or by in-vivo monitoring [51]. The deficiencies in taking account of combination effects in ecotoxicity, discussed in section may be addressed by directly testing of ecotoxicity, when the focus is on ecosystem functioning [52,53]. However it should be noted that small effects on the functioning of ecosystems may have large effects over time [35]. This necessitates large numbers of replicate tests that may well be beyond routine practice [35].

In determining combination effects on human health, direct testing on humans is an 'unethical option'. However biomarker-based monitoring of some aspects of soil pollution relevant to humans may be an option. For instance Roos et al. [54] have applied a biomarker based test to original and remediated soils that were contaminated by a variety of polycyclic aromatic hydrocarbons (PAH). They tested the expression profile of cytochromes P 450 [54]. Xiao et al. [55] have measured genotoxic risk of soil contamination using an in-vitro assay with Salmonella. Though the relation between such biomarker-based data gathered and the in-vivo risks awaits further elucidation, the application of tests based on biomarkers for soil pollution is an interesting option in dealing with combination effects on humans.

Also, estimates of risk may be derived from biomarkers which may be monitored in people exposed to soil pollution. Such biomarkers have emerged from epidemiological studies considering the combined effect of substances. An illustration thereof is the study by Lee et al. [56] which found a graded association of the concentration of blood lead and urinary cadmium concentrations with oxidative stress related markers in the US population. This suggests that oxidative stress may be useful as a biomarker for combination effects. It has furthermore been proposed to evaluate effects of exposure to nitroarenes by measuring haemoglobin adducts [57], and of mixtures of volatile organochlorines by measuring glutathione conjugative metabolites [58]. Bioassays based on aryl hydrocarbon (Ah) receptor mediated mechanisms have been proposed which will allow a better alternative to the measurement of polyhalogenated aromatic hydrocarbons [41]. Another option is to estimate risks to human health by taking into account cumulative combination effects in line with established cause-effect relations and research into the effects of actual combinations. It has been shown that risks of compounds with the same targets and the same modes of action may be estimated on the basis of concentration addition, while including toxicity equivalence factors for the compounds involved [59].

This has been shown to apply to receptor-mediated-and reactive mechanisms of toxicity, provided that no chemical reactions occur between the components of the mixture considered [60]. Currently this approach is applied to halogenated dioxins, benzofurans and planar polybiphenyls, though non-linear interactions are not completely absent in this category of compounds [61], and neurodevelopment effects may be underestimated, as pointed out before [41]. Extension of this approach is possible to e.g. polycyclic aromatics, including heterocyclic polycyclic aromatics [18,62] organophosphates that inhibit the enzyme cholinesterase [44,63], compounds that bind to estrogen receptors [64-66], carcinogens [67], a variety of petroleum products [68] and compounds that inhibit the MXR efflux pump [69].

3. ECOLOGICAL RISK ASSESSMENT

Ecological risk assessment (ERA) is a process of collecting, organizing, and analyzing environmental data to estimate the risk or probability of undesired effects on organisms, populations, or ecosystems caused by various stressors associated with human activities. The basic principles of ecological risk assessment are described in numerous papers [70-72]. All varieties of ERA are associated with uncertainties. The value or usefulness of the different ERA methodologies depends on the uncertainty, predictability, utility, and costs. There are typically two major types of ERA. The first is predictive and is often associated with the authorization and handling of hazardous substances such as pesticides or new and existing chemicals in the European Union. This kind of ERA is ideally done before environmental release. The second type of ERA could be described as an impact assessment rather than a risk assessment, as it is the assessment of changes in populations or ecosystems in sites or areas already polluted. The predictive method is based on more or less generic extrapolations from laboratory or controlled and manipulated semi field studies to real-world situations. The descriptive method is more site specific as it tries to monitor ecosystem changes in historically contaminated soils such as old dumpsites or gas facilities or in field plots after amendment with pesticides or sewage sludge, for example.

Often ERA is performed in phases or tiers, which may include predictive as well as descriptive methods. The successive tiers require, as a rule of thumb, more time, effort, and money. The paradigm or schemes for ERA may vary considerable from country to country, but often consist of an initial problem formulation based on a preliminary site characterization, and a screening assessment, a characterization of exposure, a characterization of effects, and a risk characterization followed by risk management. Although exposure assessment is often just as or even more important, this chapter primarily considers effect assessment. In most European countries, ERA of contaminated soils consists of rather simplified approaches including soil screening levels (SSL) (a.k.a. quality objectives, quality criteria, benchmarks, guideline values) and simple bioassays for a first screening of risk [73-75]. National research or remediation programs have led to the development of a large variety of guideline values.

Although hard to categorize, most fall into two categories: generic or site specific. While the site-specific guidelines require a characterization of pH, organic matter, etc., at the site, generic guideline values are more independent of modifying factors and hence straightforward to legislate. Three major classes of tools for assessing ecological effects may be identified: standardized ecotoxicity experiments with single species exposed under controlled conditions to single chemicals spiked to soil; *ex situ* bioassays, here defined as simple laboratory assays where single species are exposed to historically contaminated soils collected in the field; and finally monitoring, analyzing, and mapping of population or community structures in the field. Furthermore, mesocosm, lysometer, or terrestrial model ecosystems (TME) may be useful; these may be considered as large (multispecies) bioassays or ecotoxicity tests [76-79]. TMEs have the advantage that they operate with the (relatively) undisturbed intrinsic soil populations that make up a small food web. TME hence allow the assessment of effects of toxicants that are mediated through changes in food supply or competition and predation.

One of the keystones in deriving environmental quality criteria is the use of standardized terrestrial test procedures. The emphasis of these prognostic tests is on reproducibility, standardization, international acceptance, and site independence. Although increasing in numbers, relatively few terrestrial tests are still approved by the International Standardisation Organisation (ISO) or Organization for Economic Cooperation & Development (OECD). However, other tests have shown promising results and are likely to be prepared for standardization in the future [80].

However, the major problem in using simple laboratory tests to extrapolate to contaminated land may not be the limitations of test species and the natural variation in species sensitivity. The problems associated with extrapolating from one or a few species, exposed under controlled and typically optimal conditions, to the complex interaction of species and chemicals found in most contaminated ecosystems should also cause concern. Although single-species laboratory tests with spiked materials have their obvious benefits, e.g., they measure direct toxicity of chemicals and interpretation is therefore simple, supplementary tools are often needed. Bioassays, as defined in this context are one of the more frequently used higher-tier alternatives. Basically the same test species may be used in bioassays for assessing the risk of a specific contaminated soil as in standard laboratory tests. However, bioassays have the advantage, compared to the use of spiked soil samples, that the exact toxicity of a specific soil may be accessed directly: this includes the combined and site-specific toxicological effect of the mixture of contaminants and their metabolites. Furthermore, the *in situ* bioavailability of that specific soil is (at least almost) maintained in the laboratory during the exposure period. Several studies have shown a reduction in bioavailability and/or toxicity of soils with an old history of contamination [80-85].

Bioassays are therefore often considered a more realistic tool than generic soil screening levels based on spiked laboratory soils. However, a number of uncertainties or problems may be associated with the use of bioassays and the interpretation of their results. First, the test species are still exposed to the contaminants in a relatively short period compared to the permanent exposure condition found at contaminated sites. Furthermore, they are exposed under more or less optimal conditions, in that stressors such as predation inter- and interspecies competition, drought, frost, and food depletion are eliminated during exposure. Finally, typically only a few species are tested individually.

To compensate for some of the limitations just described, contaminated soil may be assessed using multispecies mesocosms, lysometers, or TME. In these, species interactions may be evaluated by manually introducing several species to the systems or monitoring the intrinsic populations of the soil. Natural climatic conditions may be included if the test system is kept outdoors. However, if we want to get a more realistic and large-scale picture of the impact caused by, for example, pesticide use or sewage sludge application, or to assess the environmental health at waste sites, industrial areas, or gas works, it is often necessary to conduct some kind of field observations. Several case studies exist in which field studies have successfully elucidated the ecological risk of specific activities or the ecological impact at specific sites [85-87]. The small single-species bioassay, large multispecies TME, and field surveys have some drawbacks in common. First of all, it may be difficult to actually link the observed effect to a specific toxic component in the soil. Which of the many substances is actually causing the majority of the observed effects, or is it perhaps a combination of effects? For a hazard classification of soils or a ranking of soils this may not be so important.

However, to evaluate potential risk-reduction measures or risk management procedures it may be important to identify the most problematic substances. A comparison of soil screening values with measured concentrations for each chemical present at a site may be helpful to identify the most likely group of substances causing the observed effect. Other possible tools may include a toxicity identification evaluation (TIE) approach [88]. The TIE approach is a relatively new method, which aims to identify groups of toxicants in soils with mixed pollution. Potentially toxic components present in the soil are fractionated and determined, and the toxicity of each individual fraction is determined by a *Lux* bacteria-based bioassay or the Microtox bioassay. Although perhaps promising, TIE is a time-consuming and hence costly procedure not yet used routinely. Another crucial issue when analyzing the result of bioassays, TME, and field studies is the presence or absence of a proper reference site or soil. The control soil should in principle resemble the contaminated soil in all relevant parameters, e.g., texture, pH, organic matter, waterholding capacity, and nutrient content, a practical problem that very often is difficult to solve. The lack of adequate control or reference sites may, however, be conquered at least partially by the use of multivariate techniques [89], which relate the species composition and abundance to gradients of pollutants. It is not the intention of this chapter to present a review of statistical tools for ecological risk assessment, and hence a detailed discussion about the use of these is not given.

However, it is obvious that increased computer power and the presence of new easy-to-use software tools have increased the possibility to move away from more conventional univariate statistics such as analysis of variance (ANOVA) to more powerful multivariate statistics that use all collected data to evaluate effects at a higher level of organization. Statistical methods such as the power analysis may also be very useful in planning and designing large-scale ecotoxicity studies such as mesocosms,TME, or field surveys.

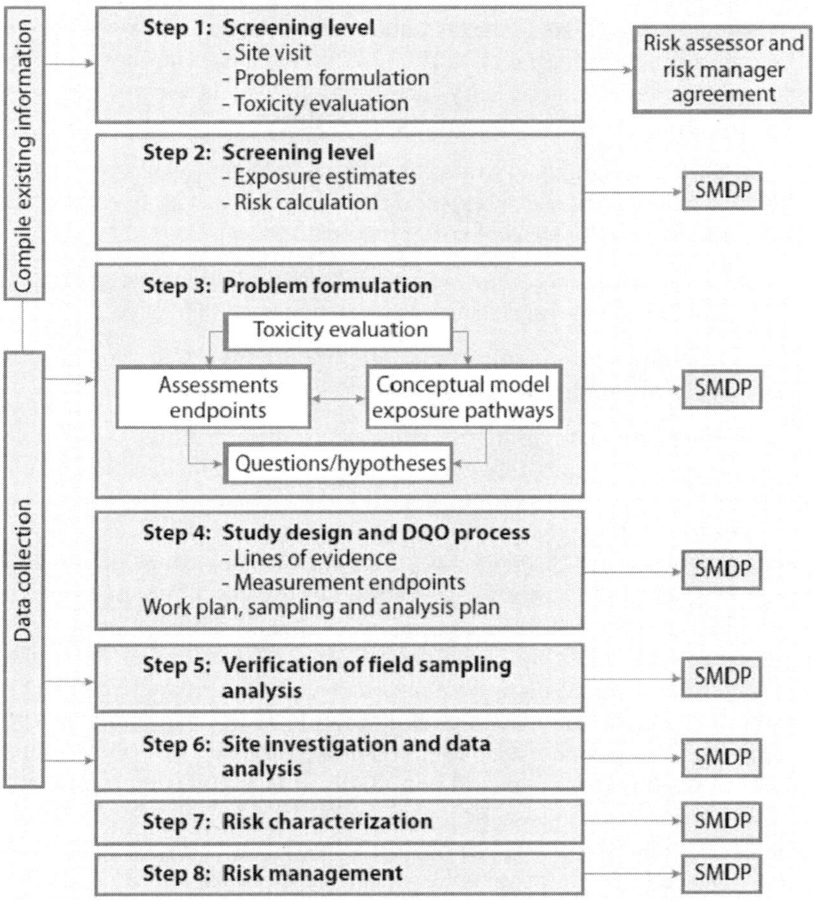

FIGURE 2. The eight steps in the US-EPA framework for risk assessment of contaminated Superfund sites. DQO = data quality objectives [89]

The US-EPA has published an Ecological Risk Assessment Guidance, which should be followed when assessing risks at Superfund sites. As all sites are considered unique this should always be done in a site-specific manner. The ERA process suggested by the US-EPA for Superfund sites follows an eight step process, which can be broken down into four categories, i.e. 1) planning and scoping, 2) problem formulation, 3) stressor response and exposure analysis and 4) risk characterisation. Essential for all steps are a negotiation and agreement of the need for further action between the risk assessor, the risk manager and other stakeholders, the so-called scientific-management decision points (SMDP).

SMDP made at the end of the screening-level assessment will not set an initial cleanup goal. Instead, hazard quotients, derived in this step, are used to help determine potential risk. Thus, requiring a cleanup based solely on those values would not be very likely, although it is technically feasible. There are three possible decisions at the SMDP:

1. There is enough information to conclude that ecological risks are very low or non-existent, and therefore there is no need to clean up the site on the basis of ecological risk.
2. The information is not adequate to make a decision at this point, and the ecological risk assessment process will proceed.
3. The information indicates a potential for adverse ecological effects, and a more thorough study is necessary.

In the Netherlands contaminated sites are first determined using a set of soil screening levels called target and intervention values, which take both human and ecological risks into account. At seriously contaminated sites remediation or other soil management decisions are required if the risks cannot be neglected based on a site-specific ecological and human risk assessment, and the chance for dispersion of the contaminants. Until now, the ecological risk assessment has been based on chemical analysis, including a Decision Table harbouring critical dimensions of the impacted area. The United Kingdom and Canada have also developed framework for ecological risk assessment of contamination land. A cornerstone in the UK framework of ERA is the connection to the statutory regime for identification and control of land potentially affected by contamination. The UK framework is based on schemes found in e.g. USA, Canada and the Netherlands. Like these it is a based on a tiered approach where the initial Tier 0 aims to determine whether a site falls under the Part IIA of the legislation. It involves the development of a Conceptual Site Model (CSM), which described what is already (historically) known about the site, e.g. whether there is a likely source-pathway-receptor linkage. The conceptual site model is followed by an initial screening phase (Tier 1) and an actual site-specific characterisation (Tier 2). Tier 1 is a simple deterministic comparison of chemical residue data and the soil quality guideline values supplemented with simple soil-specific toxicity testing. The final step (Tier 3) involves more detailed *in-situ* studies and for example ecological modelling based on a more advanced ecological theory. Tier 3 is not likely to be conducted at many sites.

3.1. Descision Support System for Ecological Risk Assessment

Ecological Risk Assessment is often a complex process with many variables to take into account. ERA involves many stakeholders and all have to be dealt with in a clear and consistent way. A stepwise or tiered approach is therefore useful to overcome the complexity of an ERA. In order to structure all the information collected, a Decision Support System (DSS) can be used. Each tier will lead to a decision to proceed or to stop. A number of decisions supporting systems or frameworks have already been developed in other countries, e.g. UK, the Netherlands and the USA. The DSS presented here is based on basic principles also common in the methodologies used in the USA and UK. However, in the present DSS measures of bioavailability and the use of the Triad approach may be built into the system more systematically. This chapter introduces the overall framework of a novel DSS including the Triad approach and the challenge to weight and scale results used in that process.

3.2. Framework for Ecological Risk Assessment

Rutgers et al. (2000) developed a basic flowchart for Ecological Risk Assessment [90], which is used as the backbone of the decision support system (DSS) presented in the Figure 5.

The DSS is separated in three different stages, i.e.

- Stage I. Site characterization and description of land-use.
- Stage II. Determination of ecological aspects.
- Stage III. Site-specific tiered assessment (the Triad):
- – Tier 1. Simple screening
- – Tier 2. Refined screening
- – Tier 3. Detailed assessment
- – Tier 4. Final assessment.

Each of these four tiers is based on a weight of evidence (WoE) approach combining three lines of evidence (Chemistry, (eco)Toxicology and Ecology).

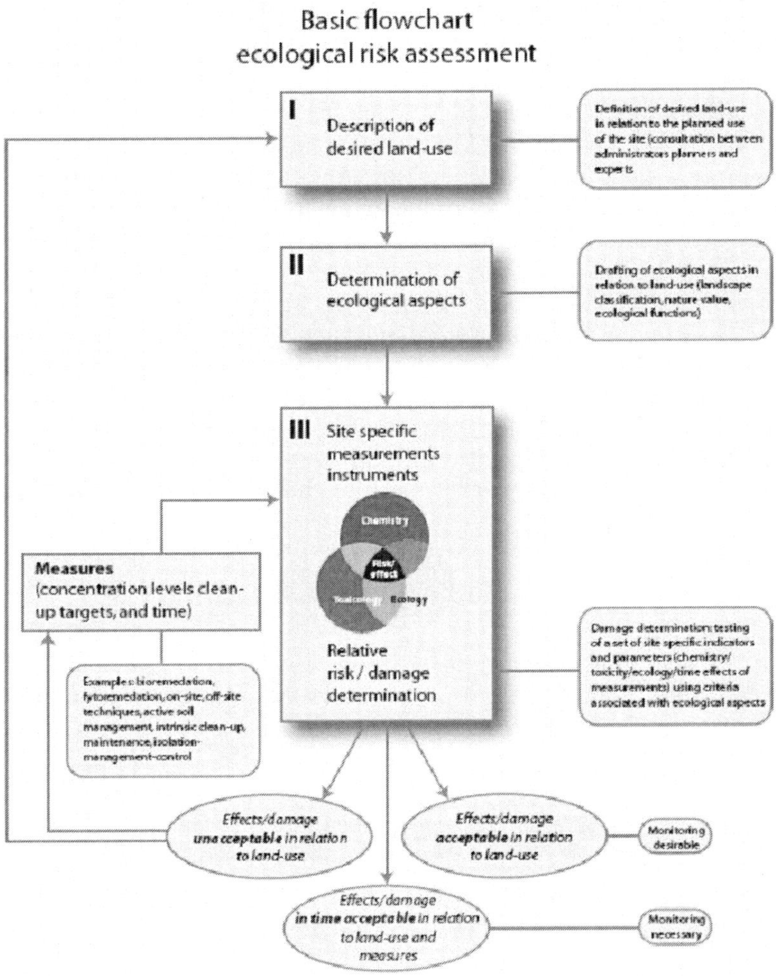

Figure 3. Basic flowchart for ecological risk assessment [90]

3.2.1. Boundaries of the dss

The DSS in this chapter is not a full and comprehensive document for managing risk of contaminated land. It focuses strongly on supporting decisions made when considering risk to the terrestrial environment. Therefore it addresses only indirectly the risk to ground water and associated (connected) fresh water systems. Nevertheless information about e.g. reduced bioavailability may be useful when assessing potential risk for leaching of contaminants to ground water or fresh water.

Furthermore, it is important to realise that the management of a contaminated site is more than assessing ecological risk. Issues like for example risk for humans, availability and cost of remediation solutions, development plans for the vicinity or the region are equally important.

3.2.2. Stage i — site characterisation and land-use definition

The first step in the DSS is to establish what is often referred to as a Conceptual Site Model. It aims at involving as many stakeholders as possible in order to describe site characteristics and to review all available information from the site, e.g. historical information about land-use, investigation of whether the site may be regulated under specific directives, obvious data gaps and urgency for reaction and data collection. The spatial borders of the site should be defined and the current and the future landuse have to be defined. Consultation between administrators, planners and experts therefore has to take place as early as possible in the process.

a. Initial requirements in the DSS

An inquiry among all stakeholders should be conducted as one of the first initiatives. The aim should be to collect as much information about soil characteristics as possible.

a. Defining land-use

One of the first actions to be taken among all stakeholders is to decide which landuse is required for the site, as this will determine the required data collection and testing. Many land-uses may be defined, but generally the four following overall categories of land-use classes are used:

- industrial area (including infrastructure and pavement).
- urban/residential area (including recreational and green areas).
- agricultural area.
- nature area.

a. When is an ecological risk assessment needed?

Most often a site specific ERA will be initiated only when soil concentrations exceed soil screening levels. However, this may not in itself be a sufficient criterion to go through the entire ERA procedure. Some boundary conditions, based on the present and future type of land-use, the level of contamination and various ecological considerations have to be met in order to rationalize an ERA. The experts and the rest of the stakeholders should answer a number of simple questions in order to conclude whether the required boundary conditions are fulfilled.

3.2.3. Stage ii — determination of ecological aspects

At stage II, site-specific ecological features and receptors relating to the land-use defined in Stage I need to be outlined. This includes aspects like key species and life support functions. The potential ecological receptors should be identified in order to determine whether potential source-pathway-receptor linkages can be established. This includes not only ecological receptors directly linked to the site but also those linked indirectly e.g. through leaching of contaminants to connected fresh water systems or (migrating) birds or mammals feeding in the area. In Table 9 some examples are given of land-use and related ecological aspects. This table can be used as a starting point for the selection of ecological aspect. Experts from ecotoxicology and ecology should be involved in the selection of ecological aspects.

3.2.4. Stage iii — site specific instruments (the triad)

If after finalising Stage I and Stage II it is still considered that there is a need for a site specific evaluation of ecological risk the process continues to Stage III using the weight of evidence approach described below.

3.2.5. Weight of evidence approaches

In order to deal with conceptual uncertainties in a pragmatic way, it has been proposed to use weight of evidence (WoE) approaches for ERA [90-93]. The rationale is, like in justice, that many independent ways to arrive at one conclusion will provide a stronger evidence for ecological effects, making ERA less uncertain.

In the sediment research area the application of WoE started at an early stage and was called the Sediment Quality Triad. For terrestrial ecosystems WoE approaches and the Triad are still in a developing stage. The Triad approach is based on the simultaneous and integrated deployment of site-specific chemical, toxicological and ecological information in the risk assessment as given in Figure 6. The major assumption is that WoE in three independent disciplines will lead to a more precise answer than an approach, which is solely based on, for example, the concentrations of pollutants at the site. A multidisciplinary approach will help to minimise the number of false positive and false negative conclusions in ERA. It also gives acknowledgement to the fact that ecosystems are too complex to analyse in one-factorial approaches.

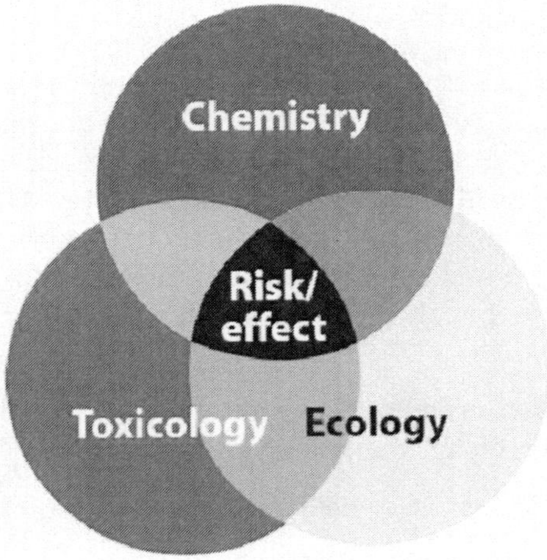

Figure 4. Schematic presentation of the integration of three fields of research according to a Triad [92]

- **Chemistry:** The concentration of contaminants in the environment (totals, bioavailable), accumulated in biota, or modelled via food-chains is used for calculation of risks on the basis of toxicity data from the literature.
- **Toxicology:** Bioassays with species across genera are carried out in order to measure the actual toxicity present in environmental samples from the site.
- **Ecology:** Field ecological observations at the contaminated site are compared to the reference site. Deviations from the reference site, which can be plausibly attributed to the contamination levels, are funnelled into the Triad.

3.3. Using The Triad In Site Specific Assessment Of Contaminated Soil

Triad is a powerful weight of evidence approach originally developed in order to evaluate sediment quality. In the terrestrial compartment less experience is available on the practical use of the Triad. This chapter describes the use of Triad in more detail and gives an insight into some of the important decisions risk assessors have to make when conducting the Triad in practise, e.g. how to scale, weight and integrate the outcome of the various investigations.

The Triad approach exists of three lines of evidence (LoE), the so-called Triad "legs", i.e. chemistry, (eco) toxicology and ecology. The Triad approach includes a tiered system in which each consecutive tier is increasingly fine-tuned to the site-specific situation. In the first tier the research is simple, broad and generic. In later tiers more specific and complex tests and analyses may be used. For each of the LoE in the Triad there are a variety of analyses or tests that can be chosen. Some examples are:

- **Chemistry:** Measurement of total concentrations, bioavailable concentrations, bioaccumulation, etc.
- **Toxicology:** Bioassays (in field and/or in lab), biomarkers etc.
- **Ecology:** Field observations of vegetation, soil fauna, micro-organisms, etc. In Chapter 6, a number of tests or tools that are for suitable for use in each tier are presented for the chemistry, toxicology and ecology LoE.

3.4. Decision Charts In Ecological Risk Assessment Of Contaminated Sites

3.4.1. Flowcharts

This chapter is an attempt to present a decision support system, which can guide risk assessors in their assessment of site-specific ecological risk. A number of site-specific questions need to be answered before a final decision on performing an ecological risk assessment can be made. This chapter introduces a flow chart for ecological risk assessment of contaminated sites. The flowchart is presented as decision trees as shown in Figure 8 together with a more in-depth introduction to the relevant questions that needs to be addressed and answered when performing a site-specific ecological risk assessment.

3.4.2. Decision Making In Era

The assessment of ecological risk is performed stepwise in tiers. Higher tiers represent gradually more and more complex studies, but also more expensive and laborious studies. The full site-specific risk assessment covers four tiers, i.e.

- Simple screening: Tier 1.
- Refined screening: Tier 2.
- Detailed assessment: Tier 3.
- Final assessment: Tier 4.

The main principle in going from a simple screening over a more refined screening to a detailed assessment of the contaminated site is to minimize time and effort. The actual performance of the risk assessment and use of the various tiers may be very site-specific.

a. Tier 1 — Simple screening

After deciding in the two first stages of the ERA that ecological concern needs special consideration, the risk assessment starts typically with a simple evaluation at the screening level. This is done in order to minimize costs until new information indicates the need for further assessment and more sophisticated studies. Therefore, the tools used in the first screening need not only to be reasonably quick and easy, but also relatively cheap. The tools for use in Tier I are described in more detail in the toolboxes C1, T1 and E1. On the basis of the results of instruments used in Tier 1 it is decided to either stop further assessment or continue to a higher tier.

a. Tier 2 — Refined screening

Tier 2, still considered being at the screening level, aims at refining the measurement of exposure and at the same time to provide further insight into the toxicological and ecological properties of the contaminated soil. Tier 2 deviate from the conservatism normally associated with the use of total concentration in the risk assessment by taking (rough) estimations of bioavailability into consideration in the chemical LoE. A better screening of the toxicological and ecological properties of the soil compensates for the reduced conservatism in the Chemistry LoE of the Triad. The tools for use in Tier 2 are described in more details in the toolboxes C2, T2 and E2. On the basis of the results in Tier 2 a decision should be made to either stop further assessment or continue to a higher Tier.

a. Tier 3 — Detailed assessment

The tools in Tier 3 differ from the ones used in Tier 1 and Tier 2 in that they are more laborious, costly and may take longer. On the other hand they are (often) more realistic and/or ecological relevant in order to give a more comprehensive assessment of the ecological risk at the specific site. The stakeholders should beforehand negotiate a minimum set of tests. Is it for example necessary to consider all trophic levels in the toxicological and ecological LoE? Or does the land-use suggest otherwise? Is it necessary (or possible) to estimate the bioavailability of all the substances exceeding their SSL? If not, how are the non-investigated substances dealt with? The tools described for use in Tier 3 are described in more details in the toolboxes C3, T3 and E3. Depending on the results from Tier 3 a decision should be made to either stop further assessment or continue with an even more detailed assessment in Tier 4

a. Tier 4 — Final assessment

In Tier 4, the aim of the studies is to answer any remaining questions and to decrease existing uncertainties and this may often require more in-depth research. Tools in Tier 4 can be similar to tools of Tier 3, but more focus has to be on site-specific circumstances. For example bioassays should be done with organisms, which normally occur at the site. Furthermore, it may be more relevant to consider ecological effects outside the contaminated area on e.g. predators or herbivores feeding in the area or effects in adjacent fresh water systems. This Tier requires specialised knowledge and experience with ERA, which implies that costs can be high and only a limited number of people may be able to perform the tests. Generally only on a very limited number of site evaluations will include investigations at this level. If the results of Tier 4 still indicate risk there are basically two possible solutions. Accept the risk and leave the contamination or remove (parts of) the contamination.

3.4.3. Screening Tools

3.4.3.1. Triad based selection of methods

For each of the three Lines of Evidence (LoE) in the Triad various methods or tools are available. In order to facilitate the selection of appropriate tools in the right context, the tools have been compiled in subclasses or toolboxes. Each of these is a collection of tools considered to be potentially useful in the designated tiers and LoE of the Triad, i.e. chemistry, toxicology and ecology. Furthermore, the tools are arranged according to their complexity, price and practicability or in other words depending on whether they are most useful for screening or detailed assessment, i.e.

- Toolbox C1. Chemistry tools for simple screening.
- Toolbox T1. Toxicology tools for simple screening.
- Toolbox E1. Ecology tools for simple screening.
- Toolbox C2. Chemistry tools for refined screening.
- Toolbox T2. Toxicology tools for refined screening.
- Toolbox E2. Ecology tools for refined screening.

- Toolbox C3. Chemistry tools for detailed assessment.
- Toolbox T3. Toxicology tools for detailed assessment.
- Toolbox E3. Ecology tools for detailed assessment.
- Toolbox IV. Various tools for the final (Tier 4) assessments.

a. Toolbox C1 — Chemistry tools for simple screening

At the very first stage of the ERA process, total concentrations of all relevant chemicals are individually compared to soil screening levels (SSL) in order to evaluate whether there is a need for a site specific assessment of ecological risk. In the current Stage III of the ERA, this first generic evaluation of risk is followed by a more site-specific screening of risk including information from all three lines of evidence in the Triad. In the Chemistry part of the Triad more site-specific information is collected by:

Refining and targeting the comparison of soil concentrations with soil related benchmarks for site-specific purposes.

Incorporation of the accumulative risk of a mixture of contaminants by calculating the toxic pressure (TP) of a mixture and by doing so generating more site-specific insight to the potential ecological impact of a contaminated site. Each of these steps can be done separately or in combination, e.g. the TP can be calculated using existing SSL or using new developed benchmarks based on either NOEC or EC50 values or site-specific benchmarks can be compared to soil concentrations individually. The approach entirely depends on the strategy taken by the stakeholder group and the availability of data.

a. Toolbox T1 — Toxicology tools for simple screening

The main objective of the selected toxicity tests or bioassay at Tier 1 should be to screen the soil for presence of toxic compounds. This includes toxic degradation products or compounds, which are not routinely included in various national analytical programs for contaminated sites. This Tier is the first screening level of the ERA and the cost in form of manpower and money should hence be relatively low.

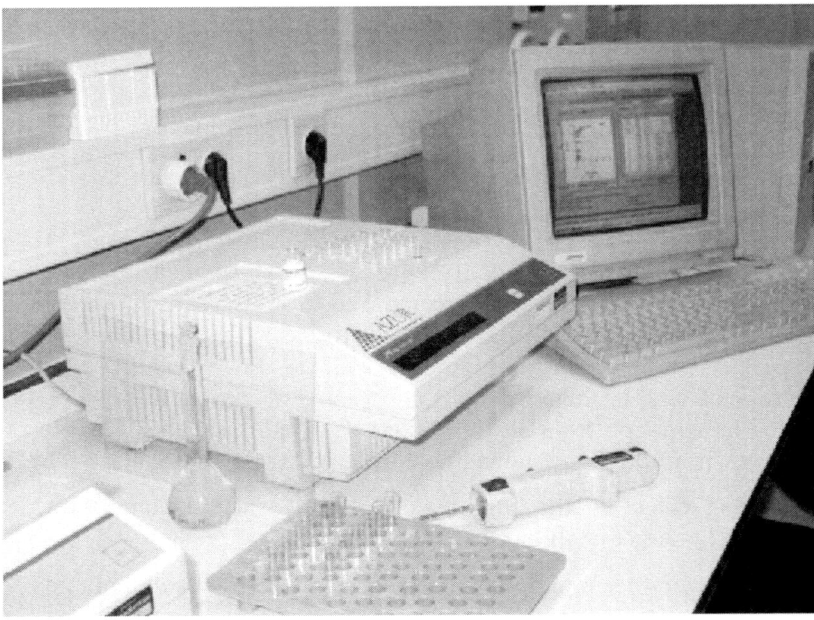

Figure 5. Equipment for measuring luminescence of Vibrio fischeri [94]

a. Toolbox E1 — Ecology tools for simple screening

Ecological surveys or monitoring studies are generally considered a time consuming effort performed by experts. This is in most cases true, wherefore detailed surveys normally take place in higher tier assessment. However, in order to ensure that also ecological information is collected and used in the Triad already in the screening phase, it is recommended to perform a limited examination of the site. A survey of the area with special focus on visible changes in e.g. plant cover or presence or absence of specific plants, trees or scrubs may indicate ecological damage, which can be associated to contaminants present at the site.

Figure 6. Simple survey of the site [95]

If any aerial pictures areavailable from the area these may give valuable information about the plant cover also historically, which may be helpful in identifying parts of the site where the impact may be highest (hot spots). At this stage the conclusion can in most cases only be indicative. Therefore if the results from the other line of evidence may cause any doubt or the survey indicated potential impact, it is recommended to either continue with a more refined screening in Tier 2 or go directly to the detailed assessment in Tier 3.

a. Toolbox C2 — Chemistry tools for refined screening

Selective solvent extraction

It may be considered useful to adjust the estimate of exposure by taking bioavailability into consideration and hereby deviating from the conservatism normally associated to the use of total concentration in the risk assessment. The principle in this refinement of the ecological risk assessment is to extract a more ecotoxicologically relevant fraction of the contamination than the total concentration. The latter generally tends to overestimate the risk of historically contaminated soils. In this screening phase no attempt is made to estimate the freely dissolved or readily bioavailable concentration of contaminants. Table 10 explains principal studies that employed chemical extractants to evaluate bioavailability.

Table 5. Outline of principal studies that employed chemical extractants to evaluate bioavailability

Contaminant (Ref)	Solvent	Bioassay	Operation	Comments
Atrazine Phenanthrene (1)	Methanol/water, n-Butanol	Earthworm uptake Degradation	25 ml extractant, 10 g solid. Shaking for 2 h	Methanol/water best predictor for atrazine, n-butanol for phenanthrene.
DDT, DDE, DDD PAH (mixture) (2)	THF Ethanol	Earthworm uptake	15-20 ml extractant, 1 g soil, 10 sec mixing	Good correlation with earthworm accumulation.
Anthracene, Fluoranthene, Pyrene (3)	n-Butanol Propanol Ethyl acetate	Plant retention Earthworm uptake Degradation	25 ml extractant, 1-2 g soil, 5 sec mixing	Reasonable correlation with bioassays.
Phenanthrene Pyrene Chrysene (4)	n-Butanol	Earthworm uptake Degradation	15 ml extractant, 5-10 g soil, Mixing: 5 sec (worm) or 120 sec (degradation)	Applicable for bioavailability prediction.

Instead the fraction of the contaminants is extracted, which can be directly compared to the existing soil screening levels. This is considered to be a relatively simple and quick method to screen for potential risk of contaminants in a more realistic way than using total concentrations. The extracted concentration (mg kg-1) is compared to the SSL and the result used in the Triad. It is therefore a prerequisite of this comparison that the extractability in the tests (with spiked soils) used for deriving SSL is close to 100% by the methods used. In most short-term tests (< four weeks) it will be reasonable to assume that only little "true" ageing or strong sequestering occurs and hence a majority of the spiked chemicals are still extractable with mild organic solvents. However, for most methods this still has to be fully validated.

Organic solvents most frequently used include methanol/water in different ratios, nbutanol, ethanol, propanpol, ethyl acetate and tetrahydrofuran (THF) given in Table 10. The method establishes preferential partitioning of hydrophobic contaminants to the extractant by increasing their solubility in the aqueous phase whilst removing pollutant compounds from soil surfaces establishing equilibrium conditions. No standard protocol has been adopted for mild chemical extractions in relation to bioavailability testing. Common methodology in literature primarily includes a soil sample to which a volume of chemical extractant is added (generally 1 – 10 g soil,15 – 25 ml extractant). This is followed by a period for mixing, e.g. vigorous mixing for10 – 120 seconds or shaking by orbital shakers for up to 2 hours. The extraction studies have mostly involved PAH and insecticides (including DDT, DDE, DDD and atrazine) [93-95]. Studies that haverelated extractability with results from bioassays have generally focused on uptake and accumulation (% taken up by earthworms or plants) and bacterial degradation (%removed). Therefore, since convincing relationships between the chemical and biological tests were found it may indicate a potential for such extraction methods to predict bioavailability.

a. Toolbox T2 — Toxicology tools for refined screening

In the first simple screening of Tier I focus was on marine bacteria and aquatic/sediment living species. In Tier 2 relatively simple tests with soil dwelling species are used for a more refined screening of the soil samples, i.e. the earthworm survival tests and avoidance tests using soil invertebrates.

Figure 7. Organisms screening in soil [98]

The habitat function of soils is often assessed using the reproduction test with *Eisenia fetida*. The avoidance test with *Eisenia fetida* is a suitable screening test, which is less cost-intensive in terms of duration and workload than the reproduction test, and at the same time (normally) more sensitive than the acute test with the same species.

a. Toolbox E2 — Ecology tools for refined screening

In Tier 2 the observations from the survey may be expanded by simple on-site assessment of the overall soil functioning or biological activity of the soils. Recommended tools include bait-lamina sticks and simple microbial tests using general endpoints like soil respiration or C/N mineralization rates.

Figure 8. Bait-lamina sticks [99]

The main principle for tests at this level is to be relatively simple and cheap but at the same time to give valuable information whether or not the soil has lost some of its main services. Bait-lamina sticks for example have been demonstrated useful for describing biological activity of the soils in a general matter.

a. Toolbox C3 — Chemistry tools for detailed assessment

The objective of the tools found in this toolbox is to assess the bioavailable and freely dissolved fraction of pollutants found in pore water of soils from contaminated sites. The methods should (in principle) be able to mimic the fraction of organic pollutants available for uptake in biota. The collection of methods includes various non-depleting and depleting pore water extractions. Very few terrestrial ecotoxicity data are yet expressed as e.g. pore water concentrations. Instead, the outcome of the methodologies in this toolbox is compared with water quality standards.

a. Toolbox T3 — Toxicology tools for detailed assessment

The objective of the tools found in this toolbox is to evaluate the potential impact of contaminated soils to fauna and plants and hereby the entire ecosystems. Some of the methods use introduced, and not intrinsic, species. The benefit of this is a higher degree of standardisation, as the species used in these bioassays is easy to maintain in laboratory cultures compared to naturally occurring species. The drawback may be that their ecological relevance is less obvious. For example the compost worm *Eisenia fetida* is used as a surrogate to evaluate risk to soil dwelling earthworms. Two sets of bioassays are presented. One for directly assessing potential risk for soil dwelling species, including micro-organisms, plants and soil invertebrates, and one for assessing indirectly risk to aquatic species through e.g. leaching of contaminants. It is often anticipated that soil organisms are exposed to pollutants mainly through uptake from pore water. Therefore it may also be possible to evaluate, or at least to compare or rank, the risk of contaminated soil samples to soil dwelling organisms on the basis of the outcome of the aquatic test using elutriate or pore water. The choice of bioassays depends on a number of variables, e.g.:

- The current and future land-use, i.e. targets of protection.
- The size of the contaminated area.
- The potential for ground water or surface water contamination.
- The need of many simple tests or fewer more complicated tests.

Figure 9. Simple plant tests [91]

a. Toolbox E3 - Ecology tools for detailed assessment

In this late tier of the Triad, the objective of the activities is community or population response analysis, typically by conducting field surveys. As these studies (most often) are time consuming, costly and dependent on ecologically, taxonomically and statistical expertise they are most frequently done on large-scale sites with a long-term- remediation perspective. In fresh water ecosystem community surveys have been widely used with relative success. The absence of species from places where they would be expected to occur could be a strong identification of unacceptable levels of contaminants. However, this type of studies has only seldom been used for the terrestrial environment. The reasons for this are many. One of the dominants may be the lack of a concentration gradient and obvious "upstream" reference sites at most contaminated areas. No world-wide accepted guideline on how to plan and perform a terrestrial field survey is available and hence no straight-forward and easy-to-follow description can be given. The decision on when, where and how to conduct field surveys depends on a number of issues, e.g. the size of the area, the land-use, the type of contaminants present, time of the year and last but not least the time and money available to perform the study. Nevertheless, a number of general considerations have to be made in the planning phase of a successful field survey. These include (but are not limited to):

- Identify the targets of concern and the species to monitor.
- Elucidate the natural temporal and spatial variation before initiating a field study.

- Use statistical (power) analyses to determine the minimum number of samples or replicates needed to emonstrate the decided difference, e.g. 25% change.
- In order to establish a cause-effect relationship, a number of confounding parameters need to be characterized both at the reference and the test site, e.g. soil type, pH, salinity, hydrology, nutrient- and organic matter content and the presence of other contaminants.

As no single description on how to perform ecological surveys for contaminated sites can be given, some general considerations and useful references for this tier of the ecological risk assessment are given below for:

- Assessing impact in the overall biological activity and organic matter breakdown.
- Assessing impact on the microbial community.
- Assessing impact on the plant community.
- Assessing impact on the invertebrate community.

Reference data from reference sites, reference samples and literature

A crucial factor in a risk assessment is the quality of reference data, because the results of the site-specific ecological measurements or calculations are compared against these data. This is true for as well chemical information (i.e. background levels in that region), toxicological data from bioassays (i.e. site relevant reference soil and well characterized control soil in order to verify the test performance) and ecological field surveys.

The reference soil should in principle resemble the contaminated soil in all relevant parameters, e.g. texture, pH, organic matter, water-holding capacity, nutrient content. In practice, these ideal spots are difficult to find. If there is no or inadequate reference information, effects can only be determined in relative terms by comparison with other sites. This is usually adequate for determining the degree of urgency and/or the need for remediation. Reference data can be obtained by including reference sites (preferably more than one) in the sampling scheme, including reference measurements in the experimental set-up, or by obtaining reference data from the literature or by expert based judgment.

Figure 10. Litter bags [94]

Higher tier assessment of the impact on biological activity and organic matter breakdown

In addition to the general information about biological activity in soils generated in Tier 2 from the bait-lamina test, other, slightly more laborious, tests may give additional information about the overall biological activity in soil, e.g. wheat straw decomposition (litter bag test) and cotton strip degradation. A review paper from Van Gestel et al. (2003) concluded that while the bait-lamina gave the best reflection of the biological activity of soil animals, e.g. earthworms, springtails and enchytraeids, the litter bag test and the cotton strip test are more indicative of the microbial activity in the soil [96]. Knacker et al. (2003) reviewed the usefulness of five different litter decomposition tests and concluded that the litter-bag test had distinct advantages over the others [97]. All of these simple tests only give insight into the overall activity in soils and the breakdown of organic material. They are hence most suitable on their own in cases of land-use with low sensitivity, e.g. industrial land. For land-uses where structural endpoints, e.g. biodiversity or specific species, are the target of protection other endpoint(s) should be monitored as well.

Higher tier assessment of the impact on the microbial community

The number of microorganisms, especially bacteria, in soil is extremely large. They differ widely in their function and sensitivity to chemicals.

Besides more classical (and simple) measures of the microbial community like total bacterial biomass, the number of colony forming units and substance induced respiration rate (SIR), more advanced methods for assessing the impact of contamination on soil microorganisms have recently been made available.

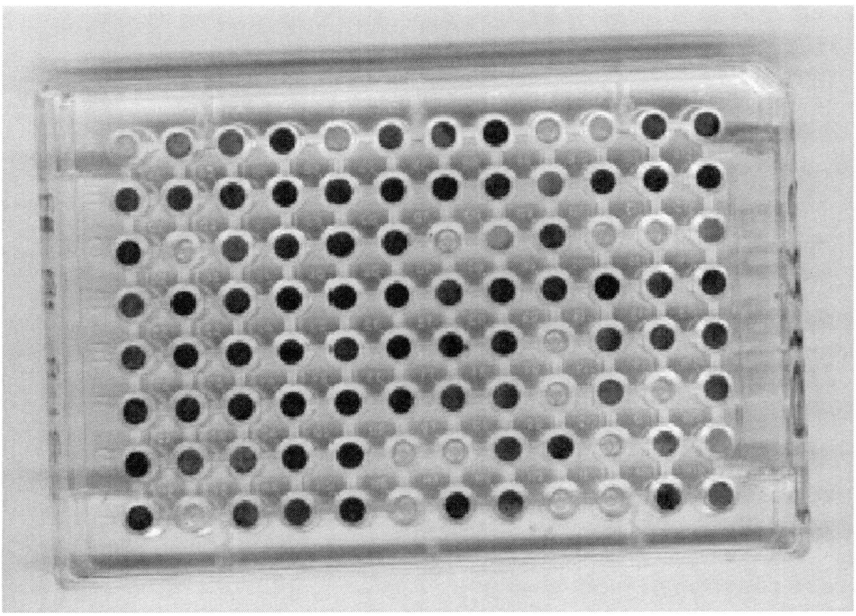

Figure 11. Microbial community [96]

This include microbial "fingerprinting" like phospholipid fatty acid analysis (PLFA) and community-level physiological-profiling (CLPP) based on metabolic response using BIOLOG plate systems, and the use of pollution induced community tolerance (PICT).

Higher tier assessment of the impact on the plant community

Plants interact dynamically with the physical and chemical characteristics of soils. Soil types and site characteristics, therefore, greatly influence the occurrence of plants and their total aboveground biomass (also called, Net Primary Production, i.e. NPP) within given climatic conditions and human management. Vegetation cover is an important indicator of soil quality and a diverse plant community is normally a good indication of essential soil functions such as the decomposition process, the mineralisation rate, and the occurrence of soil dwelling animals associated to (fresh) organic matter. Vascular plants are easy to sample. They are immobile and hence associated to soil contamination (and airborne pollution). Plant community parameters like plant cover, aboveground plant biomass, plant shoot/ root ratio, species diversity and the binary occurrence (presence/absence) of specific indicator species like metal-tolerant species may be used successfully in ERA.

Figure 12. Plant survey [95]

Higher tier assessment of the impact on the soil invertebrate community

Survey of soil biota in order to evaluate the effect of various sources of pollution on soil communities on historically contaminated sites have not yet been used on a larger scale by e.g. consultants. However, numerous (monitoring) studies by various research groups can be found in the open literature. Methods of surveying include:

- Collection of soil samples followed by extraction in the laboratory.
- Extraction or collection of organisms in the field, e.g. by hand-sorting or by the application of mustard or formalin.
- Trapping (surface dwelling) animals by the use of e.g. pit-falls. Monitoring species includes earthworms, snails, oribatid mites, nematodes, springtails, ants, ground-living beetles and spiders. Most of the studies have been done on metal contaminated sites (see references

below). A substantial amount of work has been put into the challenge of developing a soil invertebrate system for evaluating risk of pollutants. The only soil invertebrate system that is used on a regularly basis in the context of ecological risk assessment of contaminated soils is most likely the nematode Maturity Index (MI). The system is based on the evidence, that rapid colonising species dominate nematode communities in disturbed ecosystems. In the Netherlands experience with surveys of soil invertebrates from the monitoring programme Biological Indicator for Soil Quality (BISQ) has also been used in ERA..

Figure 13. Soil fauna sampling [98]

a. Toolbox for tests in Tier IV

The final assessment in the ERA process is not likely to be initiated for many contaminated sites. The choice of additional tests or monitoring at this level of the ERA is bound to be very site-specific and hence an issue for negotiation between stakeholders and experts. Accumulation in biota is included in this toolbox as the internal concentration in biota is believed, at least to some extend, to reflect uptake and then bioavailability. An alternative in this final tier could also be to model uptake in biota provided sufficient data is available [98].

Figure 14. Detailed field survey [93]

4. SOIL MANAGEMENT PROGRAM (REMEDIATION)

4.1. Reporting Norms and Standards For Contaminated Land

The aim is to investigate whether the land has been contaminated, and if contamination has occurred whether the contamination presents a significant risk of harm. The commonly encountered international practice consists of three distinct reporting phases. It progresses from Phase 1 desktop and site walkover assessments with limited investigation and testing to a Phase 2 detailed invasive investigation and testing for site characterisation to a comprehensive Phase 3 report with an evaluation of remediation objectives and a proposed remediation plan, supported by control and monitoring measures for the activities.

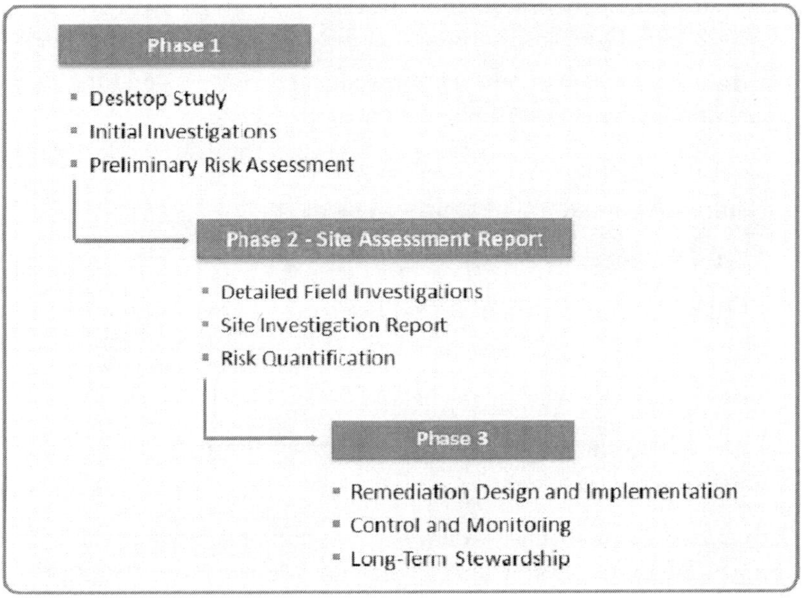

Figure 15. A phase approach for the assessment and remediation of contaminated land

The reporting system requires norms and standards of practice to be strictly applied, but also must retain flexibility to allow for decisions on the contaminated status of sites to be made in the most beneficial manner (considering ecological, social and economic aspects) also taking into account timeframes. In some cases, urgent priority works may require that the phased approach to reporting has to move forward in a concurrent single report.

4.2. Requirements for Preliminary Site Assessment Phase 1

A preliminary site assessment must consider the following elements

1. Site description-location and size
2. Nature and extent of contamination, contaminants of concern and historic activities that may be sources of contamination. List all present and past activities at the site that involves storage and

production, use, treatment or disposal of hazardous material that could contaminate the site.

3. Describe current condition of the site and contents and the results of any previous assessment report
4. Local topography and geology, drainage, surface cover, vegetation.
5. Status of ground water, approximate depth to water table
6. Proximity to surface water
7. Proximity to drinking water supplies
8. Annual rainfall and flood potential
9. Land and water use for the nearby areas and site
10. Any other regulations as Regulated by the ministry

All data may not be available, or data may vary in terms of uncertainty, it is thus important to recognize gaps in the knowledge base and to decide whether additional data must be obtained on the site characterisation. This may trigger the commencement of Phase 2 Investigations. The Phase 1 report must make clear recommendations on the status of the contamination risk posed by the site. If a complete site history clearly demonstrates that the site activities do not pose a contamination threat then no further investigation is warranted and the site should be recommended as suitable for reuse. In most cases it is likely that some level of preliminary investigation will be required to provide the level of certainty required to enable property re-development or transfer. A limited investigation of certain subsurface activities, for example, underground storage tanks, would be necessary to obtain a waiver on the contamination status of a site at a Phase 1 level of reporting.

4.3. Approval Requirements

If soil contaminants are found at concentrations that exceed the applicable standards specified in the approval, the approval holder is required to implement a soil management program.The program must first address source control to stop on-going contaminant releases. After the sources of contamination have been stopped, further assessment and delineation of the contaminated area may be necessary. When the extent of contamination is understood, remediation objectives, as described below, must be adopted for the area. Once remediation objectives have been agreed to, appropriate treatment or containment technologies can be chosen and the management plan finalized.

4.4. Developing Remediation Objectives

Remediation objectives may be developed in a variety of ways ranging from generic guidelines to site-specific risk assessment. Generic guidelines are numerical concentration limits that are applicable under a variety of site conditions. When neither Environmental Protection nor the agency has a guideline for a particular substance, four options are available. First, the remediation objective may be based on the ambient background concentration for the site. Second, the guideline development protocol may be applied. Third, a remediation objective may be adopted from another jurisdiction if the proponent can show that the remediation objective is consistent with the environmental protection goals of the approval. Finally, a remediation objective may be developed by the proponent using site-specific risk assessment procedures. A risk assessment for an approved facility will focus on human health concerns but fundamental ecological concerns must also be addressed. Site-specific risk assessment is a means of quantifying the likelihood that soil contamination will have a harmful effect under conditions found at a specific site. The essential components of human health and ecological risk assessments are similar; however, ecological risk assessments tend to be more complex than those for human health because a wide variety of receptors may have to be considered. Very briefly, a site-specific risk assessment may be described as consisting of the following steps or components as shown in Figure 20;

Problem formulation involves developing a conceptual model of the possible contaminant effects on receptors at the site. The conceptual model describes contaminant distribution and concentration in relation to the receptors and their patterns of activity on the site. The *exposure assessment* describes the pathways by which soil contaminants may be taken up by the receptor. This information is combined with receptor characteristics in order to estimate the contaminant uptake. The *toxicity assessment* describes the adverse effects that the contaminants may cause and the dose at which these effects occur.

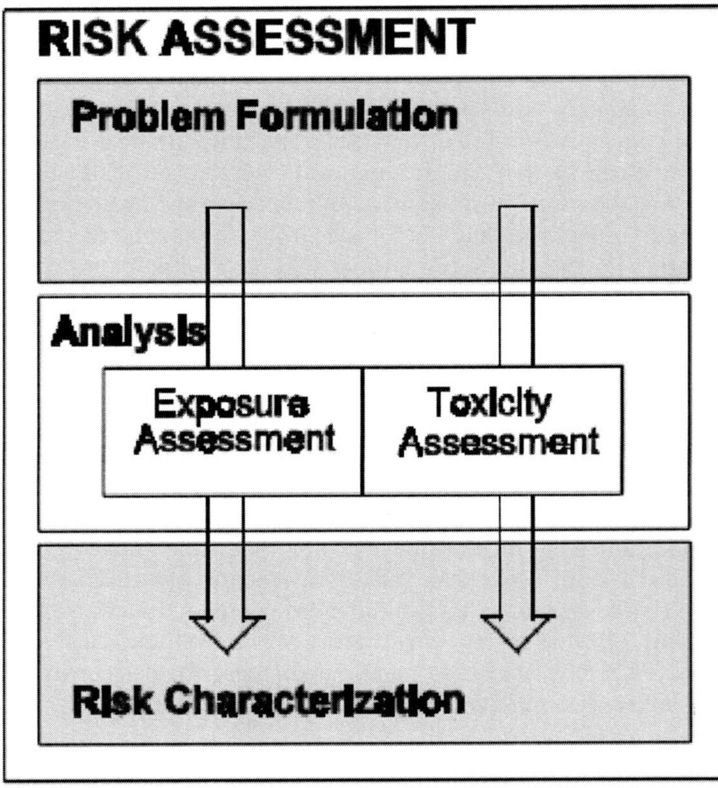

Figure 16. Risk Assessment Framework

The final step, *risk characterization*, compares this dose with the uptake rate (estimated during the exposure assessment) and determines whether or not an adverse effect is likely to occur. Risk assessment procedures can also be used to back calculate a contaminant concentration in soil at which no adverse effects are expected. At times, it may not be possible to remediate to a level compatible with industrial land-use objectives. In such cases, the approval holder must ensure that the contaminants are contained and receptor exposure does not occur. Formal risk assessment procedures are necessary to meet these requirements. Typically, an engineered containment system is required, the performance of which must be confirmed through periodic inspections and monitoring. The approval holder bears responsibility for the design, construction, assessment and maintenance of the risk management system, and any necessary emergency response should the system fail.

4.5. Enforcement Policy

Enforcement is generally the last step in a regulatory process that aims to first prevent potential problems before they arise and to resolve existing problems in a cooperative manner. When discussions between the approval holder and the Department fail to prevent or resolve a contravention of an approval or the Act, a number of abatement and enforcement tools are available to the Department including:

- environmental protection orders;
- warning letters;
- tickets;
- administrative penalties;
- enforcement orders;
- prosecutions;
- court orders; and
- cancellation of approvals or certificates.

Due to various factors, including the high cost involved in remediation interventions, it is essential that a holistic and tiered, risk-based approach be adopted that is founded on international best practice, to address remediation in a uniform manner across the country. This is relevant irrespective of the sector of occurrence to safeguard both human health and the natural environment. The Framework is based on a review of international practice in the developed countries of the world and the emergence of remediation policy from developing countries, and an assessment of alternative approaches and methodologies that may find application in the development of a remediation framework.

5. CONCLUSION

Soil pollution is a result of many activities and experiments done by mankind which end up contaminating the soil. Industrial wastes such as harmful gases and chemicals, agricultural pesticides, fertilizers and insecticides are the most common causes of soil pollution. The others are ignorance towards soil management and related systems, unfavourable and harmful irrigation practices, improper septic system and management and maintenance of the same, leakages from sanitary sewage. There is urgent need for a tiered approach in ecological risk assessment of contaminated soils. Generic soil screening levels are needed as a first tier. Higher tiers of ecological risk assessment should, however, contain some kind of site-specific assessment. It is furthermore important to organize the various studies in a framework or decision support system that is transparent and useful for all stakeholders. A weight of evidence approach may be an obvious choice to deal with these uncertainties. The TRIAD approach, which incorporates and categorizes information in a triangle – chemistry, toxicology, and ecology – is an appropriate tool for handling conceptual uncertainties. Several remedies to these shortcomings have been proposed. Regarding ecotoxicity direct testing would allow for a major improvement in risk estimates. As to human health risks: including biological availability in risk estimates, more use of up to date knowledge about exposure routes, dose-effect relations and combination effects, and biomonitoring of effects are options for improvement.

6. ACKNOWLEDGEMENTS

Authors greatly acknowledge the University Malaya Research Grant (RG257-13AFR). Moreover, the comments of two anonymous reviewers are gratefully acknowledged.

REFERENCES

1. Okrent D. 1999. On intergenerational equity and its clash with intragenerational equity and on the need for policies to guide the regulation of disposal of wastes and other activities posing very long time risks. Risk Analysis 19: 877-901.
2. Belluck, D.A., Benjamin, S.L., Baveye, P., Sampson, J., Johnson, B. 2003. Widespread arsenic contamination of soils in residential areas and public spaces: an emerging regulatory or medical crisis? International Journal of Toxicology 22: 109-128.
3. Richardson, G.M., Bright, D.A., Dodd, M. 2006. Do current standards of practice in Canada measure what is relevant to human exposure at contaminated sites? II: oral bioaccessibility of contaminants in soil. Human and Ecological Risk Assessment 12: 606-618.
4. Van Zorge, J. A. 1996. Exposure to mixtures of chemical substances: is there a need for regulations? Food and Chemical Toxicology 34, 1033-1036.
5. Swartjes, F.A. 1999. Risk-based assessment of soil and groundwater quality in the Netherlands: standards and remediation urgency. Risk Analysis 19:1235-1248.
6. Eikelboom, R.T., Ruwiel E., Goumans, J.J.M. 2001. The building materials decree: an example of a Dutch regulation based on the potential impact of materials on the environment. Waste Management 21: 295-302.
7. Crane, M. and Giddings, J.M. 2004. 'Ecologically acceptable concentrations' when assessing the environmental risks of pesticides under European Directive 91 414/EEC. Human and Ecological Risk Assessment 10: 733-747.

8. Nathanail, P., McCaffrey, C., Earl, N., Forster, N.D., Gillett, A.G., Ogden, R. 2005. A deterministic method for deriving site-specific human health assessment criteria for contaminants in soil. Human and Ecological Risk Assessment 11: 389-410.
9. Tarazona, J.V., Fernandez, M.D., Vega, M.M. 2005. Regulation of contaminated soils in Spain. Journal of Soil and Sediments 5:121-124.
10. Evans, J., Wood, G., Miller, A. 2006. The risk assessment-policy gap: An example from the UK contaminated land regime. Environment International 32: 1066-1071.
11. Huinink, J.T.M. 1998. Soil quality requirements of use in urban environments. Soil and Tillage Research 47: 157-162.
12. Urzelai, A., Vega, M., Angulo, E. 2000. Deriving ecological risk-based soil quality values in the Basque Country. Science of the Total Environment 247: 279-284.
13. Apitz, S.E. 2008. Is risk based, sustainable sediment management consistent with European policy. Journal of Soils and Sediments 8: 461-466.
14. Provoost, J., Cornelis, C., Swartjes, F. 2006. Comparison of soil clean-up standards fort race elements between countries: why do they differ? Journal of Soil and Sediments 6: 173-181.
15. Provoost, J., Reijnders, L., Swartjes, F., Bronders, J., Carlon, C., D'Allessandro, M., Cornelis, C. 2008. Parameters causing variation between soil screening values and the effect of harmonization. Journal of Soils and Sediments 8: 298-311.
16. Toccalino, P.L. and Norman, J.E. 2006. Health-based screening levels to evaluate US Geological Survey groundwater quality data. Risk Analysis 26: 1339-1348.
17. Patterson, M.M., Cohen, E., Prommer, H., Thomas, D.G., Rhodes, S., McKinley, A.I. 2007. Origin of mixed brominated ethane groundwater plume: contaminant degradation pathways and reactions. Environmental Science & Technology 41: 1352-138.
18. Leon Paumen, M. 2008. Invertebrate life cycle responses to PAC exposure. PhD thesis. Amsterdam: University of Amsterdam.
19. Nawrot, T., Plusquin, M., Hogervorst, J., Roels, H.A., Celis, H., Thijs, L., Vangronsveld, J., Van Hecke, E., Staessen, J. 2006. Environmental exposure to cadmium and risk of cancer: a prospective population-based study. The Lancet Oncology 7: 119-126.
20. Simcox, N.J., Fenske, R.A., Wolz, S.A., Lee, I.W., Kalman, D.A. 1995. Pesticides in household dust and soil: exposure pathways for children of agricultural families. Environmental Health Perspectives 103: 1126-1134.
21. Laidlaw, M.A.S., Mielke, H.W., Filippelli, G.M., Johnson, D.L., Gonzales, C.R. 2005. Seasonality and children's blood lead levels: developing a predictive model using climate variables and blood data for Indianapolis, Indiana, Syracuse, New York and New Orleans Louisiana (USA). Environmental Health Perspectives 113: 793-800.
22. Laidlaw, M.A.S., Mielke, H.W., Filippelli, G.M., Johnson, D.L. 2006. Blood lead in children: Laidlaw et al. respond. Environmental Health Perspectives 114: A 19.
23. Rothenberg, S.J. and Rothenberg, J.C. 2005. Testing the dose-response specification in epidemiology: public health and policy consequences for lead. Environmental Health Perspectives 113: 1190-1195.
24. Akesson, A., Lundh, T., Vahter, M., Bjellerup, P., Lidfeldt, J., Nerbrand, C., Samisoe, G., Strömberg, U., Skerfving, S. 2005. Tubular and glomerular kidney effects in Swedish women with low environmental cadmium exposure. Environmental Health Perspectives 113: 1627-1631.
25. Nawrot, T.S., van Hecke, E., Thijs, L., Richart, T., Kuznetsova, T., Jin, Y., Vangronsveld, J., Roels, H.A., Staessen, J.A. 2008. Cadmium-related mortality and long-term secular trends in the cadmium body burden of an environmentally exposed population. Environmental Health Perspectives 116: 1620-1628.
26. Lamphear, R.P., Dietrich, K., Auinger, P., Cox, C. 2000. Cognitive defects associated with blood lead concentrations, 10 microgram/dl in United States children and adolescents. Public Health Reports 115: 521-529.
27. Wang, C.L., Chuang, H.J., Ho, C.K., Vang, C.Y. 2002. Relationship between blood lead concentrations and learning achievement among primary schoolchildren in Taiwan. Environmental Research A 89: 167-181.

28. Wu, T., Buck, G.M., Mendola, P. 2003. Blood levels and sexual maturation in US girls. Environmental Health Perspectives 111: 737-741.
29. Von Storch, H., Costa-Cabral, M., Hagner, C., Freser, F., Pacyna, Y., Pacyna, E., Kolb, S. 2003. Four decades of gasoline lead emissions and control policies in Europe: a retroactive assessment. Science of the Total Environment 311: 151-176.
30. Mielke, H.W., Gonzales, C.R., Powell, E., Jartun, M., Mielke Jr., P.W.2007. Nonlinear association between soil lead and blood lead of children in Metropolitan New Orleans, Louisiana: 2000-2005. Science of the Total Environment 388: 43-53.
31. Provoost, J., Cornelis, C., Swartjes, F. 2006. Comparison of soil clean-up standards for trace elements between countries: why do they differ? Journal of Soils and Sediments 6: 173–181.
32. Miranda, M.L., Kim, D., Overstreet-Galeano, M.A., Paul, C.J., Hull, A.P., Morgan, S.P. 2007. The relationship between early childhood blood lead levels and performance on end-of-grade tests. Environmental Health Perspectives 115: 1242-1247.
33. Patterson, B. M., Cohen, E., Prommer, H., Thomas, D. G., Rhodes, S., McKinley, A. J. 2007. Origin of a mixed brominated ethene groundwater plume: contaminant degradation pathways and reactions. Environmental Science & Technology 41: 1352–1358.
34. Huel, G., Sahuquillo, J., Debotte, G., Oury, J., Takser, L. 2008. Hair mercury negatively correlates with calcium pump activity in human term newborns and their mothers at delivery. Environmental Health Perspectives 116: 263-267.
35. Nawrot, T., Plusquin, M., Hogervorst, J., et al., 2006. Environmental exposure to cadmium and risk of cancer: a prospective population-based study. The Lancet Oncology 7: 119–126.
36. Laidlaw, M. A. S., Mielke, H. W., Filippelli, G. M., Johnson, D. L. 2006. Blood lead in children: Laidlaw et al. respond. Environmental Health Perspectives 114: A19-A23.
37. Akesson, A., Lundh, T., Vahter, M., et al., 2005. Tubular and glomerular kidney effects in Swedish women with low environmental cadmium exposure. Environmental Health Perspectives 113: 1627–1631.
38. Mergler, D., Anderson, H.A., Chan, L.H.M., Mahaffey, K.R., Murray, M., Sakamoto, M., Stern, H.A. 2007. Methylmercury exposure and health effects in humans: a worldwide concern. Ambio 36: 3-11.
39. ten Tusscher, G.W. 2002. Later childhood effects of perinatal exposure to background levels of dioxins in the Netherlands. PhD thesis. Amsterdam: University of Amsterdam.
40. J. A. Van Zorge, "Exposure to mixtures of chemical substances: is there a need for regulations?"Food and Chemical Toxicology, vol. 34, no. 11-12, pp. 1033–1036, 1996.
41. Filser, J., Koehler, H., Ruf, A., Rombke, J., Prinzing, A., Schaefer, M. 2008. Ecological theory meets soil ecotoxicology: challenge and chance. Basic and Applied Ecology 9: 346-355.
42. Van Gestel, C.A.M. 2008. Physico-chemical; and biological parameters determine metal bioavailability in soils. Science of the Total Environment 406: 385-395.
43. Sokolik, G.A., Ovsiannikova, S.V., Ivanova, T.G., Leinova, S.L. 2004. Soil-plant transfer of plutonium and americium in contaminated regions of Belarus after the Chernobyl catastrophe. Environment International 30: 939-947.
44. Zhao, C., Ren, J., Xue, C., Lin, E., 2005. Study on the relationship between soil selenium and plant selenium uptake. Plant and Soil 277: 197-206.
45. Liste, H., Prutz, I. 2006. Plant performance, dioxygenase-expressing rhizosphere bacteria, and biodegradation of weathered hydrocarbons in contaminated soil. Chemosphere 62: 1411-1420.
46. van der Geest, H, Leon Paumen, M. 2008. Dynamics of metal availability and toxicity in historically polluted floodplain sediments. Science of the Total Environment 406: 419-425.

47. Brouwer, A., Ahlborg, U.G., van den Berg, M., Birnbaum, L.S. 1995. Functional aspects of developmental toxicity of polyhalogenated hydrocarbons in experimental animals and human infants. European Journal of Pharmacology 293: 1-40.
48. Rutgers, M., Tuinstra, J., Spijker, J., Mesman, M., Wintersen, A., Posthuma, L. 2008. Risico's voor het ecosysteem in stap twee van het saneringscriterium [Ecosystem risks in step 2 of the remediation criterion]. Bilthoven (the Netherlands): RIVM.
49. Li, J., Zhou, B., Liu, Y., Yang, Q., Cai, W. 2008. Influence of coexisting contaminants on bisphenol A sorption and desorption in soil. Journal of Hazardous Materials 151: 389-393.
50. Gennings, C., Carter Jr, W.H., Casey, M., Moser, V., Carchman, R., Simmons, J.E. 2004. Analysis of functional effects of five pesticides using a ray design. Environmental Toxicology and Pharmacology 14: 115-126.
51. Hayes, T.B, Case, P., Chui, S., Chung, D., Haeffele, C., Haston, K., Lee, M., Mai, V.P., Marjuoa, Y., Parker, J., Tsui, M. 2006. Pesticide mixtures, endocrine disruption and amphibian declines: are we underestimating the impact. Environmental Health Perspectives 114, Supplement 1: 40-50.
52. Perry, M.J., Venners, S.A., Barr, D.B., Xu, X. 2007. Environmental pyrethoid and organophosphorous insecticide exposures and sperm concentration. Reproductive Toxicology 23: 113-118.
53. Laetz, C.A., Baldwin, D.H., Clllier, T.K., Hebert, V., Stark, J.D., Scholz, N.L. 2009. The synergistic toxicity of pesticide mixtures: implications for risk assessment and the conservation of the endangered Pacific salmon. Environmental Health Perspectives 117: 348-353.
54. Roos, P. H., Tschirbs, S., Pfeifer, F., et al., 2004. Risk potentials for humans of original and remediated PAH-contaminated soils: application of biomarkers of effect. Toxicology 205: 181–194.
55. Xiao, R. Y., Wang, Z., Wang, C. X., Yu, G., Zhu, Y. G. 2006. Genotoxic risk identification of soil contamination at a major industrialized city in northeast China by a combination of in vitro and in vivo bioassays. Environmental Science & Technology 40: 6170–6175.
56. Lee, D. H., Lim, J. S., Song, K., Boo, Y., Jacobs Jr., D. R. 2006. Graded associations of blood lead and urinary cadmium concentrations with oxidative-stress-related markers in the U.S. population: results from the third National Health and Nutrition Examination Survey. Environmental Health Perspectives 114: 350–354.
57. De Zwart, D. and Posthuma L. 2005. Complex mixture toxicity for single and multiple species: proposed methodologies. Environmental Toxicology and Chemistry 24: 2665-2676.
58. Altenburger, R., Walter, H., Grote, M. 2004. What contributes to the combined effect of a complex mixture? Environmental Science & Technology 38: 6353-6362.
59. Filser, J., Wittmann, R., Lang A. 2000. Response types in Collembola towards copper in the microenvironment. Environmental Pollution 107: 71-78.
60. Drexler, J.W. and Brattin, J.W. 2007. An in vitro procedure for estimation of lead relative bioavailability: with validation. Human and Ecological Risk Assessment 13: 383-401.
61. O'Halloran, K. 2006. Toxicological considerations of contaminants in the terrestrial environment for ecological risk assessment. Human and Ecological Risk Assessment 12: 74-83.
62. Römbke, J. 2006. Tools and techniques for the assessment of ecotoxicological impacts of contaminants in the terrestrial environment. Human and Ecological Risk Assessment 12: 84-101.
63. Roos, P.H., Tschirbs, S., Pfeifer, F., Welge, P., Hack, A., Wilhelm, M., Bolt, H.M. 2004. Risk potentials for humans of original and remediated PAH- contaminated soils: application of biomarkers of effect. Toxicology 205: 181-194.
64. Xiao, R., Wang, Z., Wang, C., Yu, G., Zhu, Y. 2006. Genotoxic risk identification of soil contamination at a major industrialized city in north east China by a combination of in vitro and in vivo bioassays. Environmental Science & Technology 40. 6170-6175.

65. Lee, D., Lim, J., Song, K., Boo, Y., Jacobs, D.R. Jr. 2006. Graded associations of blood lead and urinary cadmium concentrations with oxidative- stress-related markers in the US population: Results of the Third National Health and Nutrition Examination Survey. Environmental Health Perspectives 114: 350-354.
66. Neumann, H.G. 1996. Toxic equivalence factors, problems and limitations. Food and Chemical Toxicology 34: 1045-1051.
67. Dobrev, I.D., Andersen, M.E., Yang, R.S.H. 2002. In silico toxicology: simulating interaction thresholds for human exposure to mixtures of trichloroethylene, tetrachloroethylene and 1,1,1 trichloroethane. Environmental Health Perspectives 110: 1031-1039.
68. Silva, E., Rajapakse, N., Kortenkamp, A. 2002. Something from 'nothing'-eight weak estrogenic chemicals combined at concentrations below NOECs produce significant mixture effects. Environmental Science & Technology 36: 1751-1756.
69. Richter, M. and Escher, B. 2005. Mixture toxicity of reactive chemicals by using two bacterial growth assays as indicators of protein and DNA damage. Environmental Science & Technology 39: 8753-8761.
70. Caeiro, S., Costa, M. H., Ramos, T. B. 2005. Assessing heavy metal contamination in Sado Estuary sediment: An index analysis approach. Ecological Indicators 5: 151–169.
71. Håkanson, L. 1980. An ecological risk index for aquatic pollution control: A sedimentological approach. Water Research 14: 975–1001.
72. Long, E. R., MacDonald, D. D. 1998. Perspective: Recommended uses of empirically derived, sediment quality guidelines for marine and estuarine ecosystems. Human and Ecological Risk Assessment 4: 1019–1039.
73. Pekey, H., Karakaş, D., Ayberk, S., et al., 2004. Ecological risk assessment using trace elements from surface sediments of İzmit Bay (Northeastern Marmara Sea) Turkey. Marine Pollution Bulletin 48: 946–953.
74. Roach, A. C. 2005. Assessment of metals in sediments from Lake Macquarie, New South Wales, Australia, using normalization models and sediment quality guidelines
75. Sutherland, R. A. 2000. Bed sediment-associated trace metals in an urban stream, Oahu, Hawaii. Environmental Geology 39: 611–627
76. Bhattacharya, A., Routh, J., Jacks, G., et al., 2006. Environmental assessment of abandoned mine tailings in Adak, Västerbotten District (Northern Sweden). Applied Geochemistry 21: 1760–1780
77. Usero, J., González-Regalado, E., Gracia, I. 1996. Trace metals in the Bivalve Mollusc Chamelea Gallina from the Atlantic Coast of Southern Spain. Marine Pollution Bulletin 32: 305–310.
78. Cheng, J. L., Shi, Z., Zhu, Y. W. 2007. Assessment and mapping of environmental quality in agricultural soils of Zhejiang Province, China. Journal of Environmental Sciences, 19: 50–54
79. Reeves, W.R., Barhoumi, R., Burghardt, R.C., Lemke, S.J., Mayura, K., McDonald, T.J., Phillips, T.D., Donelly, K.C.2001. Evaluation of methods for predicting the toxicity of polyaromatic hydrocarbon mixtures. Environmental Science & Technology 35: 1630-1636.
80. El Masri, H., Mumtaz, M.M., Yushak, M.L. 2004. Application of physiologically-based pharmacokinetic modeling to investigate the toxicological interaction between chlorpyrifos and parathion in rat. Environmental Toxicology and Pharmacology 16: 57 71.
81. Kortenkamp, A. and Altenburger, R. 1999. Approaches to assessing combination effects of estrogenic environmental pollutants. Science of the Total Environment 333: 131-140.
82. Sumpter, J.P. and Johnson, A.C. 2005. Lessons from endocrine disruption and their application to other issues concerning trace organics in the aquatic environment. Environmental Science & Technology 39: 4321-4332.
83. Brian, J.V, Harris, C.A., Scholze, M., Backhaus, T., Booy, P., Lamoree, M., Pojana, G., Jonkers, N., Runnalis, T., Bonfä, A., Marcomini, A., Sumpter, J. 2005. Accurate

prediction of the response of freshwater fish to a mixture of estrogenic chemical. Environmental Health Perspectives 113: 721-728.

84. Wölfe, D. 1997. Interactions between 2,3,7,8 TCDD and PCs as tumor promoters: limitations of TEFs. Teratogenesis, Carcinogenesis and Mutagenesis 17: 217-224.
85. Marchant, C.A. 1996. Prediction of rodent carcinogenicity using the DEREK system for 30 chemicals currently being tested by the National Toxicology Program. Environmental Health Perspectives 104: 1065-1073.
86. Verhaar, H.J.M., Morroni, J.R., Reardon, K.F., Hays, S,M., Gaver, D.P., Carpenter, R.L., Yang, R.S.H. 1997. A proposed approach to study the toxicology of complex mixtures of petroleum products. Environmental Health Perspectives 195 (Supplement 1): 179-195.
87. Stevenson, C.N., Macmanus-Spencer, L.A., Luckenbach, T., Luthy, R.C., Epel, D. 2006. New perspectives on perfluorochemical ecotoxicology: inhibition and induction of an efflux transporter in the marine mussel Mytilus californianus. Environmental Science & Technology 40: 5580-5585.
88. Bailer, A.J., Hughes, M.R., See, K., Noble, R., Schaefer, R. 2002. A pooled response strategy for combining multiple lines of evidence to quantitatively estimate impact. Human and Ecological Risk Assessment 8:1597-1612..
89. Chapman, P.M., McDonald, B.G., Lawrence, G.S. 2002. Weight-of-evidence issues and frameworks for sediment quality (and other) assessments. Human and Ecological Risk Assessment 8:1489-1515.
90. Rutgers, M., den Besten, P. 2005. Approaches to legislation in a global context.The Netherlands perspective – soil and sediment. In:Toxicity Testing Environmental. Thompson C, Wahhia K, Loibner AP (eds). Blackwell, CRC Press, Oxford, pp 269–289.
91. Alexander, M. 1995. How toxic are toxic chemicals in soil? Environmental Science and Technology. 29:2713–2717.
92. Alexander, M. 2000. Aging, bioavailability, and overestimation of risk from environmental pollutants. Environmental Science and Technology. 34:4259–4265.
93. Callahan, C. A., Menzie, C. A., Burmaster, D. E.,Wilborn, D. C., Ernst, T. 1991 On-site methods for assssesing chemical impact on the soil environment using earthworms: a case study at the Baird and McGuire superfund site, Holbrook, Massachusetts. Environmental Toxicology and Chemistry. 10:817–826.
94. Attrill, M. J., Depledge, M. H., 1997. Community and population indicators of ecosystem health: Targeting links between levels of biological organisation. Aquatic Toxicology 38:183–197.
95. Keddy, C., Greene, J. C., Bonnell, M. A. 1994. A review of whole organism bioassays for assessing the quality of soil, freshwater sediment and freshwater in Canada. Scientific series no. 198. Environment Canada, Ottawa, Ontario, Canada.
96. Van Gestel, C. A. M., Kruidenier, M., Berg, M. P. 2003. Suitability of wheat straw decomposition, cotton strip degradation and bait-lamina feeding tests to determine soil invertebrate activity. Biology and Fertility of Soils 37:115-123.
97. Knacker, T., Förster, B., Römbke, J., Frampton, G.K. 2003. Assessing the effects of plant protection products on organic matter breakdown in arable fields – litter decomposition test systems. Soil Biology and Biochemistry 35:1269-1287.
98. Jager, T., Fleuren, R.H.L.J., Hogendoorn, E.A., De Korte, G. 2003. Elucidating the routes of exposure for organic chemicals in the earthworm, Eisenia andrei (Oligochaeta). Environmental Science and Technology 37:3399-3404.

CHAPTER 12

Geochemical Speciation and Risk Assessment of Heavy Metals in Soils and Sediments

Santosh Kumar Sarkar[1], Paulo J.C. Favas[2, 3], Dibyendu Rakshit[1] and K.K. Satpathy[4]

[1] *Department of Marine Science, University of Calcutta, Calcutta, West Bengal, India*

[2] *Department of Geology, School of Life Sciences and the Environment, University of Trás-os-Montes e Alto Douro, Vila Real, Portugal*

[3] *IMAR-CMA Marine and Environmental Research Centre, Faculty of Sciences and Technology, University of Coimbra, Coimbra, Portugal*

[4] *Indira Gandhi Centre for Atomic Research, Environment and Safety Division, Kalpakkam, Tamil Nadu, India*

1. INTRODUCTION

Heavy metal pollution is a serious and widely environmental problem due to the persistent and non-biodegradable properties of these contaminants. Sediments serve as the ultimate sink of heavy metals in the marine environment and they play an important role in the transport and storage of potentially hazardous metals. They are introduced into the aquatic system as a result of weathering of soil and rocks, from volcanic eruptions and from a variety of human activities involving mining, dredging, processing and use of metals and/or substances containing metal contaminants. Heavy metals entering natural water become part of the water-sediment system and their distribution processes are controlled by a dynamic set of physicochemical interactions and equilibria. The properties of metals in soils and sediments depend on the physiochemical form in which they occur [1]. Heavy metals are distributed throughout soil and sediment components and associated with them in various ways, including adsorption, ion exchange, precipitation and complexation and so on [2]. Changes in environmental conditions, such as temperature, pH, redox potential and organic ligand concentrations, can cause metals to be released from solid to liquid phase and sometimes cause contamination of surrounding waters in aquatic systems [3]. They are not permanently fixed by soil or sediment. Therefore, it cannot provide sufficient information about mobility, bioavailability and toxicity of metals if their total contents are studied alone.

Natural and anthropogenic activities have the capacity to cause changes in environment conditions, such as acidification, redox potential, or organic ligand concentrations, which can remobilize contaminated soils and sediments releasing the elements from soils and sediments and pore water to the water column resulting contamination of surrounding waters. Daily tidal currents, wind energies, and storms in coastal and estuarine systems can cause periodical remobilization of surface sediments [4]. More turbulent conditions, such as seasonal flooding or storms, or bioturbation, due to feeding and movement of benthic organisms, can expose anoxic sediments to oxidant conditions. In addition, activities such as dredging result in major sediment disturbances, leading to changes in chemical properties of sediment [5].

The remediation of heavy metal pollution is often problematic due to their persistence and non-degradability in the environment. As a sink and source, soils and sediments constitute a reservoir of bioavailable heavy metals and play a significant role in the remobilization of contaminants in the aquatic systems under favorable conditions. Such potential of sediment for being a sink as well as a source of contaminant can make sediment chemistry and toxicity key components of the quality of aquatic system. Much concern has been focused on the investigation of the total element contents in soils and sediments. However, it cannot provide sufficient information about mobility, bioavailability and toxicity of elements and thus may not be able to provide information about the exact dimension of pollution. The data on total contents of metals are quite insufficient to estimate the possible risk of remobilization of total metals under changing environmental conditions and potential uptake of liberated metals by biota and thus the determination of different fractions assume great importance. This has been described as "speciation" [6]. Since each form have different bioavailability and toxicity, the environmentalists are rightly concerned about the exact forms of metal present in the aquatic environment.

The concept of speciation dates back to 1954 when Goldberg introduced the concept of speciation to improve the understanding of the biogeochemical cycling of trace elements in seawater. Kinetic and thermodynamic information together with the analytical data made it possible to differentiate between oxidized versus reduced, complexed or chelated versus free metal ions in solution and dissolved between particulate species. Florence [7] has defined the term speciation analysis as the determination of the individual physicochemical forms of the element, which together make up its total concentration in a sample. According to Lung [8], speciation analysis involves the use of analytical methods that can provide information about the physicochemical forms of the elements. Schroeder [9] distinguishes physical speciation, which involves differentiation of the physical size or the physical properties of the metal, and chemical speciation, which entails differentiation among the various chemical forms. The main objective of measuring metal species relates to their relative toxicities to aquatic biota. The second and long term aim of speciation studies is to advance an understanding of metal interactions between water and bed sediments in an aquatic ecosystem. In the last decade researchers have followed different sequential extraction techniques for the fractionation of metals in sediments of different river systems. Rauret et al. [10] studied the speciation of copper and lead in the sediments of River Tenes (Spain) while Pardo et al. [11] studies the speciation of zinc, cadmium, lead, copper, nickel and cobalt in the sediments of Pisuerga River, Spain, in order to establish the extent to which these are polluted and their capacity to remobilization. Jardo and Nickless [12] investigated the chemical association of zinc, cadmium, lead and copper in soils and sediments of England and Wales. In most samples, these four metals were associated with all the chemical fractions. Tessier et al. [13] studied speciation of cadmium, cobalt, copper, nickel, lead, zinc, iron and manganese in water and sediments of St. Fransois River, Quebec, Canada. Elsokkary and Muller [14] studied speciation of chromium, nickel, lead and cadmium in the sediments of Nile River, Egypt, reporting that a high proportion of chromium, nickel and lead is bound to organic material and sulphides, while cadmium is bound to carbonate fraction. Ure [15] and Rauret [16] have reviewed the chemical extraction procedures used for heavy metal determinations in contaminated soils and sediments. Owing to the need for validation of extraction schemes, the EC Measurement and Testing Programme (formerly BCR) has organized a project for improving the quality of determinations of extractable heavy metals, where development and validation of extraction procedures has been discussed [17, 18].

The present article aims to summarize the potentials of sequential extraction technique adopting different analytical protocols for gaining information on the mobility and dynamics of operationally determined chemical forms of heavy metals in soils and sediments. The BCR (Community Bureau of Reference, now superseded by the Standards,

Measurement and Testing Programme of the European Community) procedure has been illustrated considering the case study of Ganges (Hugli) River Estuary and adjacent Indian Sundarban mangrove wetland (a UNESCO World Heritage Site), northeastern part of the Bay of Bengal. In addition, the authors also evaluate the modified BCR sequential extraction technique as devised by various scientists, the risk assessment code (RAC) as well as assessment of toxicity comparing with sediment quality guidelines. The RAC classification is based on the strength of the bond between the metals and the different geochemical fractions in sediments or soils and the ability of metals to be released and enter into the food chain.

2. SEQUENTIAL EXTRACTION: MERITS AND DEMERITS

The sequential extraction provides more or less detailed information concerning the origin, mode of occurrence, biological and physicochemical availabilities, mobilization and transport of heavy metals. The procedure stimulates the mobilization and retention of these species in the natural environment using changes in environmental condition such as pH, redox potential and degradation of organic matter [16]. A series of reagents is applied to the sample, increasing the strength of the extraction at each step, in order to dissolve the trace metal present in different sediment phases. The extractants are inert electrolytes, weak acids, reducing agent, oxidizing agents and strong mineral acids [19].

The 3-stage sequential extraction procedure proposed by the European Community Bureau of Reference (BCR) was developed in an attempt to standardize the various schemes described in the literature [2, 20, 21], since the use of different procedures, varying in the number of steps, types of reagents and extraction condition. Hindered comparison of results obtained in the many studies of heavy metals chemical fractionation in environmental samples [22].

The BCR methods has been widely adapted by various authors, and applied to a range of type of solid sample including fresh water sediment [23-25], salt water sediment [26-28], sewage sludge and particulate matter [29-31]. This scheme enables us to associate the meals with one of the following four geochemical phases:

I. Acid-soluble phase: This phase is made up of exchangeable metals and others bound to carbonates that are able to pass easily into the water column, for example, when the pH drops. It is the fraction with the most labile bond to the soil/sediment and, therefore, the most dangerous for the environment.

II. Reducible phase: This phase consists of metals bound to iron and manganese oxides that can be released if the sediment changes from the oxic to the anoxic state, which could be caused, for example, by the activity of microorganisms present in the soils/sediments.

III. Oxidisable phase: This shows the amount of metal bound to organic matter and sulphides, which can be released under oxidizing conditions. Such conditions can occur, for example, if the sediment is resuspended (by dredging, currents, flooding, tides, etc.) and the sediment participles come into contact with oxygen-rich water.

IV. Residual phase : Lithogenous and inert (Non-bioavailable).

The heavy metals in the soils and sediments are bound to different fractions with different strengths, the value can, therefore, give a clear indication of soil and sediment reactivity, which in turn assess the risk connected with the presence of heavy metals in a terrestrial or aquatic environment. The rationale of the sequential extraction procedure is that each successive reagent dissolves a different component, which can content heavy metals within their crystalline structures. Under natural conditions, metals in minerals are unlikely to experience significant release over the time frames of interest [32, 33].

3. ANALYTICAL PROTOCOLS FOR SEQUENTIAL EXTRACTION

In recent years a great number of papers have been published on various analytical techniques proposed for the fractionation analysis of trace elements in various environmental samples (soils, sediments, etc.). An approach that has been found to be preferable is the fractionation of heavy metal into operationally defined forms under the sequential action of different extractants [2]. Selective extractants, used in sequential extraction procedures, are aimed at the simulation of natural conditions whereby metals associated with certain soil (sediment) components can be released. For example, changes in the ionic composition affecting adsorption–desorption reactions or a decrease in pH may lead to the release of metals, retained on a matrix by weak electrostatic interactions or co-precipitated with carbonates ("exchangeable" and "acid soluble" forms). Decreasing the redox potential can result in dissolution of oxides, unstable under reducing conditions, and liberation of scavenged metals ("reducible" forms). Changes in oxidizing conditions may cause the degradation of organic matter and release of complexed metals ("oxidizable" forms). Finally, the destruction of primary and secondary mineral lattice releases heavy metal retained within the crystal structure, e.g., due to isomorphous substitution ("residual" forms) [2]. The nominal "forms" determined by operational fractionation can help to estimate the amounts of total metals in different reservoirs which could be mobilized under changes in the chemical properties of the soil [34]. Since the 1970s a considerable number of extraction procedures have been proposed for determining the forms of heavy metal [2, 35-39]. Most of these procedures are based on the scheme of Tessier et al. [2]. Although most of the extracting reagents were originally used in the chemical analysis of soils, the procedures proposed have been tested on a wide variety of contaminated environmental samples—sediments, road dust, sewage sludge, etc.

Sequential extraction can be useful to have an operational classification of metals in different geochemical fractions [2] which is the most reliable criteria to quantify the potential effect of soil/sediment contamination by heavy metals. This can provide information about the identification of the main binding sites, the strength of element binding to the particulates and the phase associations of trace elements in soil/sediment. Following this basic scheme, some modified procedures with different sequences of reagents or operational conditions have been developed [40-43]. Considering the diversity of procedures and the lack of uniformity in different protocols, a European Community Bureau of Reference (BCR, now the European Community Standards Measurement and Testing Program) method was proposed [6] and was applied by a large group of researchers [31, 44-47]. In this study, we followed the sequential extraction procedure proposed by the European Union's Standards, Measurements and Testing program [3].

4. MODIFIED BCR SEQUENTIAL EXTRACTION PROCESS

As discussed above it is evident that sequential extraction provides valuable information regarding identification of main binding site, the strength of the element binding to the particulates and the phase associations of heavy metals in sediments. However, various complicated sequential extraction procedures were experimented to provide more detailed information regarding different metal phase associations [2, 48, 49]. A wide range of techniques is available whereby various extraction reagents and experimental conditions are used. These techniques involve a 5-step [2], 4-step (BCR, Bureau Commune de Reference of the European Commission), 6-step [50] and 7-step [51, 52] extraction, and are thus becoming popular methods to be used for sequential extraction [53, 54]. Following this basic scheme, some modified procedures with different sequences of reagents or operational conditions have been developed [40-43].

Several sophisticated instruments have been used for the determination of heavy metals contents in marine environments. These include; flame AAS [55, 56], atomic fluorescence spectrometry [57], anodic stripping voltametry [58, 59], ICP-AES [60] and ICP-MS [61, 62].

Heavy metal mobility and bioavailability depend strongly on their chemical and mineralogical forms in which they occur [63]. Several speciation studies have been conducted to determine study different forms of heavy metals rather their total metal content. These studies reveal the level of bioavailability of metals in harbour sediments and also confirm that sediments are indicators of heavy metal pollution in marine environment [64-67].

Since the early 1980s and 1990s sequential extraction methodology has been developed to determine speciation of metals in sediments [2, 68] due to the fact that the total concentration of metals often does not accurately represent their characteristics and toxicity. In order to overcome the above mentioned obstacles it is helpful to evaluate the individual fractions of the metals to fully understand their actual and potential environmental effects [2]. To date, strong acid digestion is used often for the determination of total heavy metals in the sediments. However, this method can be misleading when assessing environmental effects due to the potential for an overestimation of exposure risk. Moreover, in order to determine the mobility of heavy metals in sediments, various sequential extraction procedures have been developed [69-71].

Among a range of available techniques using various extraction reagents and experimental conditions to investigate the distribution of heavy metals in sediments and soils, the 5-step Tessier et al. [2] and the 6-step extraction method, Kersten and Fronstier [50] were mostly widely used. Following these two basic schemes, some modified procedures with different sequences of reagents or experimental conditions have been developed [40-43]. Considering the diversity of procedures and lack of uniformity in different protocols, a BCR, Bureau Commun de Recherche (now called the European Community (EC) Standards Measurement and Testing Programme) method was proposed [6]. It harmonized differential extraction schemes for sediment analysis. The method has been validated using a sediment certified reference material BCR-701 with certified and indicative extractable concentration of Cd, Cr, Cu, Ni, Pb and Zn [72]. This method was applied and accepted by a large group of specialists [31, 44, 45, 47, 73, 74] despite some shortcoming in the sequential extraction steps [75, 76].

Wang et al. [77] used a modified Tessier sequential extraction method to investigate the distribution and speciation of Cd, Cu, Pb, Fe, and Mn in the shallow sediments of Jinzhou Bay, Northeast China. This site was heavily contaminated by nonferrous smelting activities. They found out that the concentrations of Cd, Cu and Pb in sediments was to be 100, 73, 13 and 7 times, respectively, higher than the National guidelines (GB 18668-2002). The sequential extraction tests revealed that 39%-61% of Cd was found in exchangeable fractions. This shows that Cd in the sediments posed a high risk to the local environment. Copper and Pb were found to be at moderate risk levels. According to the relationships between percentage of metal speciation and total metal concentration, it was concluded that the distributions of Cd, Cu and Pb in some geochemical fractions were dynamic in the process of pollutants migration and stability of metals in marine sediments from Jinzhor Bay decrease in the order Pb>Cu>Cd.

Yuan et al. [78] applied BCR-sequential extraction protocol to obtain metal distribution patterns in marine sediments from the East China Sea. The results showed that both the total contents and the most dangerous non-residual fractions of Cd and Pb were extremely high. More than 90% of the total concentration of V, Cr, Mo and Sn existed in the residual fraction while more than 60% of Fe, Co, Ni, Cu, and Zn were mainly present in the residual fraction. Manganese, Pb, and Cd were dominantly present in the non-residual fractions in the top sediments.

Jones and Turki [79] worked on distribution and speciation of heavy metals in surface sediments from the Tees estuary, North East England. Tessier et al. [2] metal speciation scheme modified by Ajay and van Loon [80] was used for the study. They observed out that the sediments were largely organic-rich clayey silts in which metal concentrations exceed background levels, and which attain peak values in the upper and middle reaches of the estuary. Chromium, Pb and Zn were associated with the reducible, residual, and oxidizable fractions. Cobalt and Ni were not highly enriched while Cu is associated with the oxidizable and residual fractions. Cadmium is associated with the exchangeable fractions.

Pempkowlak et al. [81] investigated the speciation of heavy metals in sediments and their bioaccumulation by mussels. They used a 4-step sequential extraction procedure adapted from Forstner and Watmann [82]. Their investigation which was characterized by varying metal bioavailability was aimed at revealing differences in the accumulation pattern of heavy metals in mussel inhabiting that inhabit in sediments. The bioavailabilities of metals were measured using the contents of metals adsorbed to sediments and associated with Fe and Mn hydroxides. The biovailable fraction of heavy metals contents in sediments collected from Spitsbergen represented a small proportion (0.37% adsorbed metals and 0.11%, are associated with metals hydroxides). It was also revealed that the percentages of metals adsorbed and bound to hydroxides of the sediments ranged from 1 to 46% and 1 to 13%, respectively.

Wepener and Vermeulen [66] investigated on the concentration and bioavailability of selected metals in sediments of Richards Bay harbor, South Africa. Sequential extraction of sediments was carried out according to Tessier et al. [2] method. The following metals were investigated: Al, Cr, Fe, Mn, and Zn. Their studies revealed that metals concentrations in sediments samples varied only slightly between seasons, but showed significant spatial variation, which was significantly correlated to sediment particle size composition. Highest metal concentration was recorded in sites with substrates dominated by fine mud. Manganese and Zn had more than 50% of this concentration in reducible fraction while more than 70% of the Cr was associated with the inert fractions and the concentration recorded at some sites were still above action levels when considering only the bioavailable fractions. They also concluded that the concentration of Zn recorded was not elevated their results were compared with the historic data.

Coung and Obbard [54] used a modified 3-step sequential extraction procedure to investigate metal speciation in coastal marine sediments from Singapore as described by the European Community Bureau of Reference (ECBR). Highest percentages of Cr, Ni, and Pb were found in residual fractions in both Kranji (78.9%, 54.7% and 55.9% respectively) and Pulang Tokong (82.8%, 77.3% and 62.2% respectively). This means that these metals were strongly bound to sediments. In sediments from Kranji, the mobility order of heavy metals studied were Cd>Ni>Zn>Cu>Pb>Cr while sediments from Pulan Tekong showed the same order for Cd, Ni, Pb and Cr, but had a reverse order for Cu and Zn (Cu>Zn). The sum of the 4-steps (acid soluble + reducible + oxidizable + residual) was in good agreement with the total metal content, which confirmed the accuracy of the microwave extraction procedure in conjunction with the GFASS analytical method.

Fedotov et al. [83] applied a modified technique for accelerated fractionation of heavy metals in contaminated soils and sediments using rotating coiled columns. Rotating coiled columns (RCC) is valuable for the continuous-flow sequential extraction and can be successfully applied to the dynamic leaching of heavy metals from soil and sediments. This is a fluoroplastic or steel coil wound around a rigid cylindrical drum, which revolves about its axis and, at the same time, revolves around the central axis of the device called planet centrifuge. The stationary (liquid, solid, or heterogeneous) phase is retained in the column because of the centrifugal force field, and the mobile liquid phase is continuously pumped through the column. A solid sample was retrieved in the rotating column as the stationary phase under the action of centrifugal forces while different elements (aqueous solution of complexing reagents, mineral salts and acids) were continuously pumped through. This procedure developed is time saving and requires only 4-5 hr instead of the several days needed for individual sequential extraction. Losses of solid sample are minimal. Further studies are needed to better estimate the reproducibility of the technique.

Nemati et al. [84] used a modified BCR sequential extraction procedure (SEP) in combination with ICP-MS to obtain the metal distribution patterns in different depths of sediments from Sungai Buloh, Selangor, Malaysia. The results showed that heavy metal contaminations at Sungai Buloh River sediments were more severe than at other sampling sites, especially for Zn, Cu, Ni and Pb. Nevertheless, the element concentrations from top to bottom layers decreased predominantly.

Mossop et al. [85] compared of original and modified BCR sequential extraction procedures for the fractionation of Cu, Fe, Pb, Mn and Zn in soils and sediments. The procedures were applied to five soil and sediment substrates: a sewage sludge-amended soil, two different industrially contaminated soils, river sediment and intertidal sediment. Extractable Fe and Mn concentrations were measured to assess the effects of the procedural modifications on dissolution of the reducible matrix components. Statistical analyses (two-tailed t-tests at 95% confidence interval) indicated that recovery of Fe in step 2 was not markedly enhanced when the intermediate protocol was used. However, significantly greater amounts were isolated with the revised BCR scheme than with the original procedure. Copper behaved similarly to Fe. Lead recoveries were increased by use of both modified protocols, with the greatest effect occurring for the revised BCR extraction. In contrast, Mn and Zn extraction did not vary markedly between procedures. The work indicates that the revised BCR sequential extraction proves better attack on the Fe-based components of the reducible matrix for a wide range of soils and sediments.

5. SEQUENTIAL EXTRACTION OF METALS IN SEDIMENTS OF THE HUGLI RIVER ESTUARY AND INDIAN SUNDARBAN WETLAND: A CASE STUDY

5.1. Materials and Methods

5.1.1. Sample Collection and Sediment Quality Analysis

The delta region formed by Hugli (Ganges) River Estuary (HRE) and is famous for its luxuriant mangrove vegetation, known as Sundarban wetland, acclaimed as UNESCO World Heritage Site for its capacity of sustaining an excellent biodiversity. The wetland is characterized by a complex network of tidal creeks, which surrounds hundreds of tidal islands exposed to different elevations at high and low semi-diurnal tides. This is one of the most sensitive and vulnerable ecosystems in the world and suffers from environmental degradation due to rapid human settlement, tourism and port activities, operation of mechanized boats, deforestation, and increasing agricultural and aquaculture practices. The ongoing degradation is also related to huge siltation, flooding, storm runoff, atmospheric deposition, and other stresses resulting in changes in water quality, depletion of fishery resources, choking of river mouth and inlets, and overall loss of biodiversity. Moreover, the rapid economic development in this deltaic region has caused highly dense areas of human activity and led to serious contamination including heavy metals and persistent organic pollutants (POPs).

Nine sampling sites, namely Barrackpur (S_1), Dakshineswar (S_2), Babughat (S_3), Budge budge (S_4), Ulubaria (S_5), Diamond Harbor (S_6), Frezergunge (S_7), Gangasagar (S_8), and Haribhanga (S_9) were selected considering the existence of typical sediment dispersal patterns along the drainage network systems (as shown in Figure 1) and their position was fixed by a global positioning system (GPS). The stations are representative of the variable environmental and energy regimes that cover a wide range of substrate behavior, wave–tide climate, and intensity of bioturbation (animal–sediment interaction), geomorphological–hydrodynamic regimes and distances from the sea (Bay of Bengal). The sites are exposed to a variable level of heavy metal contamination mainly from anthropogenic sources as mentioned earlier. Six sampling sites (S_1 to S_6) have been chosen along the lower stretch of Hugli River Estuary, while residual three sites (S_7 to S_9) were taken into account in the coastal regions of Sundarban wetland. All sampling sites together with the main stresses to which they are subjected are presented in Table 1.

During winter months (January–March 2009) surface sediment samples weighting 10 g were randomly collected in triplicate from the top 3–5 cm of the surface at each sampling site during low tide using a grab sampler, pooled and thoroughly mixed. Immediately after collection, the samples were placed in sterilized plastic bags in the ice box and transported to the laboratory. Samples were oven dried at 50°C, most gently disaggregated, transferred into precleaned inert polypropylene bags and stored in deep freeze prior to analyses. Each sample was divided into two aliquots: one unsieved (for the determination of sediment quality parameters) and the other sieved through 63 μm metallic sieves (for elemental analyses). Organic carbon content was determined following a rapid titration method [86] and pH with the help of a deluxe pH meter (model no. 101E) using combination glass electrode manufactured by M.S. Electronics Pvt. Ltd.

(India). Mechanical analyses of sediment were done by sieving in a Ro-Tap Shaker manufactured by W.S. Tyler Company, Cleveland, Ohio.

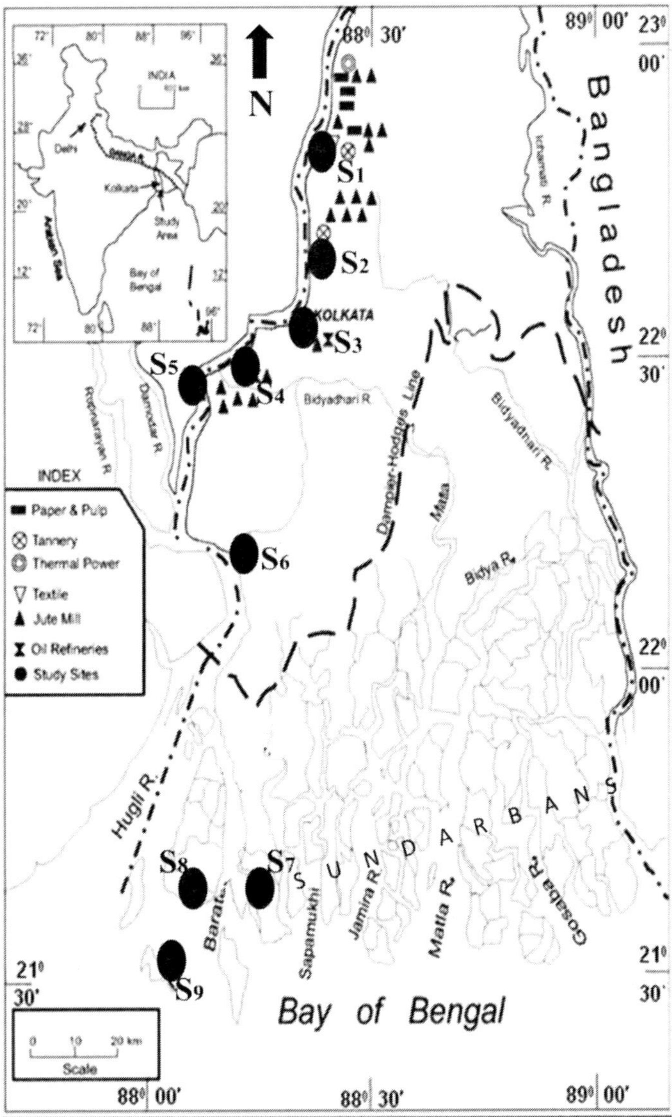

Figure 1. Map showing the location of the monitoring sites (S_1 to S_9) covering Hugli River Estuary and Sundarban mangrove wetland along with the location of the major industries.

Table 1. Details of the nine sampling sites and the main stresses to which they are subjected.

Station number	Site	Main stresses
S_1	Barrackpur	Industrial effluents, domestic sewage disposal, boating, bathing.
S_2	Dakshineswar	Industrial and domestic effluents, boating, bathing, idol immersion site.
S_3	Babughat	Power plant discharges, domestic sewage, boating, idol immersion site.
S_4	Budge budge	Domestic and industrial effluents, bathing, boating.
S_5	Ulubaria	Domestic and industrial effluents
S_6	Diamond Harbour	Boating, recreational activities, bathing, fishing, jetties for fishing trawlers
S_7	Frezergunge	Tourist activities, ferry services, fishing
S_8	Gangasagar	Boating, tourist activities, dredging, fishing, agricultural, domestic and aquaculture practices
S_9	Haribhanga	Boating, fishing and ferrying

5.1.2. Analytical Procedure

To determinate the total element concentration, sediment samples were digested in polytetrafluoroethylene vessels with aqua regia (HCl/HNO_3, 3:1) and HF neutralized with H_3BO_3 in a 650 W microwave oven (CEM MDS 2000) with a program consisting of a 20-min ramp and a 30-min hold at 100% power in pressure and temperature controlled conditions (150 psi and 180°C). The digested samples were filtered, transferred to polyethylene containers and stored at +4°C until analysis. All reagents were Suprapur® grade (Merck). Reagent blank was processed with the samples and did not show any significant contamination. Accuracy of the procedure was checked using two different certified reference materials (CRM): MESS-2 and PACS-2, which are both marine sediments certified by the National Research Council of Canada for the element content. The MESS-2 recovery ranged between 91% and 116% for all the elements (Table 2). Precision, calculated as relative standard deviation (RSD%), resulted always lower than 5%.

Table 2. Results of certified reference materials MESS-2 and PACS-2 as well as the observed values. All the values are expressed in μg/g of dry weight. MESS-2 and PACS-2 recovery rates are also reported.

	Al	As	Cd	Co	Cr	Cu	Fe	Mn	Ni	Pb	Zn
Found MESS-2	86,613± 17,773	24.0± 2.4	0.230± 0.010	15.6± 1.5	112± 12	38.4± 6.1	47,385± 3,668	372± 42	54.1± 3.8	20.0± 1.4	170±12
Found PACS-2	70,190± 3,784	29.2± 1.3	1.59± 0.80	12.8± 0.6	94.9± 4.4	307± 22	46,630± 1,411	465± 23	44.3± 2.5	184± 10	398±16
Certified MESS-2	85,698± 2,600	20.7± 0.8	0.240± 0.010	13.8± 1.4	106± 8	39.3± 2.0	43,504± 2,266	365± 21	49.3± 1.8	21.9± 1.2	172±16
Certified PACS-2	66,125± 3,184	26.2± 1.5	2.11± 0.15	11.5± 0.3	90.7± 4.6	310± 12	43,738± 585	440± 19	39.5± 2.3	183±8	364±23
Recovery MESS-2	101%	116%	95.8%	113%	106%	97.8%	109%	102%	110%	91.2%	99.0%
Recovery PACS-2	106%	111%	75.4%	111%	105%	99.1%	107%	106%	112%	100%	109%

In this study, we followed the sequential extraction procedure proposed by the European Union's Standards, Measurements and Testing program [3]. Selective extraction is based on the procedure used by Tessier et al. [2] with improvements made according to the BCR, which examined and finally eliminated irreproducibility sources. It is made up of three steps, which dissolve the following phases, respectively: exchangeable and bound to carbonate, bound to Fe and Mn oxides and hydroxides, bound to organic matter and sulphides. Exchangeable and bound to carbonate phase (phase 1) is extracted with 0.11 M acetic acid, while the fraction bound to Fe–Mn oxides (phase 2) with 0.5 M hydroxylamine hydrochloride, adjusted to pH 2 with nitric acid (65%). The phase bound to organic and sulphides (phase 3) is extracted with 8.8 M hydrogen peroxide (stabilized at a pH included between 2 and 3), treated at 80°C in a microwave oven using a program consisting of a 30-min ramp and a 60-min hold at 50% power in pressure and temperature controlled conditions (80 psi and 85°C), and 2 M ammonium acetate adjusted to pH 2 with nitric acid (65%). Each extraction was carried out overnight (16 h) at room temperature. All the reagents employed were Tracepur® grade (Merck Eurolab, Italy). After each extraction, the samples were separated from the aqueous phase by centrifuging at 4, 000 rpm for 15 min. The sediments were washed with Milli-Q water and centrifuged again. The wash water was added to supernatants. The element content of the residual phase was obtained from the difference between the total content and the sum of phases 1, 2 and 3, according to Ianni et al. [37, 38], Ramirez et al. [39], and Mester et al. [27]. Sequential extraction reagent blanks showed no detectable contamination. Accuracy of the procedure was checked with BCR-701 (SM&T). The recovery rates for trace elements in the standard reference material ranged between 77% and 118% (Table 3). Precision, calculated as RSD%, resulted generally lower than 5%, except As and Cr in the phase 1 (~20%).

Table 3. Results of certified reference materials BCR-701 as well as the observed values (expressed in μg/g of dry weight) together with recovery rates for each step. n.a.= not available.

	Al	As	Cd	Co	Cr	Cu	Fe	Mn	Ni	Pb	Zn
Found BCR-701 step 1	198±1	2.57± 0.28	6.09± 0.09	2.06± 0.08	2.41± 0.51	47.7± 1.7	43.8± 5.8	180±1	14.5± 0.3	3.38± 0.35	185±4
Found BCR-701 step 2	3,451± 46	16.5± 0.3	3.37± 0.08	3.22± 0.03	39.2± 0.4	100±2	7,042± 106	128±3	24.5± 0.4	111±2	102±1
Found BCR-701 step 3	1,912± 74	3.09± 0.20	0.28± 0.01	1.86± 0.17	169±4	64.8± 1.5	1,147± 56	31.9± 2.6	17.4± 1.7	7.15± 0.12	58.4± 5.0
Certified BCR-701 step 1	n.a.	n.a.	7.34± 0.35	n.a.	2.26± 0.16	49.3± 1.7	n.a.	n.a.	15.4± 0.9	3.18± 0.21	205±6
Certified BCR-701 step 2	n.a.	n.a.	3.77± 0.28	n.a.	45.7±2	124±3	n.a.	n.a.	26.6± 1.3	126±3	114±5
Certified BCR-701 step 3	n.a.	n.a.	0.27± 0.06	n.a.	143±7	55.2± 4.0	n.a.	n.a.	15.3± 0.9	9.3±2.0	54.2± 2.0
Recovery step 1	n.a.	n.a.	83.0%	n.a.	107%	96.8%	n.a.	n.a.	94.4%	106%	90.3%
Recovery step 2	n.a.	n.a.	89.4%	n.a.	85.9%	80.6%	n.a.	n.a.	92.3%	87.9%	89.4%
Recovery step 3	n.a.	n.a.	104%	n.a.	118%	117%	n.a.	n.a.	114%	76.9%	108%

The elemental concentrations were determined with an inductively coupled plasma atomic emission spectrometer Vista Pro (Varian), with the internal standard method. Cadmium was determined by electrothermal atomization atomic absorption spectrometry. A Varian Spectra A300 spectrometer with Zeeman effect background correction and autosampler Varian Model 96 was used employing the standard addition method for calibration. All the metal analyses were performed at the Department of Chemistry and Industrial Chemistry of the University of Genoa (Genoa, Italy).

5.1.3. Statistical Analyses

Principal component analysis (PCA) was used to characterize the metal composition in sediments, and cluster analysis was used for grouping the sampling stations. Principal component analysis (PCA) is a multivariate statistical technique used for data reduction and for deciphering patterns within large sets of data. With PCA, a large data matrix is reduced to two smaller ones that consist of principal component (PC) scores and loadings. PC loadings are eigenvectors of the correlation or covariance matrix depending on which is used for the analysis. The PC scores contain information on all of the variables combined into a single number, with the loadings indicating the relative contribution of each variable to that score [87]. Hierarchical cluster analysis (HCA) characterizes similarities among samples by examining interpoint distances representing all possible sample pairs in high-dimensional space. The sample similarities are represented on two dimensional diagrams call dendrograms [88]. All statistical analyses were performed using the computer software STATISTICA (StatSoft, Inc. 2001).

5.2. Results and Discussion

5.2.1. Sediment Geochemistry

Table 4 shows values of pH; organic carbon (%); and percentage of sand, silt, and clay in sediments of the nine sampling sites. Organic carbon values, ranging from 0.22% (in station S_8) to 1.02% (in station S_2), are low in comparison with values found in sediments from other Indian coastal areas, such as Gulf of Mannar [89], Cochin [90], and Muthupet mangroves [91]. The low organic carbon values might be related with the poor absorbability of organics on negatively charged quartz grains, which predominate in sediments in this estuarine environment [92]. In addition, the constant flushing activity by tides along with the impact of waves can support the low percentage of organic carbon in the sediments. The sediments of the studied stations are characterized by slightly basic pH (7.50–8.36) with maximum values recorded in the stations closest to the sea (stations S_6, S_8, and S_9) and minimum in station S_7.

Table 4. Geographical position, physicochemical and textural properties of sediment samples of 9 sampling sites.

Stations	Latitude and Longitude	Salinity	pH	Organic carbon (%)	Sand (%)	Silt (%)	Clay (%)
S1	22°43′ 16″ N 88°21′ 20″ E	0	7.86	0.35	4	87.1	8.9
S2	22°39′ 17″ N 88°12′ 25″ E	0	7.80	1.02	1	76.5	22.5
S3	22°33′ 53″ N 88°20′ 19″ E	0	7.90	0.52	2.24	41.97	55.79
S4	22°30′ 10″ N 88°11′ 48″ E	0–2.5	7.60	0.74	18.25	47.42	34.33
S5	22°28′ 06″ N 88°06′ 54″ E	0–1	7.90	0.91	16.7	69.6	13.7
S6	22°11′ 14″ N 88°11′ 15″ E	0–5.6	8.36	0.56	3.15	41.13	55.71
S7	21°34′ 44″ N 88°15′ 03″ E	30–34.3	7.50	0.36	98.02	0.18	0
S8	21°38′ 15″ N 88°03′ 53″ E	32–35	8.14	0.22	32.85	58.45	8.7
S9	21°34′ 20″ N 88°01′ 25″ E	35	8.10	0.46	39.3	44.25	16.45

These were different from the low pH values in most of the mangrove swamps in Hong Kong [93], where sediments were not frequently flooded by the tide and become acidic in reducible conditions. With respect to texture, the sediment samples show a variable admixture of sand, silt, and clay. Clay fractions dominate in low-energy areas of suspensional deposits. On the contrary, silt, and sand dominates where the energy level is high. Sediments from station S_7 contain higher percentage of sand (98%) compared to the others, while sediments from S_1, S_2, S_5, and S_8 contains higher percentage of silt (more than 50%) compared to the others. A variable mixture of sand, silt, and clay is present in the other stations and reflect a variable amount of erosion and deposition.

5.2.2. Total Element Concentrations

Total element concentrations in the investigated stations varied in a narrow range of values (Table 5) and were comparable with data obtained for other Indian coastal areas [94, 95]. Datta and Subramanian [96] found very similar trace element concentrations throughout the Bengal Basin, where anthropogenic perturbation is low and river channel may receive a several centimeter-thick sediment layer in a single event during peak flow, preventing to bear the signature of an accumulation of trace elements. The highest concentrations for As, Cu, Fe, Mn, and Ni were measured at station S_9 while for Cd and Pb at station S_3, close to Calcutta city (about 4.5 million residents, but about 14.2 million including suburbs). An anthropogenic input from vehicular traffic and in-dustrial activities may cause high Cd and Pb con-centrations measured in samples collected in the Calcutta urban area. The lowest element concentrations were found at station S_5. Very low (close to the detection limit) Cd concentration was found in the coastal stations (S_7, S_8, and S_9).

Table 5. Total element concentrations (μg/g) in sediments of 9 sampling sites (instrumental precision, calculated as RSD%, resulted lower than 5% for each element in all samples).

Stations	Al	As	Cd	Co	Cr	Cu	Fe	Mn	Ni	Pb	Zn
S1	70,289	8.81	0.165	13.0	67.6	27.8	37,737	591	31.9	20.4	86.6
S2	70,879	8.44	0.452	14.0	74.8	36.8	39,405	625	34.2	22.3	90.7
S3	72,134	8.65	1.79	14.5	73.5	32.3	40,070	712	35.0	33.2	83.1
S4	72,613	8.49	0.492	14.9	76.8	27.9	40,303	726	35.0	19.6	80.4
S5	62,044	6.41	0.220	12.1	58.2	21.1	33,428	597	27.5	17.0	64.1
S6	64,325	6.79	0.106	12.0	64.8	32.0	34,273	613	31.3	17.9	69.6
S7	77,529	7.77	0.044	14.0	75.1	28.4	40,084	389	38.1	20.5	74.4
S8	68,146	8.08	0.027	12.7	62.5	22.2	36,786	511	34.3	19.7	61.4
S9	72,666	9.40	0.044	14.1	74.2	36.6	40,838	785	40.1	22.9	74.9

The geoaccumulation index (I_{geo}) of Muller [97] has been calculated for the analyzed elements, by comparing current concentrations with pre-industrial levels, in order to estimate the metal contamination in sediments. The equation used for the calculation of I_{geo} is: $\log_2 (C_n/1.5\ B_n)$, where C_n is the measured content of element "n" and B_n the element's content in "average shale" [98]. Factor 1.5 is used because of possible variations in background values for a given element in the environment, as well as very small anthropogenic influences [99]. As shown in Figure 2, all sediments fall in class 0 for Al, Co, Cr, Cu, Fe, Mn, Ni, Pb, and Zn, therefore the area is not contaminated for these elements. Unlike the Hugli river, in other rivers of the Bengal Basin, such as Meghna and Brahmaputra, Cr exhibits higher I_{geo} values respect to the other elements [96]. For Cd, two stations fall in class 1 and three in class 2 exhibiting a moderate contamination for this element. In all stations, As falls in class 2 (moderate pollution). In this area, As contamination was already observed in previous studies and it is probably due to groundwater contamination [100]. This contamination can have natural origin, such as coal seams in Rajmahal basin and arsenic mineral in mineral rocks in the upper reaches of the Ganges river system. The highly reducing nature of groundwater would reduce As, causing the possible desorption of As [101].

5.2.3. Speciation Patterns

The potential environmental risk of trace elements in sediments is associated with both their total content and their speciation. The chemical partitioning of the considered elements (Al, As, Cd, Co, Cr, Cu, Fe, Mn, Ni, Pb, and Zn) from each extraction step has been described. Aluminum, Cr, and Fe are present mainly in the residual phase, representing 95.8–96.8%, 88.9–91%, and 83.0–94.7% of the total concentration, respectively, which implies that these elements are strongly linked to the inert fraction of the sediments. This result was in good agreement with data reported by several studies carried out worldwide in marine coastal areas [45, 46, 78, 102]. The high percentage of Fe in the residual phase indicates that most of the Fe exists as crystalline Fe peroxides (goethite, limonite, magnetite, etc.). The remaining Fe is associated with the reducible phase (mean, 11.25%). Large amounts of Fe accumulate in the residual phase probably because it is basically of natural origin (it is the most common element in the earth's crust).

Concentrations of Al, Fe, and Cr are very low in exchangeable phase (0.08%, 0.26%, and 1.72% as mean values, respectively), limiting their potential toxicity as pollutants. It should be noted that sediments always act as reservoir for elements; therefore, their potential risk of pollution to environment has always to be considered.

Arsenic, Co, Ni, and to some extent Zn, are found mainly in the residue (~50% of the total concentration). Nickel and Co are associated to the residue respectively for 56% and 74% of the total concentration, with a speciation similar in all the samples. A mean of 23% of Co is present in the phase 2. The highest percentage of labile Co (~13%) was measured in S_6 (Diamond Harbour) and S_8(Gangasagar) and can be due to a recent input of this element. The dominant proportion of Ni in the residual phase is in agreement with the results of other studies [27, 46]. Nickel is present, apart from the residue, in phases 2 and 3 (about 10% in each phase). Arsenic is distributed mainly between the residual (mean 47%), the reducible, and the oxidizable phases (mean 19% and 22%, respectively). Acharyya et al. [101] observed that As is adsorbed to iron-hydroxide-coated sand grains and to clay minerals in the sediments of the Ganges delta from West Bengal. Among the studied elements, As is found with the greatest proportion in the oxidizable phase coinciding with organic and sulfur compounds. Arsenic is present in the phase 1 for about 10% of the total content, in station S_7 phase 1 percentage rises up to 16%. The lower land alluvial basin of the Ganges River is recognized as an arsenic-affected area. Arsenic in solution probably is easily entrapped in the fine grained organic-rich sediments deposited in the Ganges delta [101]. The percentage of silt (lower than 70% except in $S_1$and S_2) may have contributed to a low retention of dissolved As since coarse sediments are less efficient at retaining As.

Cadmium was mainly present in the labile phase (more than 60%) in all the stations with the exception of station S_7, where the Cd labile percentage represents only 25% of the total concentration. Cadmium concentrations were negligible in phases 2 and 3. The highest labile Cd concentration was measured at station S_3, the closest to the city of Calcutta. Datta and Subramanian [96] found that the concentrations of elements in the non-detrital phases were higher in stations sampled in the Hugli river around Calcutta than in samples collected along Brahmaputra and Meghna rivers. The petroleum refinery, industrial, and mining effluents carried by the Hugli river may be responsible for this higher concentrations of non detrital fractions.

About 40% of the total Cu concentration is associated to the residue, while 33% of Cu is bound to Fe-Mn oxide and hydroxide (phase 2). The high percentage of Cu in the residue is likely due to the fact that Cu is easily chemisorbed on or incorporated in clay minerals [103]. All the samples showed lower Cu concentrations in exchangeable phase, with percentage ranging from 7% (S_7) to 22% (S_5), with a mean of 15%. Copper is characterized by high complex constant with organic matter thus it can be hypothesized that Cu is bound to labile organic matter such as lipids, proteins, and carbohydrates. On the other hand, high-element concentration in labile phase could be related to recent coastal input [39].

Manganese was found in all the four sediment phases, as observed by other researchers [45, 104]. Manganese is the most mobile element since it is present with the highest percentage (a mean of 42%) in the labile phase. This is probably because of the known close association of Mn with carbonates [105] as endorsed by other workers [69, 106]. In this phase, weakly sorbed Mn retained on sediment surface by relatively weak electrostatic interactions may be released by ion exchange processes and dissociation of Mn-carbonate phase [2]. The result indicates that considerable amount of Mn may be released into environment if conditions become more acidic [107]. The highest Mn labile percentage was measured in S_6 (57%). Differently, in S_7, Mn in the residue represents 65% of the total concentration, while the labile Mn is only 15%. A substantial Mn percentage was also found in the residue (mean 37.8%), followed by the reducible phase (14.7%), in which Mn exists as oxides and may be released if the sediment is subjected to more reducing conditions [108].

The major geochemical phase for Pb in these sediments was the Fe-Mn oxides phase (mean 55.7%) followed by the residual phase (mean 30.2%) while lower percentage of the total Pb are bound to exchangeable-labile (mean 5.3%) and oxidizable phases (mean 6.8%). At S_3 (Babughat), the reducible part is as high as 65% and only 19.9% of the total is associated with the residue. Atmospheric input as fallout from vehicular emission can be probably the major input of Pb for this station. The relatively high percentage of Pb in reducible phase is in agreement with the known ability of amorphous Fe–Mn oxides to scavenge Pb from solution [109, 110]. Caille et al. [111] observed that resuspension of anoxic sediment results in a rapid desorption of Pb and Cu adsorbed to sulphides. Thus, a high element percentage in the reducible fraction is a hazard for the aquatic environment because Fe and Mn species can be reduced into the porewaters during early diagenesis by microbially mediated redox reactions [112]. Dissolution will also release Pb associated with oxide phases to the porewater possibly to the overlying water column [113] and to benthic biota [79]. The major sources of Pb are from intensive human activities, including agriculture in the drainage basin [114], auto exhaust emission together with atmospheric deposition [115]. In addition, a substantial contribution from the factories located in the upstream of the Hugli river dealing with Pb producing lead ingots and lead alloys play a vital role as referred by Sarkar et al. [116].

The percentage of Zn in residue is highly variable (38.5–70%) and the distribution pattern in each fraction showed the following order: residual>reducible>oxidizable> exchangeable and bound to carbonates. There was some difference in Zn speciation among the sampling sites: in stations S_1, S_2, and S_3 about 40% of Zn is present in the residue, while in the other stations this percentage increases to more than 60%. In station S_1, the exchangeable and oxidisable phases shared over 22% of the total Zn, whereas labile Zn was as low as 4.6% at S_7. A major part of Zn (16.3%) is associated with Fe–Mn oxide phase, because of the high stability constants of Zn oxides. Iron oxides adsorb considerable quantities of Zn and these oxides may also occlude Zn in the lattice structures [117].

The BCR procedure as discussed above showed satisfactory recoveries, detection limits, and standard deviations for determinations of heavy metals/metalloid in the sediments. It is evident from the present results of the fractionation studies that the metals/metalloids in the sediments are bound to different fractions with different strengths leading to variations in mobility and availability and some of them show significant spatial variations subject to diverse environmental stresses. This type of association between metals and the sediments can be understood in detail by sequential extraction techniques. Hence the application of sequential extraction is fully justified as the quantification of different forms of metal is more meaningful than the estimation of its total metal concentrations. The strength values can, therefore, give a clear indication of sediment reactivity, which in turn assess the risk connected with the presence of metals in this wetland environment. The results obtained suggest the need for corrective remediation measures due to the higher accumulation of potentially dangerous metals/metalloids, which in most cases exceed the limits established by certain legislation.

5.2.4. Comparison With Sediment Quality Guidelines

Results obtained after total and sequential extraction are compared with Sediment Quality Guidelines (SQGs). Table 6 reports consensus-based values, such as TEC (concentration below which harmful effects on sediment-dwelling organisms were not expected) and PEC (concentration above which harmful effects on sediment-dwelling organisms were expected to occur frequently), and effect range-low and range-medium, such as ERL (concentrations below which adverse biological effects were observed in less than 10% of studies) and ERM (concentrations above which effects were more frequently observed in more than 75% of studies).

Table 6. Sediment Quality Guidelines concentrations with respect to total and labile element concentrations found in the analyzed samples (expressed as μg/g of dry weight).

Element	Phase	Si<TEC	TEC	Si<TEC<PEC	PEC	Si<ERL	ERL	ERL<Si<ERM	ERM
As	Total	All	9.79	None	33	S5,S6,S7,S8	8.2	S1,S2,S3,S4,S9	70
	Labile	All		None		All		None	
Cd	Total	S1,S2,S4,S5, S6,S7,S8,S9	0.99	S3	4.98	S1,S2,S4,S5, S6,S7,S8,S9	1.2	S3	9.6
	Labile	S1,S2,S4,S5, S6,S7,S8,S9		S3		S1,S2,S4,S5, S6,S7,S8,S9		S3	
Cr	Total	None	43.4	All	111	All	81	None	370
	Labile	All		None		All		None	
Cu	Total	S1,S4,S5,S7,S8	31.6	S2,S3,S6,S9	149	S1,S3,S4,S5, S6,S7,S8	34	S2,S9	270
	Labile	All		None		All		None	
Ni	Total	None	22.7	All	48.6	None	20.9	All	51.6
	Labile	All		None		All		None	
Pb	Total	All	35.8	None	128	All	46.7	None	218
	Labile	All		None		All		None	
Zn	Total	All	121	None	459	All	150	None	410
	Labile	All		None		All		None	

Comparing our results with the SQGs, it is revealed that for Pb and Zn in all the stations the measured concentrations are lower than both TEC and ERL. As regards Cd, concentration measured in station $S_3$is higher than TEC and ERL but lower than PEC and ERM both in term of total and labile concentration. For this station, some possible toxic effect on benthic organism can be hypothesized, in particular because of the large amount of element bound to the most labile phase of the sediment. Considering Cu, some stations (S_2, S_3, S_6, and S_9) exhibit total concentrations higher than TEC but lower than PEC. Concentrations of Cu are higher than ERL but lower than ERM only in stations $S_2$and S_9. Since only 7–22% of total Cu is bound to the labile phase, in all stations Cu labile concentrations are lower than TEC and ERL. Total As concentrations in stations S_1, S_2, S_3, S_4, and $S_9$are higher than ERL value but lower than TEC value. Since more than 50% of total As is not found in the residue, attention should be paid to a change in the environment conditions which could induce a release of As from the sediments. Total Ni and Cr concentrations are higher than TEC (Ni is also higher than ERL) but lower than PEC (and ERM in the case of Ni) in all the stations. Nevertheless, more than 70% of Ni as well as 90% of Cr are present in the residual fractions, therefore adverse impacts on organisms is very much negligible.

Mean sediment quality guidelines quotients (mSQGQ) have been developed for assessing the potential effects of contaminant mixtures in sediments [118]: they are determined by calculating the arithmetic mean of the quotients derived by dividing the concentrations of chemicals in sediments by their respective SQGs. The probability of observing sediment toxicity can be estimated by comparing the mSQGQ in a sample to previously published probability tables. It is important to keep in mind that mSQGQs cannot be used to accurately predict the uptake and bioaccumulation of sediment-bound chemicals by fish, wildlife, and humans, even if there is considerable evidence that this assessment tool can be predictive of the presence or absence of toxic effects [118].

SQGQs are calculated for seven elements considering ERM as sediment quality guidelines (Table 7). The mean quotient values ranges from 0.16 in station S_5 to 0.24 in station S_3. Using PEC values instead of ERM, the mean SQGQ ranges from 0.25 in station S_5 to 0.38 in station S_3 (Table 7).

Table 7. Mean Sediment Quality Guidelines Quotients calculated for the nine stations using PEC and ERM as SQGs.

Stations	$SQGQ_{PEC}$	$SQGQ_{ERM}$
S1	0.30	0.19
S2	0.33	0.21
S3	0.38	0.24
S4	0.33	0.21
S5	0.25	0.16
S6	0.28	0.18
S7	0.32	0.21
S8	0.28	0.18
S9	0.34	0.22

Compiled data from multiple data sets reporting 10-day toxicity test conducted on amphipod species in saltwater showed that the incidence of toxicity for a range of SQGQ of 0.25–0.5 is ~35%, while for a mean SQGQ range from 0.1 to 0.25, the incidence of toxicity lowers to ~20%. Measures recorded in a survey of Biscayne Bay (port of Miami and the adjoining saltwater reaches of the lower Miami River, FL, USA) showed that the average amphipod survival (*Ampelisca abdita*) decreased slightly from the least contaminated (ERMQ <0.03) to the intermediate category, (ERMQ included in 0.03–0.2 range) then decreased greatly in the most contaminated sediments (ERMQ included in 0.2–2 range). Therefore, we can presume a low toxicity of sediments sampled in the nine stations for benthic organisms. It is important to note that the benthic response to contaminants covaried among stations with both the mean ERM quotients and the effect of natural factors, such as the sediment texture, TOC, and salinity [118].

5.2.5. Statistical Analyses

The relationships between variables and the differences between stations were evaluated by PCA. The analysis was performed on 36 objects (four sediment phases for nine stations) and 11 variables (Al, As, Cd, Co, Cr, Cu, Fe, Mn, Ni, Pb, Zn). Two significant components were identified explaining 68.3% and 14.5% of the total variance, respectively. By studying the loadings of the variables (Figure 2a) on the components it can be seen that all the elements except Cd, Mn, and Pb are significantly correlated.

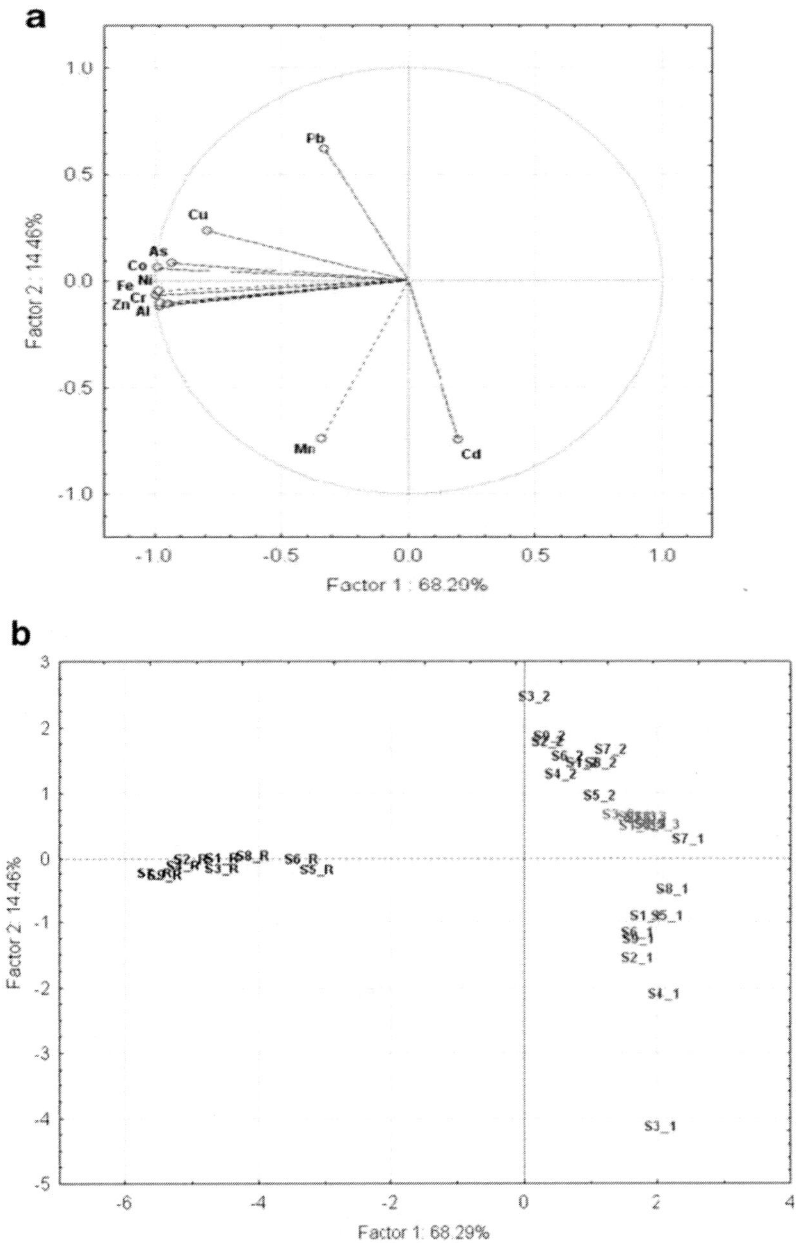

Figure 2. Principal component analysis: a) variable plot; b) score plot (phase 1, 2, 3 and 4, corresponding to labile, reducible, oxidizable and residue, are identifiable by 1, 2, 3 and 4 suffix and different colors in the score plot).

Unlike the other elements, most of Cd and Mn is present in the first phase: labile Cd and Mn represent more than 60% and 40% of the total concentration, respectively, except in station S_7. Cadmium and Mn speciation can be ascribed to their considerable affinity for carbonates. Lead is the only element which is bound to the reducible phase for more than 50%. Lead is a very reactive element in water column and, having scavenging type behavior, is easily bound to hydroxy- and oxyligands. Copper is positively and significantly correlated with all elements except Cd and Mn, but with lower correlation coefficients (0.66–0.81).

In the score plot (Figure 2b) phases 1, 2, 3, and 4 (corresponding to labile, reducible, oxidizable, and residual phases respectively) are identifiable by 1, 2, 3, and 4 suffix, respectively. In all stations, residue concentrations were characterized by negative values of PC1 and consequently by high concentrations of Al, As, Co, Cr, Fe, Ni, and Zn. Conversely, in the positive PC1 semi-axis labile and oxidizable metal concentrations, which represent a small percentage of the total elements, are distributed. For all stations, reducible concentrations are distributed along the positive PC2 semi-axis, i.e., high Pb concentrations, with a maximum for station S_3 and a minimum for S_5. The group formed by elements bound to organic matter and sulphides (phase 3) is characterized by low values of both PC1 and PC2. Therefore, a low percentage of elements (higher than 20% exclusively for all As data and for Zn in stations S_1 and S_3) is bound to the oxidizable phase, suggesting the presence of an oxidant environment. High Mn and Cd concentrations are associated with negative values of PC2, therefore a relatively high concentration of labile Mn and Cd is present in all samples (in particular in S_3), except station S_7. Samples are prevalently grouped in relation to the sediment geochemical phase, suggesting a similar element speciation among the stations. Station S_7 represents an exception, in fact the labile fraction is closely associated to the oxidizable phase group.

A HCA was carried out by applying Euclidean distances to quantitatively identify specific groups of similar stations. In the dendrogram of the sampling stations (Figure 3), we can note two main clusters: the first represented by station S_7, characterized by the highest element percentage bound to residue, and the second constituted by all the remainder stations. In the second group, a subgroup formed by station S_5 and S_6 can be individuated.

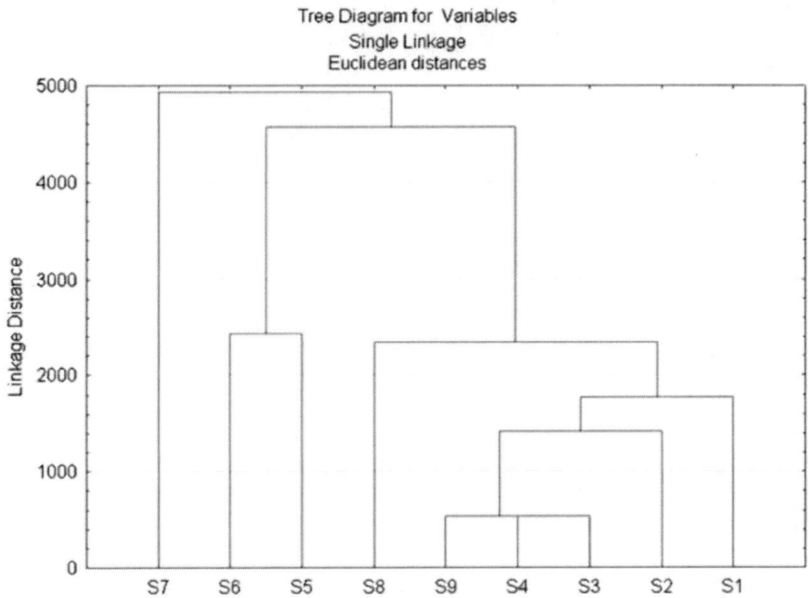

Figure 3. Dendrogram indicating linkage of sites on the basis of element concentrations.

Station S_7 was sampled in a marine coastal environment; it is characterized by a peculiar grain size percentage respect to the other stations, being the sand percentage as high as 98.6%. In general, the concentrations of elements are much higher in fine than in coarse fraction because the fine fraction larger specific surface facilitates absorption processes. As previously noted by Ramirez et al. [39], this pattern is particularly evident for Cd. It is interesting to note that the marine coastal stations S_8 and $S_9$are more similar to river stations than to station S_7. Both the stations S_8 and S_9 are, in fact, located in front of the Hugli river runoff, while station S_7 is located easternmost and probably is less influenced by the Hugli river discharge.

5.3. Conclusion and Recommendation

The study provides valuable information on the potential mobility of trace elements in sediments collected along the stretch of Hugli River and in the Sundarban mangrove wetland (northeastern part of the Bay of Bengal). The results obtained adopting BCR sequential extraction method provided the following important information: (i) Al, Cr, and Fe were found mostly in the residual phase while the other elements were found in the four phases sharing different proportions; (ii) the dominant Cd, Mn, and Pb proportion was found in the non-residual fractions and (iii) Mn had the highest percentage in the labile phase. This is worthwhile to mention that coastal environment of West Bengal is considerably constrained due to implementation of dredging, construction of port/ harbor and other industrial activities. The authors strongly recommend for periodical monitoring on the bioavailability and mobility of trace elements, control the mixing of effluent of the concentration of heavy metals in the region, environmental remediation, treatment of industrial effluent and municipal wastewater for effective management of this estuarine system. It is wisely suggested that an environmental recovery framework should be urgently implemented to avoid extension of heavy metal contamination (especially As).

6. CASE STUDIES FROM OTHER COASTAL REGIONS IN INDIA

Although the importance of metal speciation and fractionation has been realized in developed countries, the subject has not really taken off in India and only few references are available on the speciation of metals in Indian rivers. Speciation of selected heavy metals geochemistry in surface sediments (n=10 was studied by Venkatramanan et. al. [119] from Tirumalairajan river estuary, east coast of India. The results obtained from sequential extraction showed that a larger portion of the metals were associated with the residual phase, although they are available in other fractions too.

Trace metal fractionation in the Pichavaram mangrove–estuarine sediments in southeast coast of India was studied by Ranjan et. al. [120] considering the pronounced changes due to occurrence of tsunami (2004). A 5-step sequential extraction procedure was applied to assess the effects of tsunami on mobility and redistribution of selected elements (Cd, Cr, Cu, Fe, Mn, Ni, Pb, and Zn) in coastal sediments revealed that metals in the residual fraction (lattice bound) had the highest concentration suggesting their non-availability and limited biological uptake in the system. Majority of the metals (except Mn) do not constitute a risk based on the different geochemical indices.

Fractionation of selected metals in the sediments of Cochin estuary and Periyar River (southwest coast of India) was studied by Mohan et. al. [121]. The results reveal that remobilization potential of metals bound is in the range of low to medium risk to various sedimentary phases is different and is based on bond strength. Therefore, the strength values can give a clear indication of sediment reactivity that can be used to assess the risk related with metals to the aquatic organisms.

7. RISK ASSESSMENT CODE (RAC)

The risk assessment code (RAC) mainly applies the sum of the exchangeable and carbonate bound fractions for assessing the availability of metals in sediments. These fractions are considered to be weakly bonded metals which may equilibrate with the aqueous phase and thus become more rapidly bioavailable [11, 33]. This is important because the fractions introduced by anthropogenic activities, such as agricultural runoff and tourism, are typified by the adsorptive, exchangeable, and bound to carbonate fractions, which are weakly bonded metals that could equilibrate with the aqueous phase and thus become more rapidly bioavailable [122]. According to RAC guideline (Table 8), for any metal, soil/sediment which can release in exchangeable and carbonate fractions, less than 1% of the total metal will be considered safe for the environment and soil/sediment with 11-30% carbonate and exchangeable fractions will be at medium risk to the environment. On the contrary, soil/sediment releasing in the above fractions more than 50% of the total metal has considered being highly dangerous, which can be easily enter the food chain [123].

Table 8. Criteria of Risk Assessment Code [123].

Grade	Exchangeable and bounded to carbonate metal (%)	Risk
I	<1	No risk
II	1 – 10	Low risk
III	11 – 30	Medium risk
IV	31 – 50	High risk
V	>50	Very high risk

Heavy-Metal Fractionation in surface sediments was studied by Dhanakumar et. al. [124] in the Cauvery river estuarine region, southeastern coast of India. The results revealed that most of the samples fall under the category from low- to high-risk class and from low-risk to very high-risk class in terms of labile fractions of Pb as well as Zn and Cu, respectively.

8. CONCLUSION

From the above discussion it is revealed that geochemical fractionation approach to the chemical speciation has provided a useful tool and opens a new dimension in assessing the potential mobility/bioavailability of heavy metals and metalloids in soils/sediments and opens a new dimension in the field of ecology and environmental chemistry. More efficient, non-laborious and time saving processes techniques in this field of chemical speciation are also coming up to get valid information regarding geochemical behavior of soils/sediments. Besides geochemical fractionation, Dezileau et al. [125] opined that total Fe or Fe/Al may be used to infer millennial-scales climate changes in the south eastern pacific while performing sequential extraction of Fe in marine sediments from the Chileau continental margin. However, the chemical partitioning should be carefully used in the assessment of environmental pollution as large amount of metals may naturally occur as anthropogenic fractions (including loosely bonded ions, sulfide ions and metals associated with sediments).

9. ACKNOWLEDGEMENTS

The authors gratefully acknowledge full support and cooperation of the Springer press, UK for extending permission in publishing the research paper of the journal Environmental Monitoring & Assessment, vol. 184(12), pp:7561-77, 2012.

This study was partially supported by the European Fund for Economic and Regional Development (FEDER) through the Program Operational Factors of Competitiveness (COMPETE) and National Funds through the Portuguese Foundation for Science and Technology (PEST-C/MAR/UI 0284/2011, FCOMP 01 0124 FEDER 022689).

REFERENCES

1. Gleyzes C, Tellier S, Astruc M. Fractionation studies of trace elements in contaminated soils and sediments: a review of sequential extraction procedures. Trends in Analytical Chemistry 2002; 21 (6) 451-467.
2. Tessier A, Campbell PGC, Bisson M. Sequential extraction procedure for the speciation of particulate trace metals. Analytical Chemistry 1979; 51 844-851.
3. Sahuquillo A, Rigol A, Rauret G. Overview of the use of leaching/extraction tests for risk assessment of trace metals in contaminated soils and sediments. Trends in Analytical Chemistry 2003; 22(3) 152-159.

4. Calmano W, Hong J, Förstner U. Binding and mobilization of heavy metals in contaminated sediments affected by pH and redox potential. Water Science and Technology 1993; 28(8/9) 53-58.
5. Eggleton J, Thomas KV. A review of factors affecting the release and bioavailability of contaminants during sediment disturbance events. Environment International 2004; 30 973-980.
6. Ure AM, Quevauviller P, Muntau H, Griepink B. Speciation of heavy metals in solids and harmonization of extraction techniques undertaken under the auspices of the BCR of the Commission of the European Communities. International Journal of Environmental Analytical Chemistry 1993; 51 135-151.
7. Florence TM. The speciation of trace elements in waters. Talanta 1982; 29 345-69.
8. Lung W. Speciation analysis – why and how? Fresenius' Journal of Analytical Chemistry 1990; 337 557-564.
9. Schroeder WH. Development in the speciation of mercury in natural waters. Trends in Analytical Chemistry 1989; 8 339-342.
10. Rauret G, Rubio R, Lopez-Sanchez JF, Cassassas E. Determination and speciation of copper and lead in sediments of a Mediterranean River (River Tenes, Catalonia, Spain). Water Research 1988; 22(4) 449-51.
11. Pardo R, Barrado E, Perez L, Vega M. Determination and association of heavy metals in sediments of the Pisuerga River. Water Research 1990; 24(3) 373-379.
12. Jardo CP, Nickless G. Chemical association of Zn, Cd, Pb and Cu in soils and sediments determined by the sequential extraction technique. Environmental science & Technology Letters 1989; 10 743-52.
13. Tessier A, Campbell PGC, Bisson M. Heavy metal speciation in the Yamaska and St. Francois rivers (Quebec). Canadian Journal of Earth Sciences 1980; 17 90-105.
14. Elsokkary LH, Muller G. Assessment and speciation of chromium, nickel, lead and chromium in the sediments of the river Nile, Egypt. Science of the Total Environment 1990; 97/98 455-63.
15. Ure AM. Single extraction schemes for soil analysis and related applications. Science of the Total Environment 1996; 178(1/3) 3-10.
16. Rauret G. Extraction procedures for the determination of heavy metals in contaminated soil and sediment. Talanta 1998; 46(3) 449-455.
17. Quevauviller PH, Lachica M, Barahona E, Rauret G, Ure A, Gomez A, Muntau H. Interlaboratory comparision of EDTA and DTPA procedures prior to certification of extractable trace elements in calcareous soil. Science of the Total Environment 1996; 178(1/3) 127-132.
18. Quevauviller PH, van der Slootr HA, Ure A, Muntau H, Gomez A, Rauret G. Conclusions of the workshop: harmonization of leaching/extraction tests for environmental risk assessment. Science of the Total Environment 1996; 178(1/3) 133-139.
19. Bacon JR, Davidson CM. Is there a future for Sequential chemical extraction? Analyst 2008; 133 25-46
20. Förstner U, Lechsber R, Davis RA, Hermitte PL. (eds.), Chemical methods for assessing bioavailable Metals in Sludges. Elsevier, London, 1985.
21. Meguellati M, Robbe D, Marchandise P, Astruc M, Proceedings International Conference on Heavy Metals in the Environment, Heidelberg CEP Consultants, Edinburgh, 1983, p. 1090.
22. Filgueiras AV, Lavilla I, Bendicho C. Chemextraction for metal partitioning in environmental solid samples, Journal of Environmental Monitoring 2002; 4 832-857.
23. Lopez-sanchez JF, Rubio R, Rauret G. Comparison of two sequential extraction procedures for trace metal partitioning in sediments. International Journal of Environmental Analytical Chemistry 1993; 51 113-121.
24. Hlavay J, Polyak K. Chemical speciation of elements in sediment samples collected at Lake Balaton. Microchemical Journal 1998; 58 281-290.

25. Tokalioglu S, Kartal S, Elçi L. Determination of heavy metals and their speciation in lake sediments by flame atomic absorption spectrometry after a four-stage sequential extraction procedure, Analytical Chimica Acta 2000; 413 33-40.
26. Belazi AU, Davidson CM, Keating GE, Littlejohn D. Determination and speciation of heavy metals in sediments from the Cumbrian coast, NW England, UK. Journal of Analytical Atomic Spectrometry 1995; 10 233-240.
27. Mester Z, Cremisini C, Ghiara E, Morabito R. Comparison od two sequential extraction procedures for metal fractionation in sediment samples. Analytical Chimica Acta 1998; 259 133-142.
28. Gomez-Ariza JL, Giraldez I, Sanchez-Rodas D, Morales E. Comparison of the feasibility of three extraction procedures for trace metal partitioning in sediments from southwest Spain, Science of Total Environment 2000; 246 271-283.
29. Zhang T, Shan X, Li F. Comparison of two sequential extraction procedures for speciation analysis of metals in soils and plant availability. Communications in Soil Science and Plant Analysis 1998; 29 1023-1034.
30. Albores AF, Cid BP, Gomez P, Lopez EF. Comparison between sequential extraction procedures and single extrations for metal partitioning in sewage sludge samples. Analyst 2000; 125 1353-1357.
31. Ho MD, Evans GJ. Operational speciation of cadmium, copper, lead and zinc in the NIST standard reference materials 2710 nad 2711 (Montana soil) by the BCR sequential extraction procedure and flame atomic spectrometry. Analytical Communications 1997; 34 353-364.
32. Zemberyova M, Bartekov J, Hagarov I. The utilization of modified BCR three-step sequential extration procedure for the fractionation of Cd, Cr, Cu, Ni, Pb and Zn in soil reference materials of different origins, Talanta 2006; 70 973-978.
33. Singh KP, Mohan D, Singh VK, Malik A. Studies on distribution and fractionation of heavy metals in Gomti river sediments- a tributary of the Ganges India. Journal of Hydrology 2005; 312 14-27.
34. Davidson CM, Duncan AL, Littlejohn D, Ure AM, Garden LM. A critical evaluation of the three-stage BCR sequential extraction procedure to assess the potential mobility and toxicity of heavy metals in industrially-contaminated land. Analytica Chimica Acta 1998; 363(1) 45-55.
35. McLaren RG, Crawford D. Studies on soil copper I. The fractionation of copper in soils. Journal of Soil Science 1973; 24(2), 172-181.
36. Kersten M, Förstner U. Speciation of trace elements in of water-soluble organic components. sediments and combustion waste.In Ure AM, Davidson CM (ed.) Chemical speciation in the environment. Blackie Academic and Professional, Glasgow, UK. 1995; 234-275.
37. Ianni C, Magi E, Rivaro P, Ruggieri N. Trace metals in Adriatic coastal sediments: distribution and speciation pattern. Toxicological and Environmental Chemistry 2000; 78 73-92.
38. Ianni C, Ruggieri N, Rivaro P, Frache R. Evaluation and comparison of two selective extraction procedures for heavy metal speciation in sediments. Analytical Sciences 2001; 17 1273-1278.
39. Ramirez M, Massolo S, Frache R, Correa J. Metal speciation and environmental impact on sandy beaches due to El Salvador copper mine, Chile. Marine Pollution Bulletin 2005; 50 62-72.
40. Borovec Z, Tolar V, Mraz L. Distribution of some metals in sediments of the central part of the Labe (Elbe) River, Czech Republic. Ambio 1993; 22 200-205.
41. Campanella L, D'Orazio D, Petronio BM, Pietrantonio E. Proposal for a metal speciation study in sediments. Analytica Chimica Acta 1995; 309 387-393.
42. Zdenek B. Evaluation of the concentrations of trace elements in stream sediments by factor and cluster analysis and the sequential extraction procedure. Science of the Total Environment 1996; 177 237-250.

43. Gomez-Ariza JL, Giraldez I, Sanchez-Rodas D, Moralesm E. Metal sequential extraction procedure optimized for heavily polluted and iron Oxide rich sediments. Analytica Chimica Acta 2000; 414 151-164.
44. Lopez-Sanchez JF, Sahuquillo A, Fiedler HD, Rubio R, Rauret G, Muntau H, Quevauviller P. CRM 601, a stable material for its extractable content of heavy metals. Analyst 1998; 123 1675-1677.
45. Usero J, Gamero M, Morillo J, Gracia I. Comparative study of three sequential extraction procedures for metals in marine sediments. Environment International 1998; 24 478-496.
46. Martin R, Sanchez DM, Gutierrez AM. Sequential extraction of U, Th, Ce, La and some heavy metals in sediments from Ortigas River, Spain. Talanta 1998; 46 1115-1121.
47. Agnieszka S, Wieslaw Z. Application of sequential extraction and the ICPAES method for study of the partitioning of metals in fly ashes. Microchemical Journal 2002; 72 9-16.
48. Templeton DM, Ariese F, Cornels R. IUPAC guidelines for terms related to chemical Speciation and Fractionation of elements. Pure and Applied Chemistry 2001; 72 1453-1470.
49. Bordas F, Bourg ACM. A critical evaluation of sample for storage of Contaminated Sediments to Be investigated for the potential mobility of their heavy metal load. Water. Air. Soil. Pollution 1998; 103 137-149.
50. Kersten M, Frostner U. Chemical fractionation of heavy metals in anoxic estuarine and coastal sediments. Water Science and Technology 1986; 18 121-130.
51. Dold B. Speciation of the most soluble phases in a sequential extraction procedure adapted for geochemical studies of copper sulfide mine waste. Journal of Geochemical Exploration 2003; 80 55-68.
52. Favas PJC, Pratas J, Gomes MEP, Cala V. Selective chemical extraction of heavy metals in tailings and soils contaminated by mining activity: Environmental implications. Journal of Geochemical Exploration 2011; 111 160-171.
53. Pardo R, Vega M, Debán L, Cazurro C, Carretero C. Modelling of chemical fractionation patterns of metals in soils by two-way and three-way principal component analysis. Analytica Chimica Acta 2008; 606 26-36.
54. Cuong TD, Obbard JP. Metal speciation in coastal marine sediments from Singapore from Singapore using a modified BCR-sequential extraction procedure. Applied Geochemistry 2006; 21 1335-1346.
55. Dapaah RK, Takano N, Ayame A. Solvent extraction of Pb (II) from acid medium with zinc Hexamethylenedithiocarbamate followed by back-extraction and subsequent determination by FAAS. Analytica Chimica Acta 1999; 386 281-286.
56. Gomez-Ariza JL, Giraldez I, Sanchez-Rhodes D, Morales E. Metal readsorption and re-distribution during analytical fractionation of trace elements in toxic estuarine sediments. Analytica Chimica Acta 1999; 399 295-307.
57. Cheam V, Lechner J, Sekerka I, Desrosiers R, Nriagu J. Development of laser- excited atomic fluorescence spectrometer and a method for the direct determination of lead in Great Lake waters. Analytica Chimica Acta 1992; 269 129-136.
58. Fischer E, Van D, Berg CMG. Anodic Stripping Voltammetry of Pb and Cd using a Hg film electrode and thiocyanate. Analytica Chimica Acta 1999; 385 273-280.
59. Morales MM, Mart P, Llopis A, Compos L, Sagrado S. An environmental study by factor Analysis of surface sea waters in the Gulf of Valencia (Western Mediterranean). Analytica Chimica Acta 1999; 394 109-117.
60. Hirade M, Chen Z, Sugimoto K, Kawaguchi H. Co precipitation with tin (IV) hydroxide followed by removal of tin carrier for the Determination of trace heavy metals by graphite-furnace atomic absorption Spectrometry. Analytica Chimica Acta 1980; 302 103-107.
61. Ridout PS, Jones HR, Williams JG. Determination of trace elements in a marine reference material of lobster hepatopancreas (TORT-1) using inductively coupled plasma mass spectrometry. Analyst 1988; 113 1383-1386.

62. Sakao SY, OgawaY, Uchida H. Determination of trace elements in seaweed samples by inductively coupled plasma mass spectrometry. Analytica Chimica Acta 1999; 355 121-127.
63. Baeyens W, Monteny F, Leermakers M, Bouillon S. Evalution of sequential extractions on dry and wet sediments. Analytical and Bioanalytical Chemistry 2003; 376 890- 901.
64. Guevara-Riba A, Sahuquillo A, Rubio R, Rauret G. Assessment of metal mobility in dredged harbour sediments from Barcelona, Spain. Science of the Total Environment 2004; 321 241-255.
65. Idris AM, Eltayeb MAH, Potgieter-Vermaak SS, Grieken R, Potgieter JH. Assessment of heavy metal pollution in Sudanese harbours along the Red Sea coast. Microchemical. Journal 2007; 87 104-112.
66. Wepener V, Vermeulen LA. A note on the concentrations and bioavailability of selected metals in sediments of Richards Bay Harbour, South Africa. Water SA 2005; 31 589-595.
67. Esslemont G. Heavy metals in seawater, marine sediments and corals from the Townsville section, Great Barrier Reef Marine Park, Queensland. Marine Chemistry 2000; 71 215- 231.
68. Coetzee PP. Determination and speciation of heavy metals in sediments of the Hartebeespoort Dam By sequential extraction. Water SA 1993; 19 291-300.
69. Salmons W, Förstner U. Trace metal analysis on polluted sediments. Part II:evaluation of Environmental impact. Environmental science and Technology letters 1980; 1 14-24.
70. Li X, Thornton I. Chemical partitioning of trace and major elements in soils contaminated by mining and smelting activities. Applied geochemistry 2000; 16 1693-1706.
71. Kiratli N, Ergin M. Partitioning of heavy metals in surface Black Sea sediments. Applied Geochemistry 1996; 11 775-788.
72. Rauret G, Lopez-Sanchez JF. New sediment and soil CRMs for extractable Trace metal content. International Journal of Environmental Analytical Chemistry 2001; 79 81-95.
73. Salmons W. Adoption of common schemes for single and sequential extractions of Trace metals in soil and sediments. International Journal of Environmental Analytical Chemistry 1993; 51 3-4.
74. Fiedler HD, Lopez-Sanchez JF, Rubio R, Rauret G, Quevauviller PH. Study of the stability of extractable trace metal contents in a river sediment using Sequential extraction. Analyst 1994; 119 1109-1114.
75. Ramos L, Hernandez LM, Gonzalez MJ. Sequential fraction of copper, lead, copper, Cadmium and zinc in soils from or near Donana National Park. Journal of Environmental Quality 1994; 23 7-50.
76. Tu Q, Shan XZ, Ni Z. Evaluation of a sequential extraction procedure for the Fractionationation of amorphous iron and manganese oxides and organic matter in soils. The Science of The total Environment 1994; 151 159-165.
77. Wang S, Jia Y, Wang S, Wang X, Wang H. Fractionation of heavy metals in shallow marine sediments from Jinzhou Bay, China. Journal of Environmental Science (China) 2010 22 23-31.
78. Yuan CG, Shi JB, He B, Liu JF, Liang LN. Speciation of heavy metals in marine sediments from the East China Sea by ICP-MS with sequential extraction. Environment International 2004; 30 769-783.
79. Jones B, Turki A. Distribution and Speciation of heavy metals in surficial sediments from the Tees Estuary, North – East England. Marine Pollution Bulletin 1997; 34 768-779.
80. Ajay SO, Van Loon GW. Studies on redistribution during the analytical fractionation of metals in sediments. The Science of the Total Environment 1989; 87 171-187.
81. Pempkowiak J, Sikora A, Biernacka E. Speciation of heavy metals in marine sediments vs their bioaccumulation by mussels. Chemosphere 1999; 39 313- 321.

82. Forstner U, Wittmann GTW. Metal Pollution in the Aquatic Environment, Springer-Verlag, Berlin. Springer-Verlag, Heidelberg; 1981.

83. Fedotov PS, Zavarzina, a AG, Spivakov BYa, Wennrich, b R, Mattusch J, De K, Titzeb PC, Demin VV. Accelerated fractionation of heavy metals in contaminated soils and sediments using rotating coiled columns, Journal of Environmental Monitoring 2002; 4 318-324.

84. Nemati K, Kartini N, Bakar A, Abas MR, Sobhanzadeh E. Speciation of heavy metals by modified BCR sequential extraction procedure in different depths of sediments from Sungai Buloh, Selangor, Malaysia, Journal of Hazardous Materials 2011; 192(1) 402-410.

85. Mossop KF, Davidson CM. Comparison of original and modified BCR sequential extraction procedures for the fractionation of copper, iron, lead, manganese and zinc in soils and sediments. Analytica Chimica Acta 2003; 478 (1) 111-118.

86. Walkey A, Black TA. An examination of the Dugtijaraff method for determining soil organic matter and proposed modification of the chronic and titration method. Soil Science 1934; 37 23-38.

87. Farnham IM, Johannesson KH, Singh AK, Hodge VF, Stetzenbach KJ. Factor analytical approaches for evaluating groundwater trace element chemistry data. Analytica Chimica Acta 2003; 490 123-138.

88. Ragno G, De Luca M, Ioele G. An application of cluster analysis and multivariate cassification methods to spring water monitoring data. Microchemical Journal 2007; 87 119-127.

89. Jonathan MP, Ram Mohan V. Heavy metals in sediments of the inner shelf off the Gulf of Mannar, Southeast coast of India. Marine Pollution Bulletin 2003; 46 263-268.

90. Sunil Kumar R. Distribution of organic carbon in the sediments of Cochin mangroves, south west coast of India. Indian Journal of Marine Science 1996; 25 274-276.

91. Janaki-Raman D, Jonathan MP, Srinivasalu S, Armstrong-Altrin J S, Mohan SP, Ram-Mohan V. Trace metal enrichments in core sediments in Muthupet mangroves, SE coast of India: Application of acid leachable technique. Environmental Pollution 2007; 145 245-257.

92. Sarkar SK, Bilinski SF, Bhattacharya A, Saha M, Bilinski H. Levels of elements in the surficial estuarine sediments of the Hugli river, northeast India and their environmental implications. Environment International 2004; 30 1089-1098.

93. Tam NFY, Wrong YS. Spatial variation of heavy metalsin surfe sediments of Hong, Kong mangrove swamps. Environmental Pollution 2002; 110 195-205.

94. Subramanian V, Mohanachandran G. Heavy metals distribution and enrichment in the sediments of southern east coast of India. Marine Pollution Bulletin 1990; 21 324-330.

95. Chatterjee M, Massolo S, Sarkar SK, Bhattacharya AK, Bhattacharya BD, Satpathy KK, Saha S. An assessment of trace element contamination in intertidal sediment cores of Sunderban mangrove wetland, India for evaluating sediment quality guidelines. Environmental Monitoring and Assessment 2009; 150 307-322.

96. Datta DK, Subramanian V. Distribution and fractionation of heavy metals in the surface sediments of the Ganges–Brahmaputra–Meghna river system in the Bengal Basin. Environmental Geology 1998; 36 93-101.

97. Müller G. Schwermetalle in den sedimenten des Rheins-Veranderungen seit. Umschan Verlag 1979; 79 133-149.

98. Salomon W, Förstner U. Metals in the hydrocycle. Berlin: Springer; 1984.

99. Buccolieri A, Buccolieri G, Cardellicchio N, Dell'atti A, Leo AD, Maci A. Heavy metals in marine sediments of Taranto Gulf (Ionian Sea, Southern Italy). Marine Chemistry 2006; 99 227-235.

100. Dowling CB, Poreda RJ, Basu AR, Aggarwal PK. Geochemical study of arsenic release mechanisms in the Bengal Basin groundwater. Water Resources Research 2002; 38(9) 1173-1190.

101. Acharyya SK, Lahiri S, Raymahashay BC, Bhowmilk A. Arsenic toxicity of groundwater in parts of the Bengal basin in India and Bangladesh: the role of Quaternary stratigraphy and Holocene sea-level fluctuation. Environmental Geology 2000; 39 231-238.
102. Takarina ND, Browne DR, Risk MJ. Speciation of heavy metals in coastal sediments of Semarang, Indonesia. Marine Pollution Bulletin 2004; 49 854-874.
103. Pickering WF. Metal ion speciation—soil and sediments (a review). Ore Geology Reviews 1986; 1 83-146.
104. Ngiam LS, Lim PE. Speciation patterns of heavy metals in tropical estuarine anoxic and oxidized sediments by different sequential extraction schemes. Science of the Total Environment, 2001; 275 53-61.
105. Dassenakis M, Adrianos H, Depiazi G, Konstantas A, Karabela M, Sakellari A, Scoullos M. The use of various methods for the study of metal pollution in marine sediments, the case of Euvoikos Gulf, Greece. Applied Geochemistry 2003; 18 781-794.
106. Morillo J, Usero J, Gracia I. Heavy metal distri-bution in marine sediments from the southwest coast of Spain. Chemosphere 2004; 55 431-442.
107. Thomas RP, Ure AM, Davidson CM, Littlejoh D, Rauret G, Rubio R, López-Sánchez JF. Three-stage sequential extraction procedure for the deter-mination of metals in river sediments. Analytica Chimica Acta 1994; 286 423-429.
108. Panda D, Subramanian V, Panigrahy RC. Geochemical fractionation of heavy metals in Chilka Lake (east coast of India) – a tropical coastal lagoon. Environmental Geology 1995; 26 199-210.
109. Dawson EJ, Macklin MG. Speciation of heavy metals in floodplain and flood sediments: a reconnaissance survey of the Aire Valley, West Yorkshire, Great Britain Environmental Geochemistry and Health 1998; 20 67-76.
110. Ramos L, González M, Hernández L. Sequential extraction of copper, lead, cadmium, and zinc in sediments from Ebro River (Spain): relationship with levels detected in earthworms. Bulletin of Environmental Contamination and Toxicology 1999; 62 301-308.
111. Caille N, Tiffreau C, Leyval C, Morel JL. Solubility of metals in an anoxic sediment during prolonged aeration. Science of the Total Environment 2003; 301 239-250.
112. Canfield DE. Reactive iron in marine sediments. Geochimica et Cosmochimica Acta 1989; 53 619-632.
113. Petersen W, Wallman K, Li PL, Schroeder F, Knauth HD. Exchange of trace elements at the sediment– water interface during early diagenesis processes. Marine and Freshwater Research 1995; 46 19-26.
114. Monbet P. Mass balance of lead through a small macrotidal estuary: the Morlaix River estuary (Brittany, France). Marine Chemistry 2006; 98 59-80.
115. Adriano DC. Trace elements in terrestrial environments. New York: Springer; 1986.
116. Sarkar SK, Saha M, Takada H, Bhattacharya A, Mishra P, Bhattacharya B. Water quality management in the lower stretch of the river Ganges, east coast of India: an approach through environmental education. Journal of Cleaner Production 2007; 15 1559-1567.
117. Banerjee ADK. Heavy metal levels and solid phase speciation in street dusts of Delhi, India. Environmental Pollution 2003; 123(1) 95-105.
118. Long ER, Ingersoll CG, MacDonald DD. Calculation and uses of mean sediment quality guideline quotients: a critical review. Environmental Science and Technology 2006; 40 1726-1736.
119. Venkatramanan S, Ramkumar T, Anithamary I, Jonathan MP. Speciation of selected heavy metals geochemistry in surface sediments from Tirumalairajan river estuary, east coast of India. Environmental Monitoring and Assessment 2013; 185(8) 6563-6578.
120. Ranjan RK, Singh G, Routh J, Ramanathan AL. Trace metal fractionation in the Pichavaram mangrove–estuarine sediments in southeast India after the tsunami of 2004. Environmental Monitoring and Assessment 2013 (article in press).

121. Mohan M, Augustine T, Jayasooryan KK, Chandran MSS, Ramasamy EV. Fractionation of selected metals in the sediments of Cochin estuary and Periyar River, southwest coast of India, Environmentalist 2012; 32 383-393.
122. Hseu ZY. Extractability and bioavailability of zinc over time in three tropical soils incubated with biosolids. Chemosphere 2006; 63 762-771.
123. Perin G, Craboledda L, Lucchese M, Cirillo R, Dotta L, Zanette ML, Orio AA. Heavy metal speciation in the sediments of Northern Adriatic Sea- a new approach for environmental toxicity determination, in: T.D. Lekkas (Ed.), Heavy Metal in the Environment 1985; 2 454-456.
124. Dhanakumar S, Murthy KR, Solaraj G, Mohanraj R. Heavy-Metal Fractionation in Surface Sediments of the Cauvery River Estuarine Region, Southeastern Coast of India, Arch Environmental Contamination Toxicology 2013; 65 14-23.
125. Dezileau L, Pizarro C Rubio MA. Sequential extraction f iron in marine sediments from the Chilen continental margin. Marine Geology 2007; 241 111-116.

CHAPTER 13

Soil Organic Carbon Dynamics in Eroding and Depositional Landscapes

Kenneth R. Olson[1], Mahdi Al-Kaisi[2], Rattan Lal[3], Larry Cihacek[4]

[1]*Professor Emeritus, Department of NRES, College of ACES, University of Illinois, Champaign, IL, USA*

[2]*Department of Agronomy, College of Agriculture, Iowa State University, Ames, IA, USA*

[3]*School of Environment and Natural Resources, The Ohio State University, Columbus, OH, USA*

[4]*School of Natural Resources Science, North Dakota State University, Fargo, ND, USA*

ABSTRACT

As a requisite to determining management practice effects on stored soil organic carbon (SOC) stock in a landscape unit, the baseline SOC stock with depth must be determined and the land use, management practices and erosion-induced changes measured periodically or over a period of time. The SOC loss and additions due to soil erosion, transport and deposition must be accounted for or be quantified when determining the real impact of the management practices on net SOC stock over time. Quantifying the SOC loss due to erosion will help avoid over estimation of the management practice performances. Appropriate soil sampling designs and sampling procedures are needed to establish a SOC stock baseline and to monitor and verify new SOC storage or sequestration as a result of a management practice. The Dinesen Prairie landscape in western Iowa, USA was sampled to provide a SOC stock baseline and then the adjacent cropland was sampled to determine the past impact of land use change, management practices and erosion on SOC stock retention. After 100 to 150 years of farming, the entire cropland landscape retained only 49% of the baseline prairie SOC stock. Only the cropland toe-slope (TS) retained more SOC stock than the prairie TS as a result of the erosion, transport and deposition of SOC rich sediment on the TS.

Keywords: Soil Organic Carbon, Erosion, Monitoring

1. INTRODUCTION

The conversion of prairie and forest to cultivated cropland has decreased the land's capacity to produce and maintain the soil's original organic carbon (SOC) stocks and created a paradox in attempts to achieve sustainable agricultural intensification, human food security and ecosystem resilience [1] , [2] . Soil degradation processes, including soil erosion, in cropping systems have affected SOC stocks and are limiting agricultural productivity. SOC stocks are critical to the production of food, fresh water, air quality, erosion prevention, nutrient cycling, and support of wildlife habitat. The degradation of soil can be a result of soil erosion, transport and deposition of soil sediments and loss of SOC stocks. Cihacek et al. [3] proposed using universal standards for methods and procedures for measuring and verifying SOC stored in a terrestrial system.

De Gruijter et al. [4] provided a comprehensive overview of monitoring land resources. These researchers describe the many and varied decisions that must be made in the design and implementation of a monitoring strategy. As prerequisite to demonstrating the stored SOC stock in a landscape unit, Chappell et al. [5] suggested measuring changes in SOC stock in a landscape unit (entire or part of a field) over a period of time. In this case, the measurements represent the net balance between the storage and loss of SOC caused by management practices [6] and the loss and addition due to soil erosion. The impact of soil erosion must be documented to determine (quantify) the real impact of the management practices on SOC stocks [7] , [8] . When net removal of SOC by soil erosion exceeds net addition by management practices, no real SOC gain would be documented even when management practices proven to enhance SOC stocks are implemented [9] , [10] . Quantifying SOC loss due to erosion can avoid overestimation of the management practice performances. Appropriate soil sampling designs and sampling procedures are needed to monitor and verify SOC net storage or sequestration [3] , [11] .

Areas of permanent vegetation near wind-eroding soils can capture substantial quantities of SOC containing sediments [12] . Direct deposition of fresh or aged plant materials occurs where the materials move along with soil sediments in wind erosion events, sliding across the soil surface until a leeward face of a drift is reached. At this point, the plant materials drop out of the wind stream and are buried within the sediment deposit enriching the SOC as the plant materials decompose at a later time. Soil erosion is caused by energy from water, wind, gravity and chemical reactions. However, water erosion, receives most attention because its after-effects are more noticeable, and it is linked to other negative outcomes, such as impacted water quality and loss of productivity that resonate with the general populations of the US through its link with food production and security. Transport of SOC and nutrients by wind-blown soil particles is of potential relevance to water quality [13] .

Most soil sampling techniques use distance from the soil surface as a primary metric. The soil surface, however, is a reliable datum only for measurement of C concentration characteristics directly related to distance from the soil surface at the time of sampling especially when evaluating near surface soil samples [14] . Deep SOC in soil profiles differ between the tillage treatments and various land uses. Deeper sampling will not completely overcome a bias caused by bulk density variations and resultant change in soil surface elevation except when the SOC constituent is universally absent at lower depths. Equivalent soil mass (mass-depth) instead of linear depth can be used to correct for tillage and land use treatment differences in soil bulk density, allowing more precise and accurate quantitative comparison of SOC constituents [14] , [15] . It is recommended to sample to a lower soils depth, where SOC concentration is negligible. In most soils SOC concentration is uniformly absent at depths ranging from 0.5 to 1.0 m [3] . The SOC data should be expressed on an equal soil mass per unit area to the appropriate depth where SOC concentration is absent. An example of such units would be $Mg \cdot C \cdot ha^{-1}$ to a 1 m depth [16] .

SOC losses can occur from land use change, water erosion, conventional tillage use, oxidation and mineralization of organic matter, disturbance during planting, and disturbance when fertilizers or manure injection take place. This loss can be offset by the SOC gains from plant and root growth and residue decomposition and returned to the humus or soil organic matter (SOM). Previously eroded soils with low SOC content were identified as having significant potential to sequester SOC [17] . Olson [18] studied previously eroded soils on 6% slopes with low SOC content in an attempt to quantify the amount and rates of SOC storage and retention as a result of a conversion to a NT system.

Pre-treatment SOC baseline of the landscape unit, tract, field or plot should be determined prior to the study [6] . No SOC sequestration can actually occur in a landscape unit if SOC level was higher at the start of the experiment than at the end of the study. A pre-treatment SOC baseline is essential in all landscape studies to determine the amount and rate of SOC sequestration, steady state or loss. A pre-treatment SOC baseline is needed when determining the amount and rate of SOC sequestration, storage, retention or loss, especially on sloping and eroding landscape units (Figure 1) with more intensive cropping rotations (more row crops and fewer years of forages) during the study than in the previous years.

The primary objective of this paper is to present strategies to evaluate changes in SOC stock in a landscape unit over a period of time. A prairie grassland landscape was selected and sampled to establish the SOC stock baseline and the adjacent cropland landscape SOC stock was measured with depth to determine the effects of land use change, management practices and erosion on SOC stock retention. The impact of soil erosion on SOC stocks must be documented to determine the true impact of management practices. In this study, we examine the SOC stock monitoring protocols and the impact of soil erosion, transport and deposition on SOC stocks. There is a great deal of variation in crop rotations and management practices among land uses and cropping systems. It is these variations which influence whether SOC is retained, enhanced, and potentially sequestered or the SOC stock is simply lost to the atmosphere and aquatic ecosystems. Soil degradation within intensified agricultural systems will deplete SOC stocks and nutrients and result in a decline in soil structure, loss of soil resilience as soil functions and services decline, a loss of soil, a loss of soil biodiversity, and disruption to key biotic and a biotic processes necessary for intensive cropping system productivity [19] .

2. MATERIAL AND METHODS

Site Description and Sampling Procedure

The Dinesen Prairie study site is located in western Iowa near the city of Harlan [2] (Figure 1). The mean annual temperature for the area is 8.6 degrees C and the mean annual rainfall is 84.7 cm. Two treatments were used, prairie and agriculture, both on sloping landscapes. A monument marker at the prairie site suggests the location has not been farmed since at

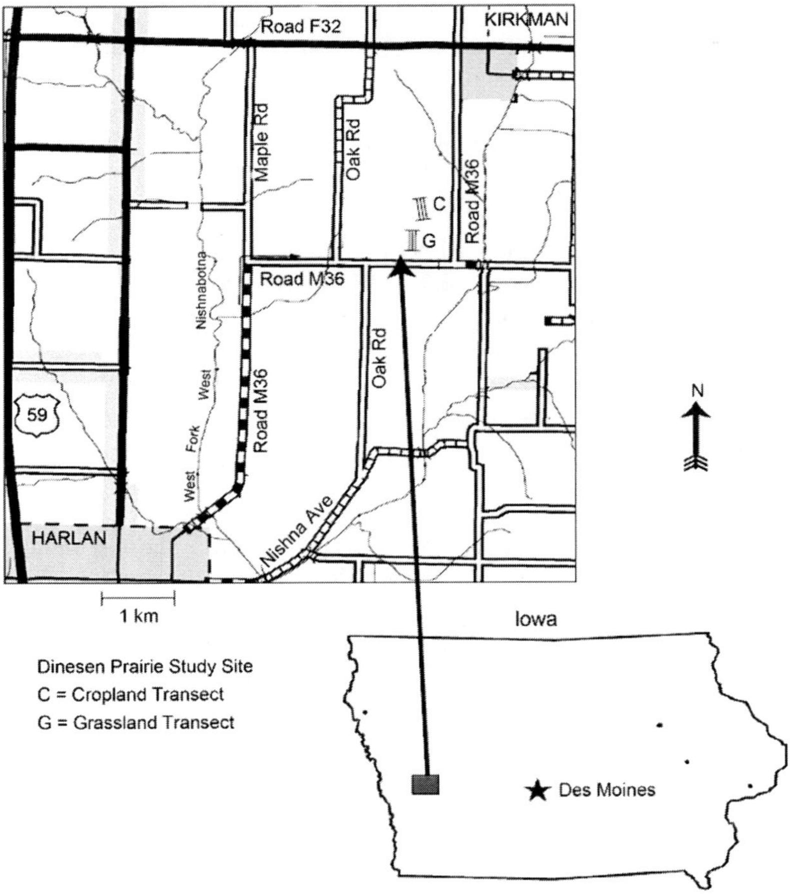

Figure 1. Map of the Dinesen Prairie study sites near Harlan in western Iowa (US) [2] (modified from Salemme 2015).

least 1946 and also states that the area has never been plowed. However, the area was seeded with legumes at some point and is believed to have been used for grazing in the past. Presently, controlled burns are used to manage the prairie in years when weather permits and the area is generally mowed in the fall. In the crop plot a corn and soybean rotation and no-till management are used. The plots were located adjacently with a south facing aspect. The prairie plot is on Marshall silty clay loam with slopes ranging from 5% to 14%. The summit of the crop plot is on Marshall silty clay loam but the majority of the catena is on Exira silty clay loam with slopes ranging from 5% to 14%.

Soil bulk density samples (mass per unit volume) were collected from the same locations using a Model 200 Soil Core Sampler manufactured by Soil Moisture Equipment Corp. The inner dimensions of the sampling ring were 5.4 cm by 6 cm. The soil bulk density samples (mass per unit volume) were oven-dried at 110 degrees C for 48 hours prior to weighing for bulk density determination. SOC was determined using a modified version of the Walkley-Black wet oxidation method, procedure number 6A1 [20] .

Statistical analysis was done using SAS version 9.4 [21] . T-tests were used to compare the different positions and depths of the two land-use treatments. The Wilcoxon rank sum test was used for any data that were not normally distributed. In all instances an alpha of 0.05 was used to declare significance.

Two treatments were used, prairie and cropland with both sides on sloping and eroding landscapes. The sampling scheme consisted of three parallel transects spaced 10 m apart and running up and down the slope from the summit (SU) to the TS. Transects for the prairie plots were 140 m in length while the adjacent crop transects were 130 m. Each transect was sampled at the SU, SH, upper backslope (UBS), lower backslope (LBS), footslope (FS), and TS. Two soil cores were collected to a depth of 1 m at each landscape position on both land use treatments. Soil cores were divided into five intervals: 0 - 15 cm, 15 - 30 cm, 30 - 50 cm, 50 to 75 cm and 75 - 100 cm for SOC stock measurements. Soil bulk density samples (mass per unit volume) were collected from the same locations and layers.

3. RESULTS

The SOC stock for the prairie site was significantly higher than the cropland site for all landscape positions except for the TS (Table 1). The SOC mass was determined using bulk density data [2] . SOC stock was also weighted by landscape segment length to account for the different transect lengths when comparing the land use treatments. The UBS, LBS and FS of the cropland had the largest differences in SOC stock and averaged 73% less SOC than the prairie. The TS of the cropland showed a 9% increase in SOC stock over the grassland site reflecting the deposition of SOC rich sediments. The cropland at the Dinesen site retained 49% of the SOC in the landscape, and contained SOC stock in 1 m layer of soil at 230 Mg·C/ha. Aland use change to cropland with erosion reduced the SOC stock in 1 m of soil to 114 Mg·C/ha. Only the cropland TS retained 9% more SOC stock than the prairie TS as a result of erosion of the upper landscape positions and deposit of SOC rich sediment on the TS. Based on high SOC stock in the 0.50 to 0.75 m and the 0.75 to 1.00 m layers [2] of the cropland TS, there would have been significantly more SOC stock in the 1.0 to 3.0 m layer. The cropland TS SOC stock was underreported as a result of protocol of sampling to a 1 m depth on all landscape positions to get a net SOC stock value for same depth for each land use treatment. The cropland adjacent to the Dinesen Prairie was only able to retain 49% of the SOC in the entire landscape to a 1 m depth.

Table 1. SOC mass at Dinesen Prairie site by landscape segment weighted for segment length.

	Mass of SOC	Landscape	Landscape	Weighed mass of SOC	Cropland
Landscape	in 1 m layer	Segment	Segment %	in Landscape Segment	impact on SOC
Position	Mg-C/ha	Length in m	of Transect	Mg-C/ha	%
Prairie Plots					
Summit	214a	26.7a	0.15	32.6a	
Shoulder	218a	30.6a	0.17	38.0a	
U. Backslope	221a	31.3a	0.18	39.4a	
L. Backslope	242a	26.1a	0.15	36.1a	
Footslope	223a	27.2a	0.16	34.7a	
Toeslope	261a	33.4a	0.19	49.8a	
Total	1379a	175a		230a	
Cropland Plots					
Summit	109b	22.9a	0.14	14.8b	−55
Shoulder	99b	27a	0.16	15.9b	−58
U. Backslope	58b	30a	0.18	10.2b	−74
L. Backslope	35b	26.9a	0.16	5.5b	−85
Footslope	85b	26.5a	0.16	13.3b	−62
Toeslope	257a	35.8a	0.21	54.4b	9
Total	642b	169a		114b	−51

Values with the same letter for the same landscape segment, same layer depth, and different land uses are not significantly different at the 0.05 probability level.

4. DISCUSSION

4.1. Impact of Soil Erosion Processes on SOC Cycling and Accounting Models

Landscape position affects the A horizon, root zone thicknesses and depth to the parent material. The A horizon and root zone were the thickest on the SU and TS. If one had sampled only the 20-cm layer, the SOC located below that depth on the SU, SH, FS and TS (Figure 2) would not have been included. Landscape position affects original soil formation processes with thicker development on higher, drier positions and thicker development on the lower more moist positions. Landscape position also affects natural SOC levels. In addition, the root zone below the 20-cm layer also contained significant SOC stock for all landscape positions [22] , [23] .

There are also erosion, transport and depositional effects on FS, TS and depressional sites through deposition of sediment and burial of topsoil, pollution by agrochemicals, inundation, and anaerobiosis leading to methanogenesis and nitrification/denitrification that lead to land degradation. An important strategy for sloping lands is to reduce risks of soil erosion and sedimentation by adoption of best management practices (BMPs) and reducing risks of soil erosion, transport and deposition. The outcomes of soil erosion are processes destructive to soil resilience and productivity, which are represented by SOC loss through increase in GHGs emissions [24] , pollution of water, and contamination of coastal ecosystems by sedimentation and other non-point source pollutants (Figure 3) [25] .

The redistribution and loss of soil through different pathways within eroding, depositional and landscape units and water bodies creates new soil conditions, where SOC-rich sediments are transported from the upper landscape to lower positions. This will create changes in soil hydrological conditions where low depressional areas can lead to both anaerobic conditions conducive to N_2O and CH_4 gas production and subsequent emissions while highly erodible uplands result with more oxidative potential for SOC and low productivity [26] .

These changes in agricultural practices have been documented by several studies [27] - [29] where land intensively cultivated is associated with medium to low agro-bio- diversity due to the conversion of natural ecosystem to cropland and the increase of soil degradation as a result of soil erosion [30] . Figure 3 shows the major pathways by which biodiversity loss services from soil erosion can occur. An increase in land degradation, due to lack of conservation of natural resources and the increase in soil erosion leads to a reduction in SOC sequestration due to loss of primary productivity and biodiversity [29] .

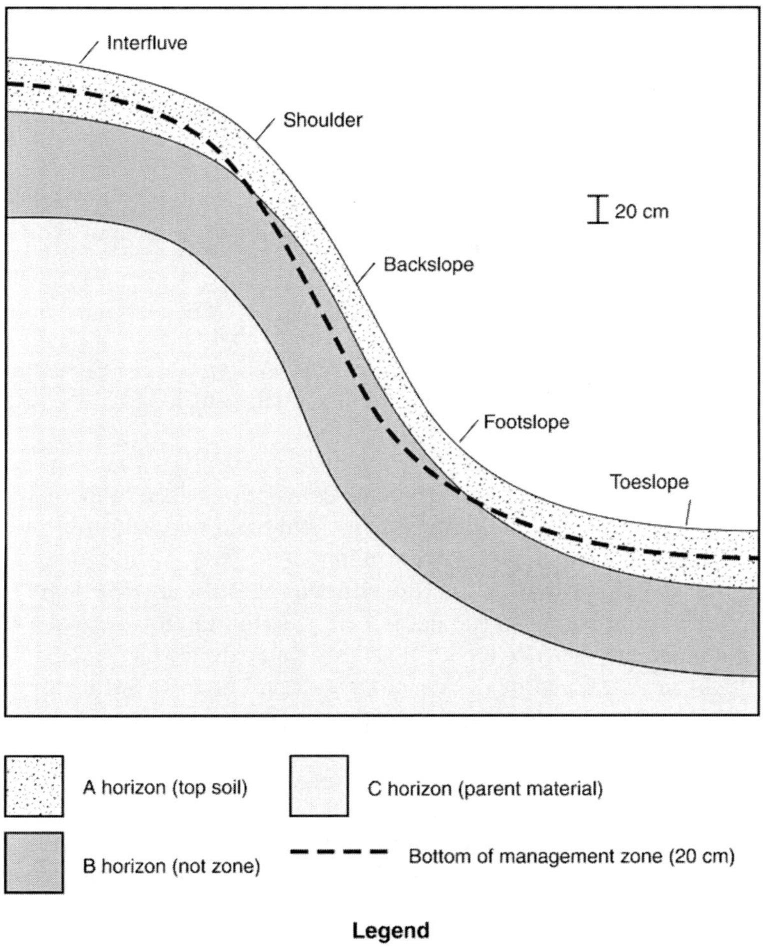

Figure 2. A horizon thickness, root zone thickness and depth to parent material is shown for a hillslope landscape [16] (modified from Olson and Al-Kaisi 2015). Published with copyright per- mission from Catena.

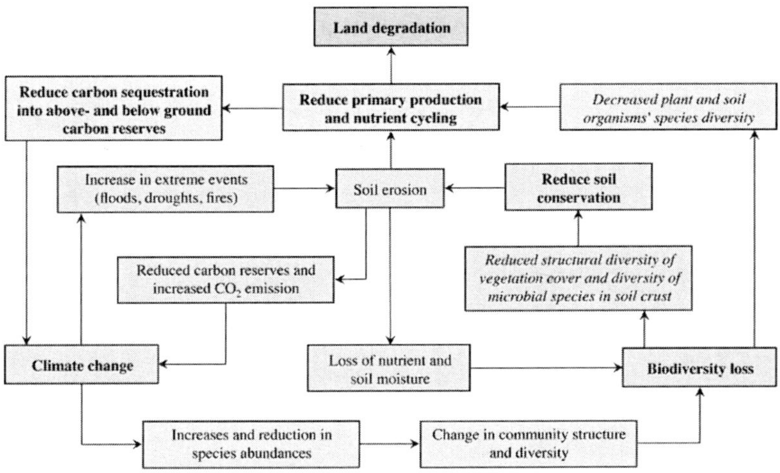

Figure 3. Major pathways by which biodiversity loss and other disservices from soil erosion occur [23] (modified from Lal et al. 2004).

Chappell et al. [5] identified the common misconceptions used to justify and perpetuate the exclusion of soil erosion, transport and deposition by wind and water from C cycling and C accounting models. These included: 1) soil erosion is no longer a threat to food production and the environment; 2) soil erosion need not be measured because it is accounted for by landscape use and management practices; and 3) soil erosion can merely redistribute SOC in a field or landscape or deliver it to a stream or to the atmosphere. The fact remains that while soil erosion processes may be related to land use and management practices, it exacerbates soil degradation and loss of productivity. Therefore, it must be accounted for when monitoring a landscape unit for accurate assessment of soil C dynamics and the pathways that affect its fate. Soil erosion contribution to SOC stock loss is driven by management practices. This highlights the coupled nature of management practices and water and wind erosion processes that ultimately lead to permanent and significant loss of SOC stock (30%) [17] .

Wind erosion, dust transport and CO_2 emission from cultivated fields are readily moved to great distances [12] and it is not uncommon for water erosion to remove soil from the terrestrial ecosystem to the ocean [5] . Water erosion removes soils on sloping land [31] and deposits material on floodplain bottomlands or in the ocean. Olson et al. [32] showed that nearly level (<1%) upland plots were subject to erosion, transport and deposition of SOC-rich sediments. Even small amounts of soil erosion can make a significant difference to the SOC stock change over time because the eroded material has higher ER (enrichment ratio) and is typically considered to be enriched relative to the soils from which the material has been removed [5]

, [18] , [31] , [32] . Because soil erosion is highly variable in space and time, it must be monitored or measured at the same time as SOC stock is measured. Otherwise, it will not be possible to unequivocally demonstrate that measured changes in SOC stocks can be attributed to land use and management practices alone.

Chappell et al. [5] reported changes in SOC stocks over time using simulation models as the difference between outputs of C and losses due to decomposition process of existing SOC stocks with rate modifiers applied to an intrinsic decay constant. Chappell and colleagues found that erosion of topsoil changed the composition and structure of the soil at the eroded location. Sediment deposition on soil from elsewhere can modify the soil micro-environment (i.e., temperature change, hydrology, SOC decomposition rates, etc.). This makes a strong case that erosion should be included in C cycling and C accounting models [33] - [36] .

Sometimes, a slow rate of erosion leads to soil formation and creation of fertile alluvial plains. When accelerated by anthropogenic activity, erosion becomes a destructive process with numerous adverse on-site and off-site effects (Figure 3). Among on-site effects is a decline in the soil quality of eroded land and reduction in agronomic productivity and decrease in plant nutrient availability [37] and water use efficiency [38] . With erosion being a selective process, soil subjected to accelerated erosion is regressively depleted of its topsoil, SOC (concentration and stock) and nutrient reserves. Decline in the root depth zone and reduction in available water capacity (AWC) increase the susceptibility of crops grown on eroded soils to pedologic and agronomic droughts [39] .

4.2. SOC Stock Measuring Challenges in Eroded Landscapes

Soil erosion causes SOC redistribution within terrestrial ecosystems, to the atmosphere and oceans. Li et al. [40] and Chappell et al. [5] reported that dust export is an essential component of the SOC and carbon dioxide (CO_2) budget because wind erosion contributes to the C cycle by selectively removing SOC from vast areas of Australia and transporting C dust quickly offshore and augmenting the net loss of C from terrestrial systems. Li et al. [40] found reduced SOC erosion and/or SOC dust accumulation by vegetation-cover crop must be induced by changes in land use and management practices. Li et al. [40] reported that C sequestration resulted from a reduction of C erosion and dust-C trapped by vegetation-cover crop. Chappell et al. [35] mapped net (1950s- 1990) SOC redistribution across Australia and estimated that water and wind erosion represent 2% of total C stock in the 0 - 10 cm soil layer. Soil erosion is not explicitly included in Australian national SOC stock accounting, which renders estimates of CO_2 flux from soils highly uncertain. The inclusion of an erosion component may reduce that uncertainty [34] , [36] , [41] and improve the accuracy for the reporting of GHG emissions. Measuring wind-eroded SOC in the dust cycle is therefore essential to quantify the release of CO_2 from SOC dust to the atmosphere and the contribution of SOC deposition to downwind C depositional sites.

Measuring SOC stocks change across a land unit, such as eroding, depositional or landscape units with different land use practices require sufficient soil samples within each landscape unit to ensure that individual temporal and spatial SOC changes are representative of the entire land unit [5] , [36] . If insufficient soil samples are taken, it is unlikely that the measured values of SOC will be representative of the landscape unit spatially and may lead to greater variability.

If variance in measured SOC stocks at the first sampling period is larger than that in the second sampling period, it will be difficult to assess change in SOC stocks over the land unit [4] , [5] . Analytical methods changes or measurements by different laboratories can affect variability. Setting up and implementing a sampling design can be a challenge when little information is available on the variability of SOC within the land unit. Consequently, it is difficult to establish what type of sampling design to use and how many samples to take in each sampling round [9] , [10] . Therefore, a consistent and standardized protocol should be followed each time samples are collected to minimize variability due to sampling protocol within each sampling time or period.

Chappell et al. [5] produced a technical document based on literature and defined the statistical justifications for recommendation of SOC monitoring in a landscape unit. The three distinct phases included: 1) setup sampling design and the assumptions/decisions necessary for implementation to determine magnitude of a detectable change in SOC over time and at an acceptable cost, 2) modify the sampling design after measurements of SOC are made in the first round of sampling (i.e., more or less SOC variance than expected), and 3) determine the actual detectable change in SOC by comparing SOC measurement from the first round of sampling with the second round of sampling (e.g., 5 years later) using a mean difference or other analytical test.

De Gruijter et al. [4] and Chappell et al. [5] restricted their consideration of sampling designs and their use over time to the most basic and easily adopted approaches. There is an expected variation in SOC if the approaches are formed from a series of decisions that were used to estimate the minimum detectable change (MDC). Assuming that spatial and temporal variations occur in land unit, the likely measurement error associated with quantifying SOC can affect the confidence in the results. Chappell et al. [5] estimated the variance of the detectable mean change in

SOC. The procedure should help address spatial variability as a basis for detecting temporal changes. One approach to reduce variability over time is to use a system of geo-referencing sampling points which remain constant on the landscape regardless of changes in land use or management [3] .

The final assumption by of this methodology relates to variability of SOC with depth [5] . The SOC stocks typically decline with increasing soil depth. However, this is not always the case. Variation of SOC across a land unit is most likely to be the largest in the surface soil layer which has the largest SOC stock depending on soil bulk density. Benefits of a sampling design that reduces the number of samples for the greatest depth can be outweighed by the need to visit more locations to collect samples near the soil surface. When sampling soil, it is typical to collect soil samples from all layers at the same time. Since the surface layer is usually the most variable in SOC stocks, a soil sampling protocol design for this layer should be adequate for collecting soil at depths where SOC is less variable.

5. CONCLUSIONS

A landscape soil sampling protocol was developed to establish the native prairie baseline SOC stock (230 Mg·C/ha) to a 1 m depth for the entire native prairie landscape. The cropland SOC stock was determined for each landscape position and the entire landscape to determine cropland retention of SOC stock (114 Mg·C/ha). The cropland landscape adjacent to Dinesen Prairie had less SOC stock than the baseline prairie landscape except for the TS which retained 109% of the SOC stock in the upper 1 m as a result of erosion from the upper landscape positions and the deposition of SOC rich sediment on the TS. For the entire landscape the cropland retained only 49% of the baseline prairie SOC stock.

If the intention of monitoring SOC stocks is to detect change and attribute the change to a particular land management practice/use then samples must be obtained over the entire landscape unit (area upon which the management practice is applied). In this case mean change and its variability determine level of detectability of such change. Consequently, this will dictate the intensity of soil sampling across the landscape unit to determine SOC stocks at different locations. When implementing, attention must be paid to the processes which maintain and form SOC stocks to ensure soil productivity and reduce the degradation processes, such as erosion, which drive SOC depletion. These processes are influenced by both natural and anthropogenic drivers. Soil erosion should be viewed as SOC redistribution not only across landscape, but also in the larger context of impacts on soil function and ecosystem sustainability and services and disservices. Appropriate soil sampling designs and sampling procedures are needed to determine SOC change and verify SOC net storage or sequestration. Quantifying the SOC loss due to erosion will help avoid over estimation of the management practices on net SOC stock over time.

ACKNOWLEDGEMENTS

Published with the approval of the Director of the Office of Research at the University of Illinois, Urbana, IL. Funded as part of Regional Research Project 367 and in cooperation with North Central Regional Project NC-1178 (Soil Carbon Sequestration).

CITE THIS PAPER

Olson, K.R., Al- Kaisi, M., Lal, R. and Cihacek, L. (2016) Soil Organic Carbon Dynamics in Eroding and Depositional Landscapes. Open Journal of Soil Science, 6, 121-134.

REFERENCES

1. Biggs, R., Schluter, M., Biggs, D., Bohensky, E.L., BurnSilver, S., Cundill, G., Dakos, V., Daw, T.M., Evans, L.S., Kotschy, K., Leitch, A.M., Meek, C., Quinlan, A., Raudsepp-Hearne, C., Robards, M.D., Schoon, M.L., Schultz, L. and West, P.C. (2012) Toward Principles of Enhancing the Resilience of Ecosystem Services. Annual Review of Environment and Resources, 37, 421-48.
2. Salemme, R. (2015) Land-Use Change Effects on Soil Organic Carbon, Total Soil Nitrogen, and Soil Erosion in a Temperate Forest and Grassland. Thesis, Graduate College of University of Illinois at Urbana-Champaign, Urbana, IL.
3. Cihacek, L.J., Botnen, B.W. and Steadman, E.N. (2010) A Sampling Protocol for Monitoring, Measurement, and Verification of Terrestrial Carbon Sequestration in Soils. Plains CO2 Reduction (PCOR) Partnership Value-Added Report, Energy and Environment Research Center, University of North Dakota, Grand Forks, ND, 9 p.
4. De Gruijter, J.J., Brus, D.J., Bierkens, M.F.P. and Knotters, M. (2006) Sampling for Natural Resource Monitoring. Springer, Berlin, Heidelberg, New York.
5. Chappell, A., Baldock, J. and Rossel, R.V. (2013) Sampling Soil Organic Carbon to Detect Change over Time. CSIRO Sustainable Agricultural Flagship, Department of the Environment, Australian Government, 39 p. [Citation Time(s):12]
6. Olson, K.R., Al-Kaisi., M.M., Lal, R. and Lowery, B. (2014) Experimental Consideration, Treatments, and Methods in Determining Soil Organic Carbon Sequestration Rates. Soil Science Society of America Journal, 78, 348-360.
7. Olson, K.R., Gennadiyev, A.N., Zhidkin, A.P. and Markelov, M.V. (2012) Impact of Land Usechange, Slope and Erosion on Soil Organic Carbon Retention and Storage USA. Soil Science, 177, 269-278.
8. Olson, K.R. (2013) Soil Organic Carbon Sequestration in U.S. Cropland: Protocol Development. Geoderma, 195-196, 201-206.

9. Allen, D.E., Pingle, M.J., Page, K.L. and Dalal, R.C. (2010) A Review of Sampling Designs for the Measurement of Soil Organic Carbon in Australian Grazing Lands. The Rangeland Journal, 32, 227-246.
10. Bishop, T.F.A. and Pringle, M.J. (2012) Paddock-Scale Soil Carbon Inventory and Monitoring: A Meta-Analysis for Sample Size Requirements. Proceedings of the Soil Science Conference, Hobart, Tasmania.
11. Olson, K.R., Al-Kaisi, M.A., Lal, R. and Cihacek, L.J. (2016) Landscape, Soil Erosion, Land use and Plot Design Considerations When Measuring Soil Organic Carbon Stocks. Journal of Soil and Water Conservation, 71, 61A-67A.
12. Cihacek, L.J. and Ulmer, M.G. (1995) Estimated Soil Organic Carbon Losses from Long-Term Crop-Fallow in Northern Great Plains of USA. In: Lal, R., Kimble, J., Levine, E. and Steward, B.A., Eds., Soil Management and Greenhouse Effect, Lewis Publishers, Boca Raton, Florida, 85-92.
13. Cihacek, L.J., Sweeney, M.D. and Deibert, E.J. (1992) Characterization of Wind Erosion Sediments in the Red River Valley of North Dakota. Journal of Environmental Quality, 22, 305-310.
14. Wuest, S.B. (2009) Correction of Bulk Density and Sampling Methods Biases Using Soil Mass per Unit Area. Soil Science Society of America Journal, 73, 312-316.
15. Lee, J., Hopmans, J.W., Rolston, D.E., Baer, S.G. and Six, J. (2009) Determining Soil Carbon Stock Changes: Simple Bulk Density Corrections Fail. Agriculture, Ecosystems, and Environment, 134, 251-256.
16. Olson, K.R. and Al-Kaisi, M.M. (2015) Impact of Soil Sampling Depth on Soil Organic Carbon Sequestration, Storage, Retention and Loss. Catena, 125, 33-37.
17. Lal, R., Kimble, J.M., Follett, R.F. and Cole, C.V. (1998) The Potential of US Cropland to Sequesters Carbon and Mitigate the Greenhouse Effect. Sleeping Bear Press Inc., Ann Arbor Press, Chelsea, 128 p.
18. Olson, K.R. (2010) Impacts of Tillage, Slope and Erosion on Soil Organic Carbon Retention. Soil Science, 175, 562-567.
19. Lal, R. (2015) Sustainable Intensification for Adaptation and Mitigation of Climate Change and Advancement of Food Security in Africa. In: Lal, R., Singh, B.R., Mwaseba, D.L., Kraybill, D., Hansen, D.O. and Eik, L.O., Eds., Sustainable Intensification to Advance Food Security and Enhance Climate Resilience in Africa, Chapter 1, Springer, Berlin, 3-17.
20. Soil Survey Staff (2004) Soil Survey Laboratory Methods Manual. Soil Survey Investigations Report No. 42 Version 4.0, United States Department of Agriculture, Natural Resources Conservation Service, Nebraska.
21. SAS Institute (2015) SAS/STAT Guide for Personal Computers. Version 9.4., SAS Institute, Cary.
22. Kreznor, W.R., Olson, K.R. and Johnson D.L. (1992) Field Evaluation of Methods to Estimate Soil Erosion. Soil Science, 153, 69-81.
23. Kreznor, W.R., Olson, K.R., Johnson, D.L. and Jones, R.L. (1990) Quantification of Post-Settlement Deposition in a Northwestern Illinois Sediment Basin. Soil Science Society of America Journal, 54, 1393-1401.
24. Olson, K.R., Gennadiyev, A.N., Zhidkin, A.P. and Markelov, M.V. (2011) Impact of Land Use Change and Soil Erosion in Upper Mississippi River Valley on Soil Organic Carbon Retention and Greenhouse Gas Emissions. Soil Science, 176, 449-458.
25. Lal, R., Griffin, M., Apt, J., Lave, L. and Morgan, G. (2004) Response to Comments on "Managing Soil Carbon". Science, 305, 1567.
26. Lal, R. (2003) Soil Erosion and the Global Carbon Budget. Environmental International, 29, 437-450.
27. Thrupp, L.A. (2000) Linking Agricultural Biodiversity and Food Security: The Valuable Role of Agrobiodiversity for Sustainable Agriculture. International Affairs, 76, 265-281.
28. Mooney, H.A., Cooper, A. and Reid, W. (2005) Confronting the Human Dilemma. Science, 434, 561-562.

29. MEA (2005) Ecosystems and Human Well-Being: Biodiversity Synthesis. World Resources Institute, Washington DC.
30. Hadgu, K.M., Rossing, W.A., Kooistra, L. and van Bruggen, A.H. (2009) Sapatial Variation in Biodiversity, Soil Degradation and Productivity in Agricultural Landscapes in the Highlands of Tigray, Northern ETHIOPIA. Food Security, 1, 83-97.
31. Loughran, R.J., Elliott, G.L., McFarlane, D.J. and Campbell, B.L. (2004) A Survey of Soil Erosion in Australia Using Caesium-137. Australian Geographical Studies, 42, 221-233.
32. Olson, K.R., Gennadiyev, A.N., Kovach, R.G. and Lang, J.M. (2013) The Use of Fly Ash to Determine the Extent of Sediment Transport on Nearly Level Western Illinois Landscapes. Soil Science, 178, 24-28.
33. Kuhn, N.J., Hoffman, T., Schwenghart, W. and Dotterwerch, M. (2009) Agricultural Soil Erosion and Global Carbon Cycle: Controversy Over. Earth Surface Processes and Landforms, 34, 1033-1038.
34. Chappell, A., Webb, N.P., Viscara Rossel, R.A. and Bui, E. (2014) Australian Net (1950s-1990) Soil Organic Carbon Erosion: Implications for CO2 Emission and Land-Atmosphere Modeling. Biogeosciences, 11, 5235-5244.
35. Chappell, A.J. Thomas, S.M., Read, A. and Leslie, C. (2012) The Dynamic of Soil Redistribution and Implications for Soil Organic Carbon Accounting in Agricultural South-Eastern Australia. Global Change Biology, 18, 2081-2088.
36. Chappell, A., Baldock, J. and Sanderman, J. (2016) The Global Significance of Omitting Soil Erosion from Soil Organic Carbon Cycling Schemes. Nature Climate Change, 6, 187-191.
37. Nizeyimana, E. and Olson, K.R. (1988) Chemical, Mineralogical and Physical Differences between Moderately and Severely Eroded Illinois Soils. Soil Science Society of America Journal, 52, 1740-1748.
38. Fenton, T.E., Kazemi, M. and Lauterbach-Barrett, M.A. (2005) Erosional Impact on Organic Matter Content and Productivity of Selected Iowa Soils. Soil & Tillage Research, 81, 163-171.
39. Er Raji, M. and Cihacek, L.J. (1994) Erosion Effects on Selected Agronomic Parameters of Hard Red Spring Wheat across a Landscape. North Dakota Farm Research, 50, 26-29.
40. Li, Y., Yu, H., Chappell, A., Zhou, N. and Funk, R. (2014) How Much Soil Organic Carbon Sequestration Is Due to Conservation Agriculture Reducing Soil Erosion. Soil Research, 52, 717-726.
41. Sanderman, J. and Chappell, A. (2012) Uncertainty in Soil Carbon Accounting Due to Unrecognized Soil Erosion. Global Change Biology, 19, 264-272.

CHAPTER 14

Organic Amendments with Chemical Fertilizers Improve Soil Fertility and Microbial Biomass in Rice-Rice-Rice Triple Crops Cropping Systems

Md. Farid Ahammed Anik[1], Md. Mizanur Rahman[2], G. K. M. Mustafizur Rahman[2], Md. Khairul Alam[1,3*], Mahammad Shariful Islam[1], Mst Fatima Khatun[1]

[1]*Soil Science Division, Bangladesh Agricultural Research Institute, Gazipur, Bangladesh*

[2]*Department of Soil Science, Bangabandhu Sheikh Mujibur Rahman Agricultural University, Gazipur, Bangladesh*

[3]*School of Veterinary and Life Sciences, Murdoch University, Perth, Western Australia*

ABSTRACT

Medium-term changes in the labile nutrient pool of microbial biomass carbon (MBC) and nitrogen (MBN) resulting from organic manure application in rice (Oryza sativa L.)-based triple cropping systems have been poorly studied. Therefore, the effects of organic materials on the soil physico-chemical properties and microbial biomass in rice fields were investigated at Bangabandhu Sheikh Mujibur Rahman Agricultural University, Bangladesh, from April 2010 to December 2012. Five treatments (control, cow dung, poultry manure, rice straw, and chemical fertilizer) were laid out in a randomized complete block design with four replications. The organic residues (2 t C ha^{-1}) were applied 7 days before transplanting and were combined with inorganic fertilizers, following integrated plant nutrition systems. This paper presents the results from the last of the five consecutive rice growing seasons. All of the organic residues increased the pH, and organic C, N, P, and K contents of the soil. However, poultry manure was more efficient in increasing soil fertility than cow dung and rice straw, resulting in a significant increase in P from 22 mg · kg^{-1} to 63 mg · kg^{-1} at crop harvest. All of the organic residues also increased the soil water holding capacity and decreased bulk density. Furthermore, poultry manure resulted in significantly higher microbial biomass C (432 mg · kg^{-1}; $P < 0.05$) and N (31.60 mg · kg^{-1}; $P < 0.05$) levels in the soil at crop harvest, followed by cow dung and rice straw. These findings indicate that the regular application of organic residues and manures will help to enhance soil fertility and production sustainability.

Keywords: Integrated Plant Nutrition Systems (IPNS), Poultry Manure, Rice Straw, Water Holding Capacity

1. BACKGROUND

The soil microbial biomass (SMB) indicates the ecological and biological quality and fertility of the soil, and is affected by organic/inorganic fertilizer application [1] . The SMB is involved in soil biogeochemical processes in soil-water-plant systems [1] [2] , not only driving the nutrient cycle in agro-ecosystems, but also playing a vital role in soil nutrient transformations and acting as a labile nutrient pool that is available to plants [3] [4] . Several studies have emphasized the effects of organic manures such as soil organic carbon (SOC) on soil physicochemical properties. However, SOC may take a very long time to increase [5] [6] and observation of SOC changes alone may not explain the effect of organic matter on nutrient availability via microbiological activities. Alternatively, soil microbial biomass, which measures changes in both SOC and total soil nitrogen (TSN), can be used as a management indicator for soil fertility and crop productivity [7] . However, the effects of the various influencing factors, such as climatic and edaphic factors, fertilizer use, and crop establishment practices, on SMB are difficult to quantify due to their interactions; and estimation of the spatial and temporal distribution of SMB on the basis of limited field observations are even more difficult [4] [8] .

Conventional agriculture on the Indo-Gangetic Plains is in a fragile state due to the overexploitation of land and water resources, and the indiscriminate use of agrochemicals [9] which reduce the diversity and quantity of microbes in the soil [10] . The use of different organic and inorganic fertilizers improves crop yields and also has direct or indirect effects on the physicochemical and biological properties of the soil. These changes are believed to have a significant effect on the quality and productive capacity of the soil on a long-term basis [11] . However, existing information on this issue is scarce and the long-term effects of organic inputs in combination with synthetic fertilizers on SMB have not been duly addressed.

Rice (Oryza sativa L.)-based cropping systems in the Indo-Gangetic Plains are very resource consumptive and input intensive [12] [13] . Though farmers use chemical fertilizers, the rates and methods of application are not justified according to either the integrated nutrient management system (INMS) or the integrated plant nutrition system (IPNS), and so most of the applied fertilizers are lost through various channels. Determination of microbial biomass carbon (MBC) and nitrogen (MBN) levels is crucial for the determination of C and N dynamics in rice systems, which regulate the SOC and N availability and loss, particularly in high-input-dependent cropping systems [14] . However, little research has investigated the effects of organic amendments on MBC and MBN in rice cropping systems, particularly in the eastern part of the Gangetic Plains, limiting our ability to generalize about soil C and N dynamics in subtropical climatic areas. Therefore, there is an urgent need to study how soil fertility and microbial biomass change with organic amendments under intensive rice-based cropping systems.

The micro-organismal growth and activities are affected by different factors depending on soil type and management practices [15] . Soil microbes are active drivers of soil nutrient cycling, being associated with the decomposition of organic matter, and the transformation and cycling of nutrients, which help to maintain crop productivity and the physical and chemical quality of the soil [16] . Thus, an improved understanding and quantification of the effect of organic residue applications on MBC and MBN, and soil properties will enhance our ability to manipulate nutrient cycling, to maintain and improve soil quality and production sustainability.

2. OBJECTIVES

Therefore, the study investigated the changes of MBC and MBN over the rice growing season; and effect of application of various organic residues in addition to chemical fertilizer on other soil characteristics after growing five consecutive rice crops.

3. MATERIALS AND METHODS

This study was conducted at the experimental farm of Bangabandhu Sheikh Mujibur Rahman Agricultural University (BSMRAU), Gazipur, Bangladesh, as part of an ongoing research project entitled "Carbon sequestration in soils of Bangladesh," which ran from April 2010 to July 2012. This period covered five consecutive seasons of transplanted aman (T. aman) and boro rice, starting with T. aman in 2010. Aman rice is grown as a rainfed crop during the monsoon season, when maximum rainfall occurs (300 - 650 mm; supplemental irrigation is occasionally required when there is insufficient rainfall), while boro rice is grown after T. aman had been harvested under fully irrigated conditions during the dry season (0 - 55 mm rainfall). This report covers the last of the five consecutive growing seasons which was T. aman season rice.

3.1. Experimental Site and Climate

The study site belongs to the Salna series and has been classified as Shallow Red-Brown Terrace soil (Inceptisol) under the US Department of Agriculture classification system [17] . The area is part of the Madhupur Tract agro-ecologi- cal zone (AEZ No. 28), in which the soils are acidic in nature and characterized by clay loam within 50 cm of the surface. The site is located at 24.09°N latitude and 90.25°E longitude, with an elevation of 8.2 meters above sea level. The experimental area experiences a subtropical climate, and is characterized by comparatively high levels of rainfall, high humidities, high temperatures, and relatively long days from April to September, and minimum levels of rainfall, low humidities, low temperatures, and short days from October to March (Figure 1).

3.2. Experimental Treatments

The experimental crop was rice and variety BRRI dhan39 was procured from Bangladesh Rice Research Institute (BRRI), Gazipur, Bangladesh and used as the experimental crop. This variety is non-photosensitive, short-duration. The grains are long slender and kernels are white. This variety takes about 130 to 145 days to mature. It attains a plant height of 95 - 100 cm and at maturity the flag leaf remains green and erect. The experiment comprised five treatments (T_1 = Control [no fertilizer]; T_2 = cow dung; T_3 = poultry manure; T_4 = rice straw; and T_5 = soil test based chemical fertilizer [STB]), which were laid out in a randomized complete block design with four replications per treatment. IPNS-based chemical fertilizers were applied in combination with treatments T_2, T_3, and T_4.The organic materials were applied 7 days before transplanting at 2 t C ha^{-1} during the Aman seasons.

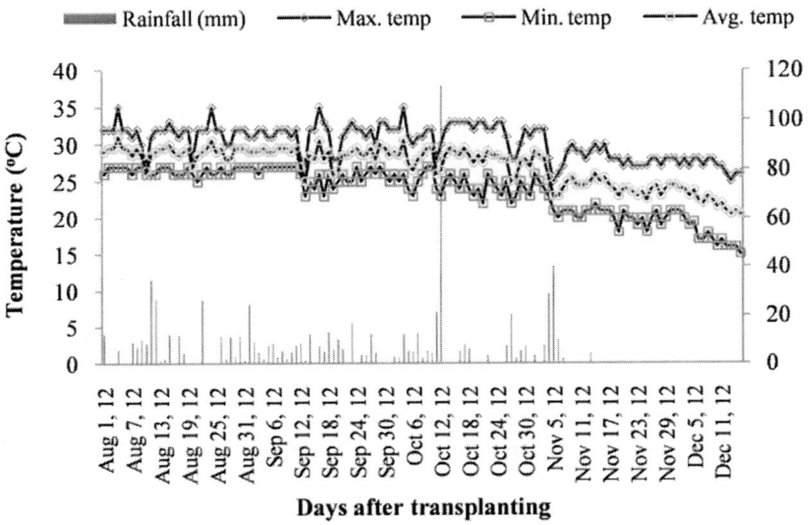

Figure 1. The maximum, minimum and average temperature (°C) with line graphs (unit on the primary axis); the bars show the precipitation (mm) in the area over the growing season (unit on the secondary axis).

On August 5, 2012, 30-day-old rice seedlings were transplanted into the experimental plots at 20 cm × 20 cm spacing. The recommended doses of chemical fertilizers were calculated based on soil test results. Urea was applied in three equal parts, whereby one-third of the urea and the entire amount of triple superphosphate (TSP), murate of potash (MOP), and gypsum were applied during final land preparation; one-third of the urea was applied at the maximum vegetative growth stage (30 days after transplanting [DAT]); and one-third of the urea was applied before the panicle initiation stage of boro season rice (45 DAT).

Irrigation was used to keep the paddy field well flooded until the rice plants reached maturity. This was applied at 1, 5, 10, 20, 30, 40, 48, 54, 63, and 78 DAT, with a total of 45 cm of irrigation water being applied during rice growth. Weeding was performed at 25 DAT. Precautionary measures were taken at every stage of crop production using natural methods of pest and disease management, such as light traps, the placement of sticks for birds to stand on, and the removal of disease-infected plants at an early stage of infection, removing any need for chemical and herbicide use.

3.3. Soil Sample Collection and Processing

Soil samples were collected from depths of 0 - 5, 5 - 10 and 10 - 15 cm separately at 0, 30, 60, 90, and 120 DAT, and were immediately placed in separate labeled ice bags to minimize microbial activity. The samples were then taken to the BSMRAU Soil Science Laboratory for analysis as quickly as possible to preserve their field properties. In the laboratory, they were spread out for partial air drying following removal of the roots, insects, worms, stones, and some small pieces of organic matter. This procedure was necessary to reduce the excess soil moisture and to allow the samples to be passed through a 2-mm mesh sieve. After sieving, part of each sample was kept in a polythene bag and incubated aerobically for 3 weeks at 25°C to remove the effect of handling on soil respiration. The remainder of each sample was fully air dried, ground, passed through a 2-mm sieve, and then stored in a clean polythene bag for physical and chemical analysis.

The soil samples were analyzed for bulk density, pH, SOC, total N, available N (mineral N), available P, exchangeable K, MBC, and MBN. The maximum water holding capacity was also determined. Initial soil samples were also collected at 0 - 15 cm depth and analyzed for pH, SOC, total N, available P, and exchangeable K.

3.4. Methods of Soil Analysis

The soil bulk density was determined using the core sampling method, as outlined by Karim et al. [18] . Soil pH was measured with a glass electrode pH meter (Horiba pH meter M-8L) using a soil water suspension ratio of 1:2.5, as described by Jackson [19] .

Maximum water holding capacity (MWHC) and adjustment of soil moisture at 60% MWHC were performed as suggested by Haider et al. [20] . Maximum water holding capacity (MWHC) of the soil was estimated by placing a quantity of soil (25 g) in a glass funnel lined with filter paper. Distilled water was then poured into the funnel and allowed to drain out for at least 1 hour. Following this, a small amount of soil was removed and placed in a pre-weighed aluminum can, and the combined weight was noted. The soil was then dried in an oven at 105°C for at least 24 hours until it attained a constant weight. The ratio of the weight of soil water to the weight of oven-dried soil was considered the MWHC of the soil. To adjust the soil samples to 60% MWHC, the difference between the initial moisture content of the soil and 60% MWHC was calculated, and the measured amount of water was then added to the samples.

Soil organic carbon (SOC) was estimated using Walkly and Black's wet oxidation method, as described by Jackson [19] . Total nitrogen was determined using the micro-Kjeldahl method [21] , available P was measured using Olsen's method [22] , and exchangeable K was estimated using the ammonium acetate extraction method [23] .

3.5. Measurements of Microbial Biomass C and N

The chloroform fumigation extraction method was used to estimate the amount of microbial biomass C in the soil samples, according to the method described by Vance et al. [24] . Biomass N was determined according to the method described by Brookes et al. [25] [26] .

3.6. Statistical Analysis

SPSS statistical software was used to analyze the data using one-way analysis of variance (ANOVA) and univariate analysis. Means were separated using the least significant difference (LSD) test at the 95% confidence level.

4. RESULTS

4.1. Soil Chemical Properties

The initial soil samples indicated that the soil in the study area was slightly acidic, and contained medium SOC, very low N, optimum P, and low K contents (Table 1), reflecting a low to medium overall fertility. After five consecutive rice growing seasons, the soil chemical properties significantly varied among treatments ($P < 0.05$), with soil pH, SOC, and available P significantly increasing following application of the different organic residues (Table 1).

Soil pH was significantly higher in the poultry manure treated plots (pH 7.17) than in the control plots (pH 5.59) and the initial soil samples (pH 5.85) ($P < 0.05$). This reflected a 23% increase compared with the initial soil pH.

Table 1. Soil properties of the experimental field under different organic residues and chemical fertilizer management at harvest.

Treatment	Soil properties				
	pH	Organic carbon (%)	Total nitrogen (%)	Available phosphorus (μg·g⁻¹)	Exchangeable potassium (meq 100 g⁻¹ soil)
Initial soil properties	5.85	1.13	0.09	22.0	0.13
Rice straw$	5.88	1.81	0.16	27.2	0.26
Poultry manure$	7.17	1.83	0.19	63.3	0.23
Cow dung$	5.89	1.45	0.19	36.0	0.22
Soil test based fertilizer	5.82	1.25	0.15	28.3	0.17
Control	5.59	1.17	0.13	18.2	0.15
SE (±)	0.52	0.10	0.82	1.22	0.83
$LSD_{0.05}$	0.16	0.32	NS	3.76	NS
CV (%)	1.70	1.70	13.6	7.9	8.5

Legends: SE (±) means standard error of means; $LSD_{0.05}$ means least significant differences at 5% level of significance and CV (%) are coefficient of variations. $-nutrient content in the organic amendment was deducted from recommended fertilizer dose.

Plots that were treated with a combination of organic residues and chemical fertilizers also had significantly higher SOC contents than the control plots ($P < 0.05$). Soil organic C concentrations in the rice straw (1.81%) and poultry manure (1.83%) treated plots were statistically similar ($P > 0.05$), but significantly higher than the cow dung treated plots (1.45%; $P < 0.05$). By contrast, the plots that were only treated with chemical fertilizer had a significantly lower SOC (1.25%) than all of the organic residue treated plots.

Available P was greatly enhanced in the poultry manure treated plots ($P < 0.05$; Table 1), increasing from an initial level of 22 mg P kg^{-1} soil to 63 mg · kg^{-1}. The cow dung treated plots also had 14 mg · kg^{-1} more available P than the initial soil, whereas the control plots experienced a 3.78 mg · kg^{-1} depletion in available P.

4.2. Soil Biomass Carbon

The MBC content of the soil varied between treatments at 0, 30, 60, 90, and 120 DAT (Table 2). The significant difference on transplanting day (day 0; $P < 0.05$) was due to the experiment starting back in 2010. This was observed not only on transplanting day, but also found in 30, 60, 90 and 120 days after transplanting (Table 2). Irrespective of residues or for all treatments, the soil MBC contents increased with advancement of the rice growing period, i.e., maximum MBC was found at crop harvest and followed the order 0 < 30 < 60 < 90 < 120 DAT. Soil MBC contents were significantly higher in plots that were treated with organic residues than in the control plots ($P < 0.05$), with poultry manure having the greatest effect and rice straw having the least effect. Thus, the maximum MBC was found in the poultry manure treated plots at harvest (432 mg · kg^{-1}), while the lowest level was found in the control plots on transplanting day (17 mg · kg^{-1}).

Table 2. Microbial biomass carbon (mg · kg?1) in soil under different organic residues and chemical fertilizer management practices at different days after transplanting (DAT).

Treatment	Biomass carbon ($mg{\cdot}kg^{-1}$) at different days after transplanting				
	0	30	60	90	120
Rice straw[$]	32.5	47.5	67.5	182	305
Poultry manure[$]	101	115	121	301	432
Cow dung[$]	47.3	71.3	91.6	178	380
Soil test based fertilizer	73.4	141	111	167	329
Control	17.0	35.5	49.3	118	200
SE (±)	0.78	12.4	4.58	9.42	14.7
$LSD_{(0.05)}$	2.41	38.2	14.1	29.0	45.5
CV (%)	2.90	30.3	10.4	10.0	9.00

Legends: SE (±) means standard error of means; $LSD_{0.05}$ means least significant differences at 5% level of significance and CV (%) are coefficient of variations. [$]-nutrient content in the organic amendment was deducted from recommended fertilizer dose.

4.3. Soil Biomass Nitrogen

The treatments also affected the MBN content of the soil at different days after transplanting the rice (Table 3) and significantly differed between each other on transplanting day ($P < 0.05$).

As observed for MBC, the maximum MBN was recorded at crop harvest and followed the order 0 < 30 < 60 < 90 < 120 DAT. Soil MBN contents were significantly higher in plots that were treated with organic residues than the control plots ($P < 0.05$), and poultry manure again had the greatest effect, followed by cow dung and then rice straw. Thus, at harvest, the maximum MBN was found in the poultry manure treated plots (32 $mg \cdot kg^{-1}$), while the lowest was found in the control plots (9.88 $mg \cdot kg^{-1}$).

4.4. Soil Water Holding Capacity

Organic amended soils had a significantly higher water holding capacity than the STB fertilizer and control treatments (Figure 2) after five consecutive rice cultivating seasons and repeated additions of the organic manures. The lowest water holding capacity was found in soil in the control plots (53%). There was no significant difference in the water holding capacity of soil between the three organic residue treatments, but cow dung treated plots had the highest level while poultry manure treated plots had the lowest.

4.5. Soil Bulk Density

The soil bulk density varied significantly following the application of different organic residues in the surface soil layer (Table 4). However, the highest bulk density was found in the control plots, while the lowest was found in the cow dung treated plots at all soil depths. Thus, the bulk density varied from 1.29 to 1.43 $g \cdot cm^{-3}$ in the top 0 - 5 cm of soil, from 1.17 to 1.47 $g \cdot cm^{-3}$ at a depth of 5 - 10 cm, and 1.20 to 1.44 $g \cdot cm^{-3}$ at a depth of 10 - 15 cm for the control and cow dung treated plots, respectively. Microbial biomass C and MBN are inversely related to the soil bulk density.

5. DISCUSSION

At all sampling dates, MBC and MBN were significantly higher in the poultry manure treated plots than in the other treatment plots. This may have been due to the higher amounts of N that were present in the poultry manure resulting in a lower or optimum C:N ratio, favoring faster microbial colonization. These findings agree with those of several other studies [27] [28] . In addition, a higher amount of MBN was found in the STB fertilizer treatment than in the rice straw and cow dung treatments. This was due to the presence of higher amounts of available nutrients, particularly N, in the fertilizer treatment, which favored rapid microbial activity, i.e., mineralization in the soils. These findings agree with those of Akmal et al. [29] who attributed higher microbial biomass to higher available C substrates and nutrient contents in soils. Thus, the availability of organic residues in the soils coupled with high moisture contents favored microbial activities, which ultimately contributed to the higher MBC and MBN under these treatments [30] [31] .

Table 3. Microbial biomass nitrogen (mg · kg?1) in soil under organic residues and chemical fertilizer management practices at different days after transplanting (DAT).

Treatment	Biomass nitrogen ($mg \cdot kg^{-1}$) at different days after transplanting				
	0DAT	30DAT	60DAT	90DAT	120DAT
Rice straw$	4.65	6.81	8.41	15.5	18.2
Poultry manure$	5.01	13.4	13.8	18.7	31.6
Cow dung$	4.70	6.99	11.6	14.4	27.2
Soil test based fertilizer	5.79	11.8	10.9	18.2	20.1
Control	3.25	6.43	6.01	11.5	9.88
SE (±)	0.17	0.65	0.53	1.48	1.30
$LSD_{(0.05)}$	0.53	2.02	1.63	4.56	4.01
CV (%)	7.40	14.5	10.6	18.9	12.2

Legends: SE (±) means standard error of means; $LSD_{0.05}$ means least significant differences at 5% level of significance and CV (%) are coefficient of variations. $-nutrient content in the organic amendment was deducted from recommended fertilizer rate.

Table 4. Bulk density of soil under different organic residues and chemical fertilizer management practices at different depths.

Treatment	Bulk density ($g \cdot cm^{-1}$) at different soil depth		
	0 - 5 cm	5 - 10 cm	10 - 15 cm
Rice straw$	1.36	1.31	1.45
Poultry manure$	1.34	1.41	1.38
Cow dung$	1.29	1.17	1.20
Soil test based fertilizer	1.38	1.47	1.44
Control	1.48	1.54	1.46
SE (±)	0.14	0.24	0.27
$LSD_{0.05}$	0.08	ns	ns
CV (%)	2.20	3.60	4.00

Legends: SE (±) means standard error of means; $LSD_{0.05}$ means least significant differences at 5% level of significance and CV (%) are coefficient of variations. $nutrient content in the organic amendment was deducted from recommended fertilizer rate.

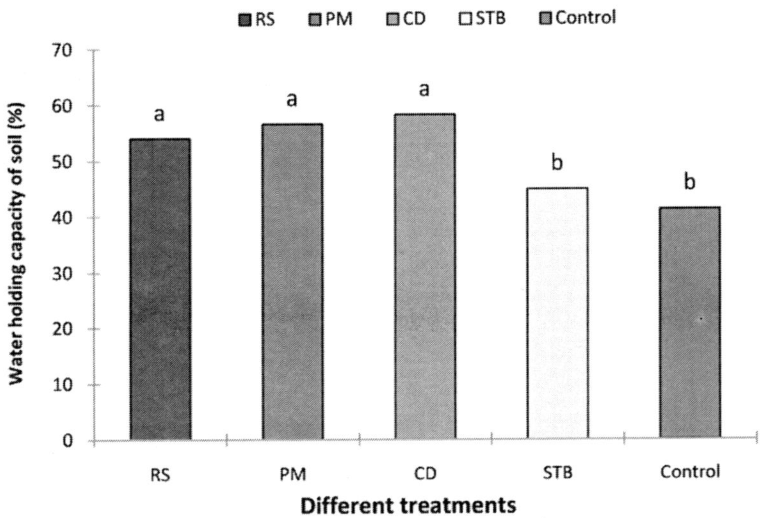

Figure 2. Water holding capacity of soil under different organic residues and chemical fertilizer management practices. RS-Rice straw + IPNS based chemical fertilizer, PM-poultry manure + IPNS based chemical fertilizer, CD-cow dung+ IPNS based chemical fertilizer, STB-soil test based chemical fertilizer and control. Vertical bars represent standard error of means.

The use of organic amendments in combination with chemical fertilizers improved the pH, and SOC, N, P, and K contents of the soils after seven consecutive rice growing seasons. The use of poultry manure led to the greatest improvement in these initially low to medium fertility soils. Organic C was applied at a rate of 2 t C ha^{-1} in all plots, irrespective of its source. However, poultry manure contains a higher amount of the stable form of C, which explains the greater level of C enrichment in these plots. The higher pH in the poultry manure treated plots can be attributed to the presence of Ca in poultry manure due to the calcium-rich diet of poultry. The accumulation of Ca in the soil will have neutralized the soil acidity and ultimately increased the soil pH to around neutral levels. Finally, it is known that poultry manure usually contains a higher concentration of P, with a previous study showing that at 30% - 35% moisture level, poultry manure, cow dung, and rice straw contained 0.70%, 0.30%, and 0.10% P, respectively [32] . Since the rate of C application in the present study was determined on an oven dry basis, the P accumulation will have been higher than these reported values. Therefore, the application of different organic residues improved soil fertility, with poultry manure being the most efficient organic amendment.

The soil water holding capacity also increased following the application of different organic residues (Figure 2), with rice straw, cow dung, and poultry manure having similar effects. The greatest improvement was recorded with poultry manure, which resulted in 38% and 30% higher levels than the control and STB fertilizer treated soils, respectively. Once again, this can be explained by the availability of organic residues in these soils coupled with their high moisture contents favoring microbial activities. These findings agree with those of Pavlova et al. [31] and Efremov et al. [30] , who reported that the total pool of microbial biomass was dependent on the soil moisture content, which predetermined the water permeability of soils.

At the soil surface (top 0 - 5 cm), organic amendment combined with chemical fertilizers significantly decreased the soil bulk density, and similar trends were observed at 5 - 10 cm and 10 - 15 cm. This resulted from the organic residues favoring microbial biomass growth [33] and matches the findings of Glover et al. [34] , who confirmed that the addition of organic residues with chemical N resulted in lower bulk densities and higher microbial biomass.

6. CONCLUSION

Poultry manure amendment in combination with chemical fertilizers under rice growing conditions resulted in the most biologically active soils with higher levels of MBC and MBN. Thus, poultry manure is a more efficient organic amend- ment than cow dung and rice straw for increasing soil fertility. The application of organic residues also increased the pH, and SOC, N, P, and K contents of the soil, further improving soil fertility. Compared with the initial soil, the poultry manure treatment increased the soil pH from an acidic (5.85) to neutral (7.17) state, SOC by 62%, available P by 187%, total N by 112%, and exchangeable K by 77%, though the latter two were similar to other organic amendments. The soil water holding capacity and bulk density increases that were observed following the application of organic residues reflected the microbial biomass improvements as a result of the same treatments, with poultry manure having the greatest effect and rice straw the least effect. The maximum MBC and MBN were found in the poultry manure treated plots at harvest (432 and 31.6 $mg \cdot kg^{-1}$, respectively). Furthermore, irrespective of treatment, the biomass carbon contents increased with advancement of the rice growing period, i.e., $0 < 30 < 60 < 90 < 120$ DAT, with a maximum level at crop harvest. These findings show that the regular application of organic materials to soils using whatever sources available will increase the microbial dynamics and nutrient pools in the soil, enhancing soil fertility and productivity.

ACKNOWLEDGEMENTS

The study was conducted under the project "Carbon sequestration in soils of Bangladesh (ID # 305)" funded by World Bank, IFAD & Government of Bangladesh through Bangladesh Agricultural Research Council (BARC), NATP, Phase 1 and implemented by the Department of Soil Science, Bangabandhu Sheikh Mujibur Rahman Agricultural University, Gazipur, Bangladesh.

DECLARATION OF CONFLICT OF INTEREST

The Authors of the manuscript are hereby declared that there will not take place any conflicts of interest for any reasons, such as personal and/or financial relationships, academic competition, and intellectual passion. No gender issue is also involved in publishing the manuscript.

AUTHORS' CONTRIBUTIONS

All persons who meet authorship criteria are listed as authors, and all authors certify that they have participated sufficiently in the work to take public responsibility for the content, including participation in the concept, design, analysis, writing, or revision of the manuscript. Furthermore, each author certifies that this material or similar material has not been and will not be submitted to or published in any other publication before its appearance in The Journal of Agricultural Science and Technology.

FUNDING/SUPPORT

The work was supported by World Bank, IFAD & Government of Bangladesh through Bangladesh Agricultural Research Council (BARC) [(ID # 305)].

CITE THIS PAPER

Anik, Md.F.A., Rahman, M., Mustafizur Rahman, G.K.M., Alam, Md.K., Islam, M.S. and Khatun, M.F. (2017) Organic Amendments with Chemical

Fertilizers Improve Soil Fertility and Microbial Biomass in Rice-Rice-Rice Triple Crops Cropping Systems. Open Journal of Soil Science, 7, 87-100.

REFERENCES

1. Cerny, J., Balik, J., Kulhanek, M. and Nedved, V. (2008) The Changes in Microbial Biomass C and N in Long-Term Field Experiments. Plant, Soil and Environment, 54, 212-218.
2. Laurie, A.T., Mary, A.A. and Ruth, D.Y. (1999) Forest Floor Microbial Biomass across a Northern Hardwood Successional Sequence. Soil Biology and Biochemistry, 31, 431-439.
3. Jenkinson, D.S., et al. (1990) The Turnover of Organic Carbon and Nitrogen in Soil. Philosophical Transactions of the Royal Society B Biological Sciences, 329, 361-368.
4. Wu, J., O'Donnel, A.G. and Syers, J.K. (1993) Microbial Growth and Sulfur Immobilization Following the Incorporation of Plant Residues into Soil. Soil Biology and Biochemistry, 25, 1567-1573.
5. Hao, X.H., Liu, S.L., Wu, J.S., Hu, R.G., Tong, C.L. and Su, Y.Y. (2008) Effect of Long-Term Application of Inorganic Fertilizer and Organic Amendments on Soil Organic Matter and Microbial Biomass in Three Subtropical Paddy Soils. Nutrient Cycling in Agroecosystems, 81, 17-24.
6. Huang, Q.R., Hu, F., Huang, S., Li, H.X., Yuan, Y.H., Pan, G.X., et al. (2009) Effect of Long-Term Fertilization on Organic Carbon and Nitrogen in a Subtropical Paddy Soil. Pedosphere, 19, 727-734.
7. Murphy, D.V., Sparling, G.P. and Fillery, I.R.P. (1998) Stratification of Microbial Biomass C and N and Gross N Mineralizsation with Soil Depth in Two Contrasting Western Australian Agricultural Soils. Australian Journal of Soil Research, 36, 45-55.
8. Liu, X.M., Zhao, K.L., Xu, J.M., Zhang, M.H., Si, B. and Wang, F. (2008) Spatial Variability of Soil Organic Matter and Nutrients in Paddy Fields at Various Scales in Southeast China. Environmental Geology, 53, 1139-1147.
9. Faroque, M.A.A., Kashem, M.A. and Bilkis, S.E. (2011) Sustainable Agriculture: A Challenge in Bangladesh. International Journal of Agricultural Research, Innovation and Technology, 1, 1-8.
10. Fox, C.A. and MacDonald, K.B. (2003) Challenges Related to Soil Biodiversity Research in Agroecosystems—Issues within the Context of Scale of Observation. Canadian Journal of Soil Science, 83, 231-244.
11. Rahman, M.M. (2013) Nutrient Use and Carbon Sequestration Efficiencies in Soils from Different Organic Wastes in Rice and Tomato Cultivation. Communications in Soil Science and Plant Analysis, 44, 1457-1471.
12. Alam, M.K., Biswas, W.K. and Bell, R.W. (2016) Greenhouse Gas Implications of Novel and Conventional Rice Production Technologies in the Eastern-Gangetic Plains. Journal of Cleaner Production, 112, 3977-3987.
13. Alam, M.K., Salahin, N., Islam, M.M. and Hasanuzzaman, M. (2014) Effect of Tillage Practices on Soil Properties and Crop Productivity of Wheat-Mungbean-Rice Cropping System under Subtropical Climatic Conditions. The Scientific World Journal, 2014, Article ID: 437283.
14. Moore, J.M., Klose, S. and Tabatabai, M.A. (2000) Soil Microbial Biomass Carbon and Nitrogen as Affected by Cropping Systems. Biology and Fertility of Soils, 31, 200-210.
15. Yang, F., Wu, J.S., Qian, X.B. and Wu, L.-J. (2006) Dynamic Changes of Soil Microbial Biomass Carbon in Phyllostachys praecox Stand with Different Fertilizers. Journal of Zhejiang Forestry College, 23, 70-74.

16. Lin, X.G., Yin, R., Zhang, H.Y., Huang, J.F., Chen, R.R. and Cao, Z.H. (2004) Changes of Soil Microbiological Properties Caused by Land Use Changing from Rice-Wheat Rotation to Vegetable Cultivation. Environmental Geochemistry and Health, 26, 119-128.
17. USDA (1978) Soil Taxonomy. Agriculture Handbook No. 436, USDA, Soil Conservation Service, Washington DC.
18. Karim, Z., Rahman, S.M., Idris, M. and Karim, A.J.M.S. (1988) A Manual for the Determination of Soil Physical Parameters. Soils and Irrigation Division, Bangladesh Agricultural Research Council, Dhaka, Bangladesh.
19. Jackson, M.L. (1973) Soil Chemical Analysis. Prentice Hall of India (pvt). Ltd., New Delhi.
20. Haider, J., Marumoto, T. and Azad, A.K. (1991) Estimation of Microbial Biomass Carbon and Nitrogen in Bangladesh Soils. Soil Science and Plant Nutrition, 37, 591-599.
21. Jackson, M.L. (1958) Soil Chemical Analysis. Prentice-Hall Inc., Englewood Cliffs, NJ, 498 p.
22. Olsen, S.R. and Sommers, L.E. (1982) Phosphorus. In: Page, A.L., et al., Eds., Methods of Soil Analysis, 2nd Edition, Part 2, Agronomy No. 9, American Society of Agronomy, Madison, WI, 403-430.
23. Baker, D.E. and Surh, N.H. (1982) Atomic Absorption and Flame Emission Spectroscopy. In: Page, A.L., Miller, R.H. and Keeney, D.R., Eds., Methods of Soil Analysis, Part 2, Chemical and Microbiological Properties, American Society of Agronomy Inc., Madison, WI, 13-26.
24. Vance, E.D., Brookes, P.C. and Jenkinson, D.S. (1987) An Extraction Method for Measuring Soil Microbial Biomass. Soil Biology and Biochemistry, 19, 703-707.
25. Brookes, P.C., Kragit, J.F., Powlson, D.S. and Jenkinson, D.S. (1985) Chloroform Fumigation and the Release of Soil Nitrogen. The Effects of Fumigation Time and Temperature. Soil Biology and Biochemistry, 17, 831-835.
26. Brookes, P.C., Landsman, A., Pruden, G. and Jenkinson, D.S. (1985) Chloroform Fumigation and the Release of Soil Nitrogen. A Rapid Direct Extraction Method to Measure Microbial Biomass Nitrogen in Soil. Soil Biology and Biochemistry, 17, 837-840.
27. Lovell, R.D., Jarvis, S.C. and Bardgett, R.D. (1995) Soil Microbial Biomass and Activity in Long-Term Grass Lands: Effects of Management Changes. Soil Biology and Biochemistry, 27, 969-975.
28. Chang, X.C. and Juma, N.G. (1996) Impacts of Crop Rotations on Microbial Faunal Populations, and Plant Carbon and Nitrogen in Gray Luvisol (Typic Cryoralf). Biology and Fertility of Soils, 22, 31-39.
29. Akmal, M., Khan, K.S. and Xu, J.M. (2004) Dynamics of Microbial Biomass in a Rain Fed Soil under Wheat Cultivation. Pedosphere, 14, 53-62.
30. Efremov, A.L. (2000) Microbiota and Biogenic Nature of Alluvial Meadow Soils in the Pripyat Poles'e Region. Pochvovedenine, 5, 579-583.
31. Pavlova, O.S., Polyanskaya, L.M., Kochkina, G.A. and Ivanushkina, N.E. (2000) Peculiarities of Microbial Succession in the Soils of Okskii Reserve. Pochvovedenie, 3, 320-328.
32. FRG Fertilizer Recommendation Guide (2005) Bangladesh Agricultural Research Council (BARC). Farmgate, Dhaka 1215, 274 p.
33. Rasiah, V. (1999) Nitrogen Immobilization and Mineralization in Legume-Amended Soils as Influenced by Texture and Compaction. Community Soil Science and Plant Analogy, 30, 829-841.
34. Glover, J.D., Reganold, I.P. and Andrews, P.K. (2000) Systematic Method for Rating Soil Quality of Conventional Organic, and Integrated Apple Orchards in Washington State. Agriculture, Ecosystems & Environment, 80, 29-43.

CHAPTER 15

Soil Contamination with Heavy Metals and Petroleum Derivates: Impact on Edaphic Fauna and Remediation Strategies

Raphael Bastão de Souza[1], Thiago Guilherme Maziviero[1], Cintya Aparecida Christofoletti[1], Tamaris Gimenez Pinheiro[1] and Carmem Silvia Fontanetti[1]

[1] Departament of Biology, São Paulo State University, Rio Claro, Brazil

1. INTRODUCTION

Soil is characterized as a complex and dynamic system. It is constituted by several layers that differ in relation to the physical, chemical, mineralogical and biological nature, which are influenced by the climate and activities of the living organisms. Besides contributing to the maintenance of all forms of life that occur in the terrestrial surface, soil plays an important role in protecting the groundwater acting as a collector filter of organic and inorganic residues, helping in sequestering possible toxic compounds [1].

During the last decades of the twentieth century there was an awareness of the importance of the soil as an environmental component and recognition of the need to maintain or improve its capacity to allow it to perform its various functions. At the same time there was a confirmation that the soil is not an inexhaustible resource and, if used improperly or poorly managed, its characteristics can be lost in a short period of time, with limited opportunities for regeneration [2].

However, the final disposal of potentially toxic residues in the soil has become a practical and inexpensive alternative and can cause alterations in the arthropod community [3, 4]. These species can present individual biological alterations (physiological, morphological and behavioural), which can be extrapolated to field studies in order to analyze ecological aspects, such as population dynamics and richness of diversity in the contaminated areas. Therefore, the gathering of biological studies, both laboratorial and field, combined with chemical analysis of the contaminants, provides a real scenario of the effects that the toxic substances can cause in the ecosystem.

Among the substances released in the soil it can be highlighted the petroleum derivatives and heavy metals [5]. In soils contaminated with petroleum and derivatives, some contaminants stand out compared to others, such as benzene, toluene, ethylbenzene and xylenes, known as BTEX, polycyclic aromatic hydrocarbons (PAH) and total petroleum hydrocarbons (TPH) [6, 7]. Pollution by heavy metals is derived from the anthropogenic activity, mainly associated to the industrial process and natural sources, such as volcanic eruptions [8].

Although researches involving soil quality are facing an important technologic challenge with several actions being taken in order to assess, correct and reduce the risks of contaminants in the soil, standardized monitoring combined with remediation strategies are still needed [5].

Thus, several researches aiming to remediate the effects of the soil contaminants have been carried out worldwide. Remediation of a contaminated area involves the application of one or more techniques aiming to remove or contain harmful substances in order to allow the reuse of the area with acceptable risk limits for human and environmental health. For this purpose, an ideal remediation process must remove all the contaminants of the soil or, at least, reduce the percentage of contamination of the environment to acceptable limits; should also avoid the migration of contaminants to other areas.

For the remediation of soils contaminated with petroleum and heavy metals, several physical, chemical and biological techniques have been developed for the removal or degradation *in situ* or *ex situ* of the pollutant [6, 9]. In this context, the chapter aims to provide a thorough revision of techniques for the removal or degradation of the pollutants as well as a discussion on the implementation of such techniques for the development of remediation strategies and policies.

2. DYNAMIC OF POLLUTANTS IN THE SOIL

Geosphere, or terrestrial layer, is that part of the earth on which the human beings live and extract the maximum of its resources. Erstwhile it was believed that the earth had unlimited capacity to absorb the impacts of humankind. Currently, the geosphere is considered very fragile and vulnerable to injuries originating from anthropogenic activities. According to Manahan [10] the definition of pollutant can be described as the increase in the concentration of a certain substance to higher levels than that they occur naturally, arising from an external source, generally related to the human activity.

There is great difficulty in predicting the behaviour of a xenobiotic in the soil, since its composition is totally complex and heterogeneous. Therefore, the knowledge of the physico-chemical characteristics of the contaminant compounds and the environment is fundamental to predict its dynamic [11].

It should be noted that several soils have the capacity to assimilate and neutralize such pollutants, since chemical and biochemical phenomena are capable of attenuating the harmful nature of the pollutants. These phenomena include processes of oxi-reduction, hydrolysis, acid-base reactions, precipitation, adsorption and biochemical degradation. Some hazardous organic chemical products can be degraded to innocuous products on the soil and the heavy metals can be sorbed, immobilized or mineralized. In general, a lot of care should be taken in the elimination of the residues, rejects and other potentially hazardous materials to the soil, particularly where there is the possibility of contaminating the existing water.

When the contaminant reaches the soil, either on purpose or accidentally, it suffers the action of geochemical and biological phenomena and is distributed by the subsurface in the vaporized, residual or adsorbed phases, free phase and dissolved phase. The distribution of such phases will depend on their physico-chemical characteristics and also on the type of the soil [12]. Thus, the mobility of the contaminants and, consequently, their toxicity are directly related to the capacity of the soil in maintaining them retained in their solid phase, making them unavailable to be absorbed by plants, eroded and/or leachate [13]. Among the factors that determine the binding of contaminants to the soil there is the available surface area of the particles (m^2/g). Moreover, the electrical charges of the particles of the soil matrix also influence in the adsorption of the contaminants to the environment. It is noteworthy that in relation to their physico-chemical properties the contaminants are classified as Dense Non-Aqueous Phase

Liquid (DNAPL), when the substance is more dense than the water and Light Non-Aqueous Phase Liquid (LNAPL), when it is less dense [14].

The main processes of interaction between the organic compounds or metals and the environment are the retention by adsorption, absorption or precipitation; biotic and abiotic transformations and transport by volatilization, leaching or runoff [15]. There are compounds highly resistant to degradation that can interact strongly in a reversible or irreversible way with the colloidal components of the soil. This process is called sorption, both for adsorption and absorption. Adsorption is characterized as an interfacial process while absorption differs for involving the penetration of the compound in the particles of the soil and can be accumulated inside the absorber system [11].

In general, the dynamic of the contaminants in the soil can be modelled by three mechanisms of mass transference, namely: advection, dispersion and attenuation.

a. *Advection* – it consists in the mechanism where the contaminants coincidentally follow the flow vectors and keep a direct relationship with the speed of percolation in the soil. It is the mechanism responsible for the formation and mobilization of the free phase of hydrocarbons.

b. *Dispersion* – Consists in the mechanism responsible for the decrease in the concentration of the contaminants in the fluid percolation and that can occur by two processes: hydrodynamic dispersion and molecular diffusion. Hydrodynamic dispersion occurs by the flow restriction in the pores of the soil that generates the reduction in the percolation velocity of the more viscous components while the molecular diffusion is, intrinsically, a phenomenon of dilution of the more soluble compounds, and is the main formation process of the dissolved phase, responsible for the greater mobility of the contaminants. In the case of emulsions, such as hydrocarbons, the dispersion can occur in a more complex mechanism, due to the phenomena of hysteresis (delay) of the entrainment of the contaminants, especially in the saturation fronts and capillary fringe. This process is associated to the formation of the adsorbed phase and also by the production of a fraction of emulsions that can compose the dissolved phase.

c. *Attenuation* – Consists in the reduction of contaminants transported by advection or dilution by chemical or physico-chemical reactions. Chemical attenuation is the more intense in soils with higher cation exchange capacity and acts reducing compounds in the free and adsorbed phase. Also in the list of reactions there are the bioconversion reactions, in which a part of the hydrocarbons is transformed or totally oxidized in organic acids. Chemical attenuation is more intense in the region with higher availability of oxygen.

Physico-chemical attenuation is responsible for the formation of the adsorbed phase and consists in the imprisonment of the contaminants that adhere to the grains of the soil, especially to the grumes of clay with higher activity. However, associated with the mechanisms of chemical attenuation, it is responsible for the formation of the dissolved phase (facilitated by the reduction of pH) [16].

3. CONTAMINATION OF SOIL AND ITS EFFECTS ON THE EDAPHIC FAUNA

Soil ecosystem harbours an enormous biodiversity and is increasingly being recognized that this diversity is essential for the maintenance of the function of other ecosystems [17], since the activities of the invertebrates have significative effects in its organization and structure, dynamics of the organic matter and in the growth of plants [18]. Despite this importance, soil has become a practical and cheap alternative for the final disposal of several toxic residues, resulting in negative consequences [4].

Contaminants can be resistant to the decomposition processes and, therefore, can be accumulated in the soil [19]. Invertebrates easily become exposed to such contaminants, which can affect their ecological function [20] and influence indirectly the ecosystem and alter the ratio predator/prey and affect the complex food chain [21]. In order to evaluate the ecological effects of this contamination it is developed tests that aim to quantify the abundance, mortality and reproduction of the organisms exposed [20].

In this sense, the following topic will address the effects caused in the edaphic fauna due to the contamination of the soil by heavy metals and petroleum derivatives.

3.1. Petroleum and Its Derivatives

According to Leblond [22], it is expected the production of 95 million barrels of petroleum per day in order to meet the growing worldwide demand of this resource. Crude petroleum is a complex mixture constituted, mainly, by hydrocarbons, organic sulphur compounds, nitrogen and oxygen [23]. Although about 80% of the total production of crude petroleum is generated from terrestrial fields, few studies about its impact on the soil are available [24].

Studies on the toxicity of petroleum have shown that some species present higher sensitivity to these contaminants. Survival of earthworms (*Eisenia andrei* and *E. fetida*) and enchytraeids (*Enchytraeus crypticus*) can be reduced in soil containing crude petroleum [25, 26], while the abundance of Isopoda and Hymenoptera in areas contaminated with complex mixtures derived from refineries can be higher in relation to uncontaminated areas [27].

Among the petroleum derivatives, the Polycyclic Aromatic Hydrocarbons (PAH) have a prominent role. Chemically, they are aromatic compounds formed by two or more benzene rings, constituted exclusively by atoms of carbon and hydrogen, arranged in a linear, angular or grouped form [28], and are residues of combustion, petroleum refinery and other industrial processes of high temperature [29]. There are thousands of these substances in the environment, each one differing in the number and position of the aromatic ring [30], but only 16 substances cause environmental concern: acenaphthene, acenaphthylene, anthracene, benzo(a)anthracene, benzo(a)pyrene, benzo(b)fluoranthene, benzo(ghi)perylene, benzo(k)fluoranthene, chrysene, dibenzo(a, h)anthracene, phenanthrene, fuoranteno, fuoreno, indeno(1,2,3-cd)pyrene, naphthalene and pyrene [31].

Although van Brummelen et al. [19] asserted that the exposure of invertebrates to PAH accumulated in the soil can affect the ecological function of these organisms, little is known about their effects [32, 33]. However, it is known that terrestrial invertebrates do not have the ability to metabolize aromatic compounds, with exception of some species that have microorganisms associated to the intestine [34], which implies in a broader problem, since it generates the bioaccumulation in the organism, enhancing the possibility of contaminating their predators via the food chain [35].

A small review performed by Souza et al. [7] discusses the main ecotoxicological assays that can be applied in soils contaminated by petroleum hydrocarbons. In this review, the authors affirm that bioassays with invertebrates have been efficient, thanks to the important role that these animals play in the ecological processes of the soil, such as cycling and decomposition.

Studies using earthworms as bioindicator organisms of contamination of the soil by PAH showed that the impact in these organisms is limited. Both the survival and reproduction rates were not altered and the concentrations of these substances in the individuals were low, suggesting low absorption by them [36]. Schaub and Achazi [apud 36] observed that PAH did not influence the survival and growth of the earthworm *E. fetida* in the concentration of 100.8 mg/kg, but the reproduction was affected in the concentration of 1.008 mg/kg. Chrysene did not alter the survival of *E. fetida* in a study carried out by Bowmer [37].

The non-toxicity of PAH for earthworms can be explained by the fact that there is a mutual interference between them [38, 39]. Earthworms are responsible for assisting the elimination of the PAH in the soil by improving the natural conditions of biodegradation, contributing to the increase of its oxygenation due to the intimate contact of the microorganisms present in their intestine with the soil [36].

In relation to Collembola, Sverdrup et al. [33] affirm that they are more sensitive to PAH when compared to other organisms, such as earthworms, being, therefore, good models of toxicity for this class of contaminants. To reach this conclusion, the authors tested 16 different PAH, from which eight affected the reproduction and survival rates of Collembola. Eom et al. [40] corroborate the fact that Collembola are more sensitive to PAH. Isopoda did not show to be more sensitive to contamination by PAH. The species *Oniscus asellus* presented a small alteration in the abundance after exposure to benzo(a)anthracene and no effect after exposure to benzo(a)pyrene. The species *Porcellio scaber* did not present alteration for any of the two substances [19].

PAH can also act indirectly on organisms and cause alterations in the populations, since the increase in the density of the soil due to their presence and their hydrophobic properties decrease the inhabitable space within the pores of the soil. Moreover, PAH can also act as fungicides, eliminating the source of food of some organisms [41].

Due to the reduced number of studies there is not a base to predict alterations in the community of invertebrates caused by contaminations of PAH. Studies with this focus does not seem to be a promising tool to assess the risks of this substance and the use of more sensitive biochemical markers (concentration of metabolites, damages in the DNA) are better strategies for this purpose [3].

In this sense, besides the traditional tests with Annelida and Collembola, studies with other terrestrial invertebrates have been developed to assess the quality of soils [42]. Diplopoda also make part of the edaphic fauna and are continuously exposed to the contaminants present in the soil. In these animals, histopathological markers have been applied [43-47].

Tissular alterations in the midgut and perivisceral fat body of the diplopod *Rhinocricus padbergi* were studied by Souza and Fontanetti [46] and Souza et al. [42], after exposure of these animals to a landfarming soil. According to the authors, the chemical analysis showed the presence of high concentrations of compounds such as PAH and metals, the authors also inferred that the histological and physiological alterations observed can be an attempt of defence of the animals exposed to this residue, in an attempt to eliminate and/or neutralize the assimilation of toxic residues [42].

3.2. HEAVY METALS

As a consequence of the technological development and global population growth, the agricultural and industrial activities have intensified, leading to a considerable increase of metals in the different compartments of the environment. Unlike organic pollutants, the toxicity of metals is intrinsic to their atomic structure and they cannot be transmuted/mineralized to a total innocuous form [48].

Pollution by heavy metals in terrestrial ecosystems has been recognized as a serious environmental concern, due to their non-biodegradability and tendency to accumulate in plants and animal tissues [49]. The extreme sensitivity of the macrofauna to the conditions of the soil make them potential indicators of the disturbance occurred in this environment [50]. For studies of this nature, the most used organisms are nematodes, earthworms, Collembola [apud 51], as well as molluscs [49] and ants, despite the last two ones be quite resistant to this type of contamination [52].

Among the most common responses of these organisms to the contamination by heavy metals it can be highlighted the decrease in the diversity of species due to changes in the composition of the community that eliminate the most sensitive species [53, 54] and promote the tolerance of opportunistic species [55] (Syrek et al., 2006) or invasive species [56].

Despite the increase in the abundance, richness and/or uniformity is not commonly found, some studies reported these types of alterations with the increase of pollution by heavy metals. It can be cited the studies developed by Russell and Alberti [57] that observed that Protura present tolerance to heavy metals, since this group was the only one found in sites highly contaminated; by Nahmani and Lavelle [51] that also found that the abundance of some groups of arthropods, such as larvae of beetles of the subfamily Hoplinae and family Staphylinidae, was positively correlated with pollution by heavy metals; by Migliorini et al. [58] that verified increase in the abundance of Protura, Diplura and Collembola with increase in the pollution by metals; by Grzés [52] that presented clear evidence of the increase of the diversity of species of ants with the increase in the pollution by metals.

Although no direct explanation for these patterns has been proposed, some of the authors point out the importance of the interactions between fauna and soil, mainly related to decrease of predation and competition between the edaphic organisms [52]. For this reason, considering all the macrofauna communities as indicator seems logical, since they have a wider range of adaptive mechanisms than a single taxonomic group [51, 59].

Pollution by metals can still influence directly the communities by the alteration of the abiotic conditions such as temperature and humidity. If the pollution decreases the density of the vegetation, the temperature of the environment will increase and this will facilitate the increase in the diversity of thermophilic organisms [52]. According to the same author, pollution by metals can favour species with affinity to humidity by reducing the microbial activity, allowing an accumulation of organic matter [52].

4. TYPES OF TREATMENT OF CONTAMINATED SOILS

Geochemical and biological processes that determine the mobilization and transformation of the compounds in the soil involve countless variables, making the remediation process a complex task. Thus, for the remediation be satisfactory and complies the environmental legislation, it is necessary to know the treatment technologies available, their advantages and disadvantages (table 1), cost-benefit relationships, applicability regarding the hydrogeology of the place and the nature of the contaminant [60].

According to Andrade et al. [6], the technique to be used depend on some factors, such as: physical, chemical and biological conditions of the contaminated site, concentration of the contaminants and time needed for the degradation/removal of the target compounds, according to the technique to be employed.

The main processes of interaction between the hydrocarbons or metals and the environment are retentions (adsorption, absorption or precipitation); biotic and abiotic transformations, transport by volatilization, leaching or runoff [15]. There are compounds highly resistant to degradation that can interact strongly in a reversible or irreversible way with the colloidal components of the soil. This process is called sorption, both for adsorption and absorption. Adsorption is characterized as an interfacial process while absorption differs for involving the penetration of the compound in the particles of the soil and can be accumulated inside the absorber system [11].

Table 1. Advantages and disadvantages in the use of different techniques in the remediation of soils contaminated by petroleum and heavy metals

TECHNOLOGY	ADVANTAGES	DISADVANTAGES
Solidification/stabilization	• Simple design • Cost-effective • Large soil volume can be treated • Very recommended for metals	• Does not promote the treatment of the contaminant, promotes only immobilization • Short-lived • Dependent on the soil characteristics and homogeneity of the mixture • Process hindered by the depth of the contaminant
Advanced Oxidative Processes (AOP)	• Cost-effective • Mineralization capacity • Recommended for soils with high permeability • Different reagents may be employed	• Mass transfer of the adsorbed phase to the aqueous phase • Risk of aquifer contamination by not recovered solvent • Limitations for large-scale application (*ex-situ* treatment) • The use of strong acids causes destruction of the basic structure of the soil
Advanced Oxidative Processes (AOP)	• *In situ* treatment • Cost-effective • Rapid process • Little or no waste is generated	• Lower efficiency for insoluble compounds • Susceptible to changes in pH • May be harmful to soil microorganisms
Thermal desorption or extraction with supercritical CO_2	• High efficiency for volatile compounds • Soil aeration can facilitate the bioremediation process • Rapid process • Low environmental impact	• Low efficiency for soils with low permeability • Not recommended in saturated areas • Treatment of the realeased vapors is required
Incineration	• High efficiency • Rapid process • Compounds mineralization • May be used where other processes are not effective	• High cost • Release of secondary compounds to the atmosphere • Periodic and rigorous monitoring are riquered • *In situ* treatment is not possible
Adsorption with clay	• Cost-effective • Simple design • Can be combined with other techniques	• Soil exchange is required • Limited by buffer capacity of the soil • Selectivity for specific ions
Electrokinetic	• High efficiency • *In situ* treatment • May be combined with bioremedition techniques	• Treatment time depends on the distance between the electrodes • pH change in areas near the electrode • Cost depends on the contaminant concentration and soil • Lower efficiency in soils with low permeability

Since 1993, information of the Environmental Protection Agency (EPA) was considered to indicate the need for innovative technologies, such as remediation, to replace conventional processes. New technologies have as objective the treatment of organic compounds, however, few alternatives are available for the removal of metals in the soil, particularly *in situ*.

Among the existing remediation processes it can be highlighted the technologies of immobilization, destruction of the contaminants and separation. Immobilization technology consists in the creation of physical barriers to avoid the migration of the contaminants, such as processes of solidification/stabilization (encapsulation of the contaminants). The processes of destruction are based, mainly, on the use of high temperatures and chemical methods, such as incineration, chemical reduction, chemical oxidation, photolysis and bioremediation; and the separation consists in retaining, isolating or extracting the contaminants to a phase of easier management or to a more concentrated phase, reducing the volume of the material to be remediated or disposed, such as processes of thermal desorption, washing the soil, extraction by solvent and supercritical extraction [61, 62].

4.1. Solidification/Stabilization

The process of solidification/stabilization, also known as immobilization, modifies the physico-chemical characteristics of the residue to contain the contaminants. Metals are commonly remediated by solidification *ex situ* by encapsulation and sometimes complexation.

The encapsulation technology has become an important alternative treatment for the disposal of hazardous residues in landfills and control of contaminated areas, since it provides an improvement of the physical and toxicological characteristics of the residue and/or soil, facilitating its management in a safe and effective form. Moreover, the cost of the encapsulation has been considered low in relation to other treatment techniques, fact that has stimulated the development of this technology in the last years. However, there is an increasing interest in more durable and safer solutions [63].

The frequently used agents for encapsulation are Portland cement and lime. In physical terms, the cement presents response in a smaller interval of time than lime, since its curing takes place in less time. Chemically, both act to alkalinize the environment, increasing the pH of the compound, decreasing the solubility of the contaminants, since it is known that the solubility is dependent on the pH [64]. Physically, it occurs the cementing of the particles, causing a decrease in the mobility of the contaminant within the soil. Therefore, the reduction in the mobility of the contaminant can be enhanced by the alkalinisation of the environment and also by the cementing effect of the particles.

After application of the encapsulation technique, some assays become necessary for analyzing the effectiveness of the method, which consist in chemical and physical analysis of the treated compound. Chemical analysis are performed based on leaching assays and chemical extraction. Physically, it is performed analysis of compressing, resistance to simple compression, permeability, durability, among others [63].

Another solidification technique involves the vitrification by the passage of an electric current between electrodes. This process results in the retention of solids and incorporation of metals in the vitrified method. This technology is being commercially evaluated and presents very promising results. Vitrification has been used for capturing mercury and other volatile metals such as lead and arsenic [65].

4.2. Washing and Extraction By Solvent And Chemical Oxidation

One technique of separation of organics in soils very used is the extraction by organic solvents. In these cases, the organic contaminant is extracted from the contaminated site and later destined to the destruction treatment. The process occurs by washing the soil using adequate solvents for each type of contaminant, such as detergents for oils or petroleum and chelators for metals. It has the disadvantage of being a process that requires specific machinery, demands specialized staff and, at the end of the process, generates great quantities of contaminated liquid residues, which must be adequately treated and disposed posteriorly [11].

Chemical oxidation or *In Situ Chemical Oxidation* (ISCO) has shown to be a promising technique for the remediation of soils contaminated by organic compounds [66]. This technique is based on the application of strong oxidant agents to degrade the organic. It has been applied both *in situ* and *ex situ*, and its application in the field is more appropriate.

ISCO also has its limitations, especially regarding the reactivity of the agent with the contaminant and mass transference between the adsorbed and aqueous phases, where generally occurs the oxidation reaction [67]. The most used agents in ISCO processes are ozone (O_3), hydrogen peroxide (H_2O_2) and potassium permanganate ($KMnO_4$). Each one has advantages and disadvantages and the application depends on the environment to be treated and the contaminant to be degraded [68].

4.3. Advanced Oxidative Processes (Aop)

The most effective processes in the destruction of organic pollutants are known as advanced oxidative processes (AOP). AOP are characterized by the generation of hydroxyl radicals (HO*), which presents high potential pattern of oxidation, superior to those of other oxidant species, such as O_3, H_2O_2 and chloride (Cl_2) [69], capable of reacting with practically all classes of organic and inorganic compounds. These processes are emerging as a promising alternative for the treatment of matrices contaminated with highly toxic and recalcitrant substances, leading them to total mineralization or formation of more biodegradable intermediates [70, 71].

Although there are more economical processes, not always the time needed to achieve the expected results allow their use, thus, AOP can be used when these limits of time and other logistics become hierarchically more important.

Fenton system is one of the most known advanced oxidative processes and consists in the combination of hydrogen peroxide and ferrous ions to form hydroxyl radicals. The oxidizing power of Fenton's reagent (H_2O_2/Fe^{2+}) is attributed to the hydroxyl radicals resulted from the catalytic decomposition of hydrogen peroxide in acid medium, whose general reaction is represented by:

$$Fe^{2+} + H_2O_2 \text{->} Fe^{3+} + OH^- + {}^*OH$$

Hydroxyl radicals generated oxidize the organic compounds of the environment, generating intermediates that are attacked again by other hydroxyl radicals and can reach the complete mineralization (CO_2 and H_2O). This system has been widely studied in the oxidation of organic compounds of high toxicity.

The reagents that compose the Fenton system present advantages over the others because they are compounds relatively inexpensive and non-toxic, besides the reaction occurs at room temperature and pressure. It is known that the hydroxyl radical oxidize effectively organic compounds in aqueous phase, including the polychlorinated biphenyls (PCB) [72].

The efficiency of the chemical oxidation in soil is influenced, mainly, by factors such as concentration of iron, concentration of peroxide, presence of other organic compounds competitive by hydroxyl and pH [62]. Moreover, some researchers have observed a strong increase in the oxidant power of the Fenton reagent when combined with radiation UV or UV-visible, called Photo-Fenton. This technique has shown to be an extremely promising alternative, especially on tropical countries, like Brazil, where the incidence of sunlight is high practically during the entire year, configuring an important source of energy, hitherto unexplored [11].

4.4. Thermal Desorption Or Extraction With Supercritical Co2

The most applied system of thermal desorption is the process of injecting steam water in the soil with a system of pumps and vacuum, i.e., it is installed in the area to be remediated a series of pipes from which will be injected in the soil steam water and other suction pipes. The steam at high temperature drags the contaminants, extracting them from the soil, which are then sucked by vacuum sites and sent to filters or condensers to receive appropriate treatment [66].

Extraction of compounds using supercritical fluid consists in making the extraction of the contaminants by passing a gas at high pressure (400 bar) and high temperature (150ºC) through the contaminated soil. In general, CO_2 is the fluid chosen due to its low toxicity and environmental acceptability and this extraction has shown to be very efficient for compounds with high solubility in CO_2, such as PAH, PCB, dioxins and organochlorine pesticides [11].

In the United States, an area with more than 170 tons of soil contaminated with benzene, arsenic, chromium and PAH was remediated using the process of thermal desorption [73].

4.5. Incineration

The use of heat to destroy toxic compounds is a very old practice. Incineration has been used for centuries to destroy or diminish the volume of domestic or agricultural residues that are unnecessary or undesirable. However, during the combustion process occurs the formation of undesirable by-products, such as dioxins and furans, highly toxic and carcinogenic. To avoid the formation of such compounds it is necessary to have strict control over the combustion conditions [74].

To remediate soils contaminated with PAH, this process is one of the most efficient and used, despite the high cost due to the need of soil excavation, transport and treatment with heat [75]. Although the treatments with high temperatures are effective in the treatment of organic residues, a serious problem occurs when the residue has metals, since a fraction of them will volatilize during the treatment and, after the gas cooling, they will condense on particles of metal [76].

4.6. Adsorption With Clay

Clays have structures in layers of lamellae that consist on sheets of silicon oxide alternating with sheets of aluminium oxide. The sheets of silicon oxide are arranged in tetrahedra in which each atom of silicon is surrounded by four atoms of oxygen, some variations can present geometrical structure in form of octahedrons. Many clays contain large quantities of sodium, potassium, magnesium, calcium and iron and other metals. Clays can attract cations such as Ca_2^+, Mg_2^+, K^+, Na^+ and $NH4^+$, retaining them between their lamellar structure in order to avoid leaching by water, but maintain then available in the soil as nutrients for the plants [10].

Thus, heavy metals and other charged species are strongly attracted and adsorbed in the clay surfaces. Heavy metals have different sorption characteristics and the mechanisms depend on the adsorbents. The sorption mechanisms include complexation of the surface (adsorption) and ion exchange. Adsorbents show difference in the sequence of selectivity for different metals. One example is lead when compared to other metals, since it is highly attracted and adsorbed by several types of clay.

4.7. Electrokinetic

Electrokinetic remediation, also called electrokinetic processing of the soil, electromigration, electrokinetic decontamination or electrocorrection, can be used to extract metals and some types of organic residues, such as PAH, of saturated or unsaturated soils, sludges and sediments [66]. This technique consists on the application of a direct current of low intensity between the electrodes located in the soil. The materials used for the construction of the electrodes can be graphite, stainless steel and platinum. Electrolysis of the water (in the disperse electrolyte) produces ions H^+ in the anodes and ions OH^- in the cathodes, generating a localized change of pH, which leads to the desorption of the contaminated ions.

Some variations of this technique involve the direct extraction of metallic ions already in the metal form and the others involve the extraction of metallic ions using a posterior process of ion exchange resins. Electrokinetic remediation can be also used to delay or prevent the migration and/or diffusion of the contaminants, directing them to specific sites and diverting them from the freatic sheets.

Currently, the application of electrokinetic process has been considered promising, especially for the remediation of low permeability contaminated soils, where the electric field generated mobilizes electrically charged species, particles and ions in the soil by the processes of electromigration, electrophoresis and eletroosmosis [66]. For the migration process in the electrodes, the contaminants can be removed by reduction in the cathode, precipitation, pumping next to the electrode, or in a more complex form with ion exchange resins.

However, the electrokinetic process is limited by the solubility of the contaminant and by desorption of the contaminants in the surface of the soil. Heavy metals in their metallic state are not being sufficiently dissolved and separated from the samples of soil. The process is also not efficient when the concentration of the ions to be removed is low and the concentration of diverse ions is high. Moreover, factors such as heterogeneity and anomalies in the local surface (boulder, large quantities of iron or iron oxides, large rocks and gravel or materials such as shells) can reduce the efficiency of removal.

The cost of remediating soils contaminated by metals, using the electrokinetic technique is strongly influenced by the soil conductivity, since the consumption of energy is directly related to the conductivity of the soil between the electrodes. The electrokinetic treatment of the soils with high ion conductivity may not be feasible due to the high cost [63].

Another method that uses the electrokinetic technology is the electroacoustic decontamination of the soil. This technology combines the eletrokinetic with the sonic vibration. The properties of the liquid contaminant in the soil can be altered in order to increase the rate of the contaminant removal by the application of a mechanical vibratory energy in the form of sonic or ultra-sonic energy. The elecroacoustic technology is technically feasible for the removal of inorganic species from the soil with clay (and partially effective for the removal of hydrocarbons) [63].

4.8. Bioremediation

According to Yeung et al. [77], biological processes are gaining increasing importance in the treatment of soils. To meet the challenges presented by environmental pollution, the objective of bioremediation (along with prevention and physical and chemical methods for remediation) is reduce the quantity and availability of hazardous chemical compounds and convert them into useful products and/or less innocuous [48]. However, biological processes, when compared to the conventional physical and chemical processes, are safer, less costly and less aggressive to the environment [78].

Bioremediation process can be defined as the use of microorganisms, such as bacteria, fungi, yeasts and algae or their enzymes to treat polluted areas or "return" them to their original condition [48, 79, 80]. In general, bioremediation is based on the biochemical degradation of contaminants

[6, 81], resulting in the transformation in metabolites or their mineralization [78].

The types of treatment involved in the remediation process can be of two types: *ex situ*, in which there is an excavation and removal of the contaminated soil to another place and the *in situ*, where the treatment is performed in the local. The *in situ* bioremediation is the most worldwide used type of process regarding the place of treatment [6].

Briefly, the main techniques involved in the bioremediation process are:

a. *Bioattenuation* (natural process) - used to described the passive remediation of the soil, which involves several natural processes, such as biodegradation, volatilization, dispersion, dilution and adsorption of the contaminants, promoted in the sub-surface by native microorganisms [6, 80].

b. *Biostimulation* (or accelerated natural attenuation) - consists in the addition of nutrients and/or descompacting agents in the contaminated soil, increasing the population of endogenous or native microorganisms [42].

c. *Biomagnification* (or bioaugmentation) - characterized by the increase of the native microbiota by the inoculation of exogenous microorganisms (allochthonous) [6, 82, 83]. In this case, according to the literature, generally, the used microorganisms are bacteria, philamentous fungi and yeasts.

d. *Landfarming* - is an *ex situ* remediation technique, based on the placement of the contaminated soil in layers with at maximum 40 cm of thickness and their processing with agricultural machines [84].

e. *Biopiles* - is an *ex situ* technology of bioremediation, which involves the stacking of contaminated soils, which stimulates the aerobic microbial activity, accelerating the degradation of the pollutant by aeration, addition of nutrients and correction of humidity.

f. *Composting* - technology that involves the addition of organic structuring agents in the contaminated soil/compounds, increasing the porosity and airflow in them. Such agents still serve as easy access source of carbon to the biomass growth. The energy released during the degradation of the organic matter result in temperature increase, which facilitates the action of different microbiological phases: mesophilic, thermophilic, cooling and maturation [85].

g. *Phytoremediation* - technique that uses plants as decontamination agent. Involves several mechanisms such as phytoextraction, phytostabilization, rhizofiltration, phytodegradation, phytostimulation, phytovolatilization, vegetative strains, artificial ponds and hydraulic barriers [86].

4.8.1. Bioremediation of sites contaminated by petroleum derivatives and heavy metals

Contaminations of soils with petroleum hydrocarbons have become a worldwide problem in the mid 80's [77]. The contamination sources by these compounds are related with exploration, production, storage, transport, distribution and final disposal of petroleum and their derivatives.

In the biological treatment of soils contaminated by petroleum, microorganisms, being bacteria the most studied, use hydrocarbons, major components of petroleum, as source of carbon and alternative energy in the production of biomass. This process involves the transformation of hydrocarbons into smaller unities and later incorporation as cellular material (biotransformation) or conversion to carbon dioxide (mineralization), resulting in the reduction of the concentration of the petroleum hydrocarbons [87].

There are, in the scientific literature, a considerable number of studies on bioremediation of soils contaminated by PAH, using different remediation methodologies such as treatment of the solid phase, landfarming/composting, phytoremediation, biostimulation among several others [85].

In the landfarming process, petroleum derivatives are removed by volatilization, biodegradation and absorption. The more volatile products, such as gasoline, are removed by volatilization during the aeration process and a small portion is degraded by the microorganism respiration. Derivatives such as diesel and kerosene have less volatile constituents than gasoline and, therefore, the biodegradation is more significative than volatilization. The heavier compounds, such as lubricating oil, are not volatile, suffering only biodegradation [88].

Composting has obtained success in the bioremediation of petroleum derivatives using different compounds, such as mushrooms [89], soot residues [90], green residues [91, 92], maple leaves and alfalfa [93] and horse manure [94]. Plants, by phytoremediation, have shown positive results in the degradation of PAH, since it stimulates the growth and microbial activity in the rhizosphere (interface soil/root) [95].

Besides the individuals use of these processes, it is possible to combine more than one technology in the bioremediaton of contaminated soils. According to Straube et al. [96], microorganisms naturally present in the

Table 2. Advantages and disadvantages in the use of different techniques in the bioremediation of soils contaminated by petroleum and heavy metals

TECHNOLOGY	ADVANTAGES	DISADVANTAGES
Landfarming	• Simple desing and implementation • Cost-effective • Large soil volumes can be treated • Favourable public opinion • Complete destruction of waste material	• Large treatment area is required • Risk of human pollutant exposure • Limited to removal of biodegradable pollutants
Phytoremediation	• Cost-effective • Easy to implement and operate • Environment-friendly • Favourable public opinion • Reduced pollutant exposure	• Slower than other methods • Soil properties, toxicity level and climate should allow plant growth • Limitations for large-scale application
Bioaugmentation	• Cost-effective • Increase the bioavailability of pollutants • Short treatment times	• Laboratory strains of microorganisms rarely grow in contaminated soil • The use of genetically modified organisms does not have public acceptance • Recent and under development • Possible environmental risk by introducing non-indigenous microorganisms
Biostimulation	• Improve the degradation potential of the inhabiting microbial population	• Dependent on the indigenous organisms • Inorganic nutrients that are injected may precipitate metals, swell clays, change redox potentials and conductivity
Bioattenuation	• Efficient and continuous process • Equipment are not required • Less impact on the environment	• Preliminary studies are required • Slow and unpredictable process • Periodic and rigorous monitoring are riquered • May be costly
Biopiles	• Rapid process	• *Ex situ* treatment • High cost • Other sites may be contaminated
Composting	• Cost-effective • Simple design • High efficiency	• Depending on the compound employed, there may be a small increase in contamination, pH, solubility and mobility of toxic elements • Poor public reception due to odour and insects

soil that degrade PAH can have their degradation capacity limited due to several environmental factors, such as low solubility and low bioavailability of PAH and limitation of nitrogen or other nutrient. Thus, it is possible to combine landfarming with biostimulation and bioaugmentation to increase the efficiency of the technique. In the bioremediation process by biopile it can be also employed procedures such as aeration, bioaugmentation, biostimulation and composting in order to increase the efficiency of the remediation of petroleum hydrocarbons [97].

Mohan et al. [85] and Megharaj et al. [80] presented a review on the main techniques used in the bioremediation of soils contaminated by organic pollutants. The different strategies of the bioremediation process have specific advantages and disadvantages (table 2), which, according to the same authors, need to be considered in several situations, since there are many factors that limit the efficiency of the microbial degradation of organic pollutants: bioavailability of the pollutant, low temperatures, anaerobic conditions, low levels of nutrients and co-substrates, presence of toxic substances and physiological potential of microorganisms.

To illustrate the difficulty and success/failure of the bioremediation of soils contaminated by petroleum and its derivatives, there are some studies performed in different parts of the world, which use different techniques of bioremediation. Bento et al. [98] assessed the efficiency of the natural attenuation, biostimulation and bioaugmentation in the degradation of TPH (Total Petroleum Hydrocarbons) in soils contaminated by diesel, in samples from California and Hong Kong. After 12 weeks of incubation, the authors observed that the three techniques employed show different effects in the degradation of light fractions (C_{12}-C_{23}) and heavy fractions (C_{23}-C_{40}) of TPH in the soil samples. However, the authors noted that the number of microorganisms that degrade diesel and the heterotrophic population were not influenced by the treatments, suggesting, therefore, that detailed studies on the characterization of the site are needed before deciding the adequate bioremediation method.

Haderlein et al. [93] studied the effects of composting or simple addition of manure in the soil, during the mineralization of pyrene and benzo[a]pyrene. It was reported that composting and addition of manure had no effect on the mineralization of benzo[a]pyrene. In contrast, the mineralization rate of pyrene increased dramatically with the amount of time that the soil was composted (more than 60% of mineralization after 20 days).

Bioremediation of metals face major obstacles in relation to the bioremediation of organic compounds, since metals introduced in the environment cannot be degraded. They persist indefinitely and can cause pollution of water, air and soil, and the main strategies in the control of their contamination are the reduction of their bioavailability, mobility and toxicity [99].

The oxidation state, solubility and association of metals with other organic and inorganic molecules can vary, however, the microorganisms, as well higher organisms can play an important role in the bioremediation of the concentration of metals, so that they become less available and less hazardous [48].

Among the main methods involved in the remediation of environments contaminated by heavy metals it is included the phytoremediation [99] and the use of microorganisms [100].

In this context, phytoremediation of heavy metals present in the soil, also called phytoextraction, is the technique that uses the capacity of the plants to absorb the metals [101]. As a general rule, metals bioavailable for absorption by plants include Cd, Ni, Zn, As, Se and Cu. Metals moderately bioavailable are Co, Mn and Fe; while the least are Pb, Cr and U [apud 101]).

Phytoremediation process can be divided into three types: phytoextraction, phytostabilization and rhizofiltration. Phytoextraction uses species of hyperaccumulator plants to transport metals to soil and concentrate them into the roots or buds, which will be later collected; in the phytostabilization the plants are used to limit the mobility and bioavailability of metals in the soil by sorption, precipitation, complexation or reduction of the valences of metals; rhizofiltration uses roots of plants in order to absorb, concentrate and precipitate metals from residual waters, which can include soil leachates [102].

Microorganisms, frequently used in the bioremediation of organic pollutants, can be also used in the bioremediation of soils contaminated with metals by biosorption (process in which metals are absorbed and/or complexed in live or dead biomass), alterations in the redox state (transformations catalyzed by enzymes) [103, 104]), by biosurfactants [105], bioleaching (immobilization of metals by excretion of organic acids or methylation reactions), biomineralization (immobilization of metals by the formation of insoluble sulphides or polymeric complexes) and intracellular accumulation [apud 100].

Since metals cannot be biodegraded in CO_2 and water, microorganisms can only modify their speciation, converting them into non-toxic forms [105]. In order to ensure the efficiency of the bioremediation process, the microorganisms added in the contaminated site must have, besides enzymes of biodegradation, resistance to the metal target [101].

5. CURRENT POLICIES FOR SOIL REMEDIATION

Bredehoeft [106] suggested that the problem of the remediation of toxic substances would be present in the society for a long time and taking into consideration the policies and expenses of the period with the issue, it would exist until mid twenty first century. Fifteen years after this statement, management of contaminated soils and waters still continue to be a current environmental issue due to the great number of areas around the world that face this problem [107].

In the countries member of the EEA (European Environment Agency), according to estimates performed in 2007, about 250,000 areas need remediation. Potentially toxic activities occurred in about 3 million areas, which are under study to determine the need for remediation and, if this tendency continues, the number of areas requiring remediation will increase in 50% until 2025. In these countries, approximately 35% of the costs with remediation were public [108].

In the USA, the report of USEPA [109] state that, despite much has already been done in the last decades of the last century, a considerable amount of work is still needed. According to the report, about 300,000 areas will still need remediation in the next three decades. The estimate cost for the remediation of these areas is around 209 billion dollars, funded by the responsible for the contamination, private or public entity.

According to Fernandes et al. [110], the resources needed, both human and economic, to overcome the challenges in the implementation of remediation programs can be great. The resources destined to this purpose will not be the same in different countries. Some countries are more prepared to deal with the costs of the remediation programs in relation to others, since they have appropriate mechanisms (technical and economic) to implement projects on a large scale. In the countries where this is not possible, the existence of contaminated areas should be a livelong problem.

New remediation technologies are under development in the physical and/or chemical areas, however most of them are still in the initial phase of elaboration [111]. However, the trend of emerging technologies are focused in methods in which the contaminants can be destroyed or carefully removed with low risk of secondary contamination [112]. The methodologies traditionally used, physical and chemical, simply transfer the contaminants, creating other sources of contamination and not eliminating the problem [113].

According to Koenigsberg et al. [114] there is the intention to use tools of molecular biology of microorganisms in contaminated areas, which can and must influence the conception and management of bioremediation engineer and open new paradigms so that the closure of a contaminated site does not occur.

Due to this, in the last years, the bioremediation methodologies have a significative portion of the remediation market [112]. According to Singh et al. [115], bioremediation entered in a new era with the use of genetically modified bacteria, however its use is still limited due to the fact that environmental factors can interfere in the process, making the results unpredictable. A study performed by Liu et al. [116] using genetically modified bacteria showed that its use is a promising strategy in the bioremediation process of environments contaminated by arsenic. Other technologies of bioremediation in development include the use of protein engineering, metabolic engineering, transcriptome and proteomics [117, 118].

For the development of tests in field using genetically modified bacteria, the major obstacle is the environmental concern and political restrictions for the use of these organisms [113]. As most of the researches on this theme are still basic, there is a growing need for regulatory and cost protocols and, thus, transform this potential technology into reality [119].

6. CONCLUSION

Contamination of the soil by petroleum and heavy metals has shown to be one of the major environmental problems that the governments and researchers must solve in the next decades. Several studies available in the literature warn about the negative effects of these substances in the living organisms, mainly in terrestrial invertebrates, since they are in direct contact with the contamination. In order to avoid that this problem become more serious, several remediation technologies have been elaborated and improved. Physical and chemical techniques are very used, however bioremediation, as it is ecologically correct, has gained great prominence, both in the remediation of petroleum and heavy metals.

Soil remediation standards are based on the protection of human health and on the protection of the ecosystem. Critical values for concentration in the soil are calculated based on human toxicology and others based on ecotoxicology. The most critical value is retained as soil remediation standard. The methodology for site specific risk assessment is based on the approach followed to derive soil remediation standards. A generic approach is followed for the derivation of soil remediation standards, while for site-specific risk assessments certain parameters, such as soil properties, can be evaluated.

REFERENCES

1. Sousa A, Pereira R, Antunes SC, Cachada A, Pereira E, Duarte AC, Gonçalves F. Validation of avoidance assays for the screening assessment of soils under different anthropogenic disturbances. Ecotoxicology and Environmental Safety 2008;71 661-670.
2. Nortcliff S. Standardisation of soil quality attributes. Agriculture, Ecosystems and Environment 2002;88 161–168.
3. Van Straalen. The use of soil invertebrates in ecological surveys of contaminated soils. Developments in Soil Science. 2004;29 159-195.
4. Fontanetti CS, Nogarol LR, Souza RB, Perez DG, Maziviero GT. Bioindicators and biomarkers in the assessment of soil toxicity. In: Pascucci S. (ed.) Soil Contamination. Rijeka: InTech; 2011. p.143-168.
5. Pascucci S., editor. Soil Contamination. Rijeka: InTech; 2011.
6. Andrade JÁ, Augusto F, Jardim ICSF. Biorremediação de solos contaminados por petróleo e seus derivados. Eclética química 2010;35(3) 17-43.
7. Souza TS, Christofoletti CA, Fontanetti CS. Ecotoxicological assays applied in soils contaminated by petroleum hydrocarbons. In: Visser J. (ed.) Ecotoxicology around the globe. New York: Nova Science Publishers; 2011. p247-26.
8. Aras S, Aydin SS, Körpe DA, Dönmez Ç. Comparative genotoxicity analysis of heavy metal contamination in higher plants. In: Begun G (ed.) Ecotoxicology. Rijeka: InTech; 2012. p.107-124.
9. Mulligan CN, Yong RN, Gibbs BF. Remediation technologies for metal-contaminated soils and groundwater: an evaluation. Engineering Geology 2001;60 193-207.
10. Manahan SE., editor. Introduction to Chemistry: Fundaments of Environmental Chemistry. Boca Raton; 2001.
11. Higarashi MM. Processos oxidativos avançados aplicados à remediação de solos brasileiros contaminados com pesticidas. PhD thesis. Universidade Estadual de Campinas; 1999.
12. Silva DN, Zagatto PJP, Guardani R, Nascimento CAO. Remediação de Solos Contaminados com Linear Alquil Benzenos Usando Reagentes de Fenton. COBEQ, 2004.
13. McBride MB., editor. Environmental chemistry of soils. New York: University Press; 1994.
14. Otten A, Alphenaar A, Pijls C, Spuij F, Wit H. In Situ Soil Remediation. Kluwer Academic Publishers; 1997.
15. Lavonreti A. Comportamento de herbicidas no meio ambiente. In: Workshop sobre bidegradação - Embrapa, Jaguariuna, São Paulo. p. 81-83; 1996.
16. Azambuja E, Costa AFU, Nanni AS. O emprego da prospecção geofísica na avaliação de plumas de contaminação por hidrocarbonetos. In: XXVII Congresso Interamericano de Engenharia Sanitária e Ambiental. ABES, Porto Alegre, Rio Grande do Sul; 2000.
17. Copley J. Ecology goes underground. Nature 2000;406 452-454.
18. Lavelle P. Diversity of soil fauna and ecosystem function. Biology International 1996; 33 3-16.
19. Van Brummelen TC, Verweij RA, Wedzinga SA, Van Gestel CAM. Polycyclic aromatic hydrocarbons in earthworms and isopods from contaminated forest soils. Chemosphere 1996;32(2) 315-341.
20. Smith R, Pollard SJT, Weeks JM, Nathanail CP. Assessing significant harm to terrestrial ecosystems from contaminated land. Soil Use and Management 2006;21 527–540.
21. Edwards CA. Assessing the effects of environmental pollutants on soil organisms, communities, processes and ecosystems. European Journal of Soil Biology 2002;38 225–231.

22. Leblond D. (2008). Total world oil output to reach 95 million b/d by 2020. Oil Gas J 2008;106 32–33.
23. Ko JY, Day JH. A review of ecological impacts of oil and gas development on coastal ecosystems in the Mississipi Delta 2004;47 597-623.
24. Wang Y, Chen H, Chen H, Wu J. Influences of chronic contamination of oil field exploitation on soil nematode communities at the Yellow River Delta of China. Front. Biol. China 2009;4(3) 376-383.
25. Filimonova ZV, Pokarzhevskii AD. Enchytraeid Enchytraeus crypticus as a Test Organism for Crude Oil Contamination of Soil. Bull. Environ. Contam. Toxicol. 2000;65 407-414.
26. Hanna SHS, Weaver RW. Earthworm survival in oil contaminated soil. Plant and Soil 2002;240 127–132.
27. Faulkner BC, Lochmiller RL. Increased Abundance of Terrestrial Isopod Populations in Terrestrial Ecosystems Contaminated with Petrochemical Wastes. Arch. Environ. Contam. Toxicol. 2000;39 86–90.
28. Netto ADP, Dias JCM, Arbilla G, Oliveira LF, Barek J. 2000. Avaliação da contaminação humana por Hidrocarbonetos Policíclicos Aromáticos e seus derivados nitratos: Uma revisão metodológica. Química Nova 2000;23(6) 765-773.
29. Bispo A, Jourdain MJ, Jauzein M. Toxicity and genotoxicity of industrial soils polluted by polycyclic aromatic hydrocarbons (PAHs). Organic Geochemistry 1999;30 947-952.
30. Eisler R. Polycyclic aromatic hydrocarbon hazards to fish, wildlife, and invertebrates: a synoptic review. U.S. Fish and Wildlife Service Patuxent - Wildlife Research Center 1987.
31. United State Environment Protection Agency - Quality criteria for water 1986. USEPA 440/5-86-001. Washington, DC. 1986.
32. Erstfeld KM, Snow-Ashbrook J. Effects of chronic low-level PAH contamination on soil invertebrate communities. Chemosphere 1999;39(12) 2117-2139.
33. Sverdrup LE, Kelley AE, Krogh PH, Nielsen T, Jensen J, Scott-Fordsmand JJ, Stenersen J. Effects of eight polycyclic aromatic compounds on the survival and reproduction of the springtail Folsomia fimetaria L. (Collembola, isotomidae).Environ. Toxicol. Chem. 2001;20(6) 1332-1338.
34. Kaplan DL, Harteinstein R. Studies on monooxygenases and dioxygenases in soil macroinvertebrates and bacterial isolates from the gut of the terrestrial isopod, Oniscus asellus L. Comp. Biochem. Physiol. 1978;60 47-50.
35. Faber JH, Heijmans JSM. Polycyclic Aromatic Hidrocarbons in Soil Detritivores: specific accumulation and effects. In:Van Straalen NM (ed.) Bioindicator Systems for Soil Pollution. Dordrecht: Kluwer Academic Publishers; 1996. p.31-44.
36. Eijsackers H, van Gestel CAM, de Jonge S, Muijs B, Slijkerman DME. Polycyclic aromatic hydrocarbon-polluted dredged peat sediments and earthworms: a mutual interference. Ecotoxicology 2001;10 35-50.
37. Bowmer CT, Roza P, Henzen L, Degeling C. The development of chronic toxicological tests for PAH contaminated soils using the earthworm Eisenia fetida and the springtail Folsomia candida. TNO report IMW-R 92/387, The Netherlands, 1992.
38. Hund K, Traunsperger W. Ecotox-evaluation strategy for soil bioremediation exemplified for a PAH-contaminated site. Chemosphere 1994;29 371-390.
39. Potter CL, Glaser JA, Chang LW, Meier JR, Dosani MA, Hermann RF. Degradation of polynuclear aromatic hydrocarbons under bench-scale compost conditions. Environ. Sci. Technol. 1999;33, 1717-1725.
40. Eom IM, Rast C, Veber AM, Vasseus P. Ecotoxicity of a polycyclic aromatic hydrocarbon (PAH)-contaminated soil. Ecotoxicology and Environmental Safety 2007;67 190-205.
41. Blakely JK, Neher DA, Spongberg AL. Soil invertebrate and microbial communities, and decomposition as indicators of polycyclic aromatic hydrocarbon contamination. Applied Soil Ecology 2002;21 71-88.

42. Souza TS; De Angelis DF, Fontanetti CS. Histological and histochemical analysis of the fat body of Rhinocricus padbergi (Diplopoda) exposed to contaminated industrial soil. Water, Air and Soil Pollution 2011; 221 235-244.
43. Triebskorn R, Köhler HR, Zanh T, Vogt G, Ludwing M, Rumpf S, Kratzmann M, Alberti G, Storch V. Invertebrate cells as targets for Hazardous substances. Zeitschfirt fuer Angewandte Zoologie 1991;78 277-287.
44. Godoy JAP, Fontanetti CS. Diplopods as bioindicators of soils: analysis of midgut of individuals maintained in substract containing sewage sludge. Water, Air and Soil Pollution 2010;210 389-398.
45. Nogarol LR, Fontanetti CS. Acute and subchronic exposure of diplopods to substrate containing sewage mud: Tissular responses of the midgut. Micron 2010;41 239-246.
46. Souza TS, Fontanetti, CS. Morphological biomarkers in the Rhinocricus padbergi midgut exposed to contaminated soil. Ecotoxicology and Environmental Safety 2011; 74 10-18.
47. Souza RB, Fontanetti CS. Alterations in the fat body cells of Rhinocricus padbergi (Diplopoda) resulting from exposure to substrate containing sewage sludge. Microscopy and Microanalysis 2012;18 317-323.
48. Bonaventura C, Johnson FM. Healthy environments for healthy people: bioremediation today and tomorrow. Environmental Health Perspectives 1997;105 5-20.
49. Otitoloju AA, Ajikobi DO, Egonmwan RI. Histopathology and Bioaccumulation of Heavy Metals (Cu & Pb) in the Giant land snail, Archachatina marginata (Swainson). The Open Environmental Pollution & Toxicology Journal 2009;1 79-88.
50. Paoletti MG., editor. Invertebrate biodiversity as bioindicators of sustainable landscapes, practical use of invertebrates to assess sustainable land use. Amsterdam: Elsevier Science BV; 1999.
51. Nahamani J, Lavelle P. Effects of heavy metal pollution on soil macrofauna in a grassland of Northern France. European Journal of Soil Biology 2002;38 297- 300.
52. Grzés, I. M. Ant species richness and evenness increase along a metal pollution gradient in the Bolesław zinc smelter area. Pedobiologia 2009;53 65 – 73.
53. Del Val C, Barea JM, Azon-Aguilar C. Diversity of arbuscular mycorrhizal fungus populations in heavy- metal-contaminated soils. Applied and Environmental Microbiology 1999;65 718–723.
54. Beyrem H, Mahmoudi E, Essid N, Hedfi A, Boufahja F, Aissa P. Individual and combined effects of cadmium and diesel on a nematode community in a laboratory microcosm experiment. Ecotoxicology and Environmental Safety 2007;68 412–418.
55. Syrek D, Weiner WM, Wojtylak M, Olszowska GY, Kwapis Z. Species abundance distribution of collembolan communities in forest soils polluted with heavy metals. Applied Soil Ecology 2006;31 239–250.
56. Piola RF, Johnston EL. Pollution reduces native diversity and increases invader dominance in marine hard-substrate communities. Diversity Distributions 2008;14 329–342.
57. Russell DJ, Alberti G. Effectsoflong-term, geogenic heavy metal contamination on soil organic matter and microarthropod communities, in particular Collembola. Applied Soil Ecology 1998;9 483–488.
58. Migliorini M, Pigino G, Bianchi N, Bernini F, Leonzio C. The effects of heavy metal contamination on the soil arthropod community of a shooting range. Environmental Pollution 2004;129 331–340.
59. Koehler HH. Soil animals and bioindication, Bioindicator systems for soil pollution. In: Van Straalen, DAKNM, Krivolutsky DA. (eds.) Dordrecht; 1996.
60. Nobre MM, Nobre RCM. Remediação de solos - Técnicas alternativas melhoram desempenho. Revista Química e Derivado 2003;417.
61. Koustas RN, Fischer D. Review of separation technologies for treating pesticide-contaminated soil. Journal of the Air & Waste Management Association 1998;48 434-440.

62. Policarpo N A. Tratamento de solos contaminados com bifenilas policloradas (PCBs). PhD thesis. Univesridade de São Paulo; 2008.
63. Pedrazzoli CD. Remediação eletrocinética de chumbo em resíduos industriais. PhD thesis. Universidade Federal do Paraná; 2004.
64. Heineck KS, ConsolI NC, Cruz RC, Knop A. Encapsulamento de um solo contaminado por óleo diesel. Revista Geotecnia 2007;110 3-17.
65. United State Environmental Protection Agency. Engineering Bulletin: Technology Alternatives for the Remediation of Soils Contaminated with Arsenic, Cadmium, Chromium, Mercury, and Lead. USEPA - Office of Emergency and Remedial Response. 1997.
66. United State Environmental Protection Agency. Engineering Bulletin: Technology Alternatives for the Remediation of Soils Contaminated with Arsenic, Cadmium, Chromium, Mercury, and Lead. USEPA - Office of Emergency and Remedial Response. 1997.
67. Watts RJ. In situ application of modified fenton's reagent spells success for BETX, MBTE site cleanup. Hazardous Waste Consulant 1998;3.
68. United State Environmental Protection Agency. In Situ Chemical Oxidation. USEPA 2005.
69. Mattos IL, Shiraishi KA, Braz AD, Fernandes JR. Peróxido de hidrogênio: importância e determinação. Química Nova 2003;26(3) 273-380.
70. Miller CM, Valentine RL, Roehl ME, Alvarez PJJ. Chemical and microbiological assessment of pendimethalin-contaminated soil after treatment with fenton's reagent. Water Research 1996;20(11) 2579-2586.
71. Muszkat L, Bir L, Feigelson L. Solar photocatalytic mineralization of pesticides in polluted waters. Journal of Photochemistry and Photobiology A: Chemistry 1995;87 85-88.
72. Yeh CK, Wu H, Chen T. Chemical oxidation of chlorinated non-aqueous phase liquid by hydrogen peroxide in natural sand systems. Journal of Hazardous Materials B. 2003;96 29-51.
73. United State Environmental Protection Agency - FRTR (Federal Remediation Technologies Roundtable): Abstracts of Remediation Case Studies and Technology Assessment – fact sheet. USEPA 2002;6.
74. Gullet TBK, Lemieux PM. Role of combustion and sorbent parameters in prevention of polychlorinated dibenzo-p-dioxin and polychlorinated dibenzofuran formation during waste combustion. Environmental Science Technology 1994;29(1) 107-118.
75. Henner P, Schiavon M, Morel JL, Lichtfouse E. Polycyclic aromatic hydrocarbons (PAH) occurrence and remediation methods. Analusis Magazine 1997;25(9-10) 1-8.
76. Ho TC, Chu HW, Hopper JR. Metal volatilization and separation during incineration. Waste Management 1993;13 455-466.
77. Yeung PY, Johnson RL, Xu JG. Biodegradation of petroleum hydrocarbons in soil as affected by heating and forced aeration. Journal of Environmental Quality 1997;26 1511–1516.
78. Molina-Barahona L, Rodrýguez-Vázquez R, Hernández-Velasco M, Vega-Jarquýn C, Zapata-Pérez O, Mendoza-Cantú A, Albores A. Diesel removal from contaminated soils by biostimulation and supplementation with crop residues. Applied Soil Ecology 2004;27 165–175.
79. Glazer AN, Nikaido H., editors. Microbial Biotechnology – Fundamentals applied of microbiology. New York: WH Freeman and Company Editors; 1995.
80. Megharaj M, Ramakrishnan B, Venkateswarlu K, Sethunathan N, Naidu R. Bioremediation approaches for organic pollutants: a critical perspective. Environmental International 2011;37 1362-1375.
81. Bernoth L, Firth I, Mcallister PE, Rhodes S. Biotechnologies For Remediation and Pollution Control in the Mining Industry. Minerals & Metallurgical Processing 2000;17(2) 105-111.
82. Sarkar D, Ferguson M, Datta R, Birnbaum S. Bioremediation of petroleum hydrocarbons in contaminated soils: Comparison of biosolids addition, carbon

supplementation, and monitored natural attenuation. Environmental Pollution 2005;136 187-195.
83. Hamdi H, Benzarti S, Manusadzianas L, Aoyama I, Jedidid N. Soil bioaugmentation and biostimulation effects on PAH dissipation and soil ecotoxicity under controlled conditions. Soil Biology and Biochemistry 2007;39 1926–1935.
84. Berger TM. Biorremediação de solos contaminados com hidrocarbonetos totais de petróleo – enfoque na aplicação do processo Terraferm. PhD Thesis. Universidade Federal do Rio Grande do Sul; 2005.
85. Mohan SV, Kisa T, Ohkuma T, Kanaly R.A, Shimizy Y. Bioremediation technologies for treatment of PAH-contaminated soil soil and strategies to enhance process efficiency. Rev Environ Sci Biotechnol. 2006;5 347-374.
86. Parrish ZD, Banks MK, Schwab AP. Effectiveness of Phytoremediation as a Secondary Treatment for Polycyclic Aromatic Hydrocarbons (PAHs) in Composted Soil. International Journal of Phytoremediation 2004;6(2) 119–137.
87. Martins A, Dinardi AL, Formagi VM, Lopes TA, Barros RM, Coneglian CMR, Brito NN, Sobrinho GD, Tonso S, Pelegrini R. Biorremediação. III Fórum de Estudos Contábeis, Faculdades Integradas Claretianas, Rio Claro, SP. www.ceset.unicamp.br /lte/artigos/3fec2401 (accessed 20 June 2012).
88. Maila MP, Cloete TE. The use of biological activities to monitor the removal of fuel contaminants – perspective for monitoring hydrocarbon contamination: a review. International Biodeterioration and Biodegradation 2005;55 1-8.
89. Lau KL, Tsang YY, Chiu SW. Use of spent mushroom compost to bioremediate PAH-contaminated samples. Chemosphere 2003;52 1539–1546.
90. Moretto LM, SIlvestri S, Ugo P, Zorzi G, Abbondanzi F, Baiocchi C, Iacondini A. Polycyclic aromatic hydrocarbons degradation by composting in a soot-contaminated alkaline soil. Journal of Hazardous Materials 2005;126(1–3) 141–148.
91. Van Gestel K, Mergaert J, Swings J, Coosemans J, Ryckeboer J. Bioremediation of diesel oil-contaminated soil by composting with biowaste. Environmental Pollution 2003;125 361–368.
92. Antizar-Ladislao B, Lopez-Real J, Beck AJ. Laboratory studies of the remediation of polycyclic aromatic hydrocarbon contaminated soil by in-vessel composting. Waste Management 2005;25(3) 281–289.
93. Haderlein A, Legros R, Ramsay BA. Pyrene mineralization capacity increases with compost maturity. Biodegradation 2005;17 293-303.
94. Kirchmann, H, Ewnetu W. Biodegradation of petroleum-based oil wastes through composting. Biodegradation 1998;9 151–156.
95. Parrish ZD, White JC, Isleyen M, Gent PNM, Iannucci-Berger W, Eitzer BD, Kelsey JW, Mattina MI. Accumulation of weathered polycyclic aromatic hydrocarbons (PAHs) by plant and earthworm species. Chemosphere 2006; 64 609-618.
96. Straube WL, Nestler CC, Hansen LD, Ringleberg D, Pritchard PH, Jones-Meehan J. Remediation of polyaromatic hydrocarbons (PHAs) trough landfarming with biostimulation and bioaugmentation. Acta Biotechnologica 2003;23 179-196.
97. Jorgensen KS, Puustinen J, Suortti AM. Bioremediation of petroleum hydrocarbon-contaminated soil by composting in biopiles. Environmental Pollution 2000;107 245-254.
98. Bento FM, Camargo FAO, Okeke BC, Frankenberger WT. Comparative bioremediation of soils contaminated with diesel oil by natural attenuation, biostimulation and bioaugmentation. Bioresource Technology 2005;96 1049-1055.
99. Mohammed AS, Kapri A, Goel R. Heavy Metal Pollution: Source, Impact, and Remedies. Earth and Environmental Science 2011;10 1-28.
100. Abioye P. Biological remediation of hydrocarbon and heavy metals contaminated soil. In: Pascucci S (ed.) Soil Contamination. Rijeka: InTech; 201. P.127-142.
101. Garbisu C, Alkorta I. Basic concepts on heavy metal soil bioremediation. The European Journal of Mineral Processing and Environmentak Protection 2003;3(1) 58-66.

102. Kumar N, Dushenkov V, Motto H, Raskin I. Phytoextraction: the use of plants to remove heavy metals from soils. Environ. Sci. Technol. 1995;29 1232-1238.
103. Lovley DR, Coates JD. Bioremediation of metal contamination. Environmental Biotechonology 1997;8 285-289.
104. Congeevaram S, Dhanarani S, Park J, Dexilin M, Thamaraiselvi K. Biosorption of chromium and nickel by heavy metal resistant fungal and bacterial isolates. Journal of Hazardous Materials 2007;146 270–277.
105. Singh P, Cameotra SS. Enhancement of metal bioremediation by use of microbial surfactants. Biochemical and Biophysical Research Communications 2004;319 291–297.
106. Bredehoeft JD. Hazardous waste remediation: a 21st century problem. Ground Water Monitoring & Remediation 1994; 95-100.
107. Lemming G. Environmental assessment of contaminated site remediation in a life cycle perspective. PhD thesis. Technical University of Denmark; 2010.
108. Europen Environmental Agency. EEA: Progress in management of contaminated sites (CSI 015). http://www.eea.europa.eu/data-and-maps/indicators/progress-in-management-of-contaminated-sites/progress-in-management-of-contaminated-1 (accessed 19 July 2012).
109. United States Environmental Protection Agency – USEPA. Cleaning up the nation's waste sites: markets and technology trends. Office of Solid Waste and Emergency Response 2004.
110. Fernandes HM, Recio MS, Forsstrom H, Carson PM. International cooperation and support in environmental remediation e is there any room for improvement? Journal of Environmental Radioactivity 2011;
111. Olexsey RA, Parker RA. Current and future in situ treatment techniques for the remediation of hazardous substances in soil, sediments, and groundwater. Soil and Water Pollution Monitoring, Protection and Remediation 2006; 3–23.
112. Lynch LJ, Moffat AJM. Bioremediation – prospects for the future application of innovative applied biological research. Annals of Applied Biology 2005;146 217–221
113. Shukla KP, Singh NK, Sharma S. Bioremediation: Developments, Current Practices and Perspectives. Genetic Engineering and Biotechnology Journal 2010; 1-20.
114. Koenigsberg SS, Hazen TC, Peacock AD. Environmental Biotechnology: A Bioremediation Perspective. Environmental Biotechnology 2005;
115. Singh S, Mulchandani A, Chen W. Highly Selective and Rapid Arsenic Removal by Metabolically Engineered Escherichia coli Cells Expressing Fucus vesiculosus Metallothionein. Applied and Environmental Microbiology 2008;74(9) 2924–2927.
116. Liu L, Zhang F, Chen J, Sun G. Arsenic removal from contaminated soil via biovolatilization by genetically engineered bacteria under laboratory conditions. Journal of Environmental Sciences 2011;23(9) 1544–1550.
117. Alcade M, Ferrer M, Plou FJ, Ballesteros A. Environmental biocatalysis: from remediation with enzymes to novel green processes. Trends in Biotechnology 2006;24 281-287.
118. Woods TK. Molecular approaches in bioremediation. Current Opinion in Biotechnology. 2008;19 572-578.
119. Singh JS, Abhilash PC, Singh HB, Singh RP, Singh DP. Genetically engineered bacteria: An emerging tool for environmental remediation and future research perspectives. Gene 2011;480 1–9.

CHAPTER 16

Metabolomics for Soil Contamination Assessment

Maria C. Hernandez-Soriano[1] and Jose C. Jimenez-Lopez[2]

[1] *School of Agriculture and Food Sciences, The University of Queensland, St Lucia QLD, Australia*

[2] *The UWA Institute of Agriculture, The University of Western Australia, Crawley, Perth WA, Australia*

1. INTRODUCTION

The evaluation of biological responses to assess and predict the impact of environmental changes in ecosystems functioning is receiving increasing attention, and current research focuses on overcoming concerns about the specificity of biomarkers (Amiard-Triquet et al., 2012). Generally, biomarkers are chemicals, metabolites, susceptibility characteristics, or physiological changes that relate to the exposure of an organism to a chemical. Accordingly, a selected biomarker, i.e. biological response, can be linked to a specific environmental exposure, being representative of the health status of the ecosystem studied. The identification of biomarker profiles has been possible upon the development of metabolomics. Those profiles allow the genuine identification of the relevant biological response/s associated to a particular exposure while the assessment of a single biomarker could only estimate the potential response of the ecosystem to a particular pollutant.

Metabolomics is the generic name assigned to a scientific field that addresses the characterization of low molecular weight organic metabolites released by living organisms in response to environmental stimuli. Morrison et al. (2007) provided an extended definition "the application of metabolomics to the investigation of both free-living organisms obtained directly from the natural environment (whether studied in that environment or transferred to a laboratory for further experimentation) and of organisms reared under laboratory conditions (whether studied in the laboratory or transferred to the environment for further experimentation), where any laboratory experiments specifically serve to mimic scenarios encountered in the natural environment".

The methodological approach of metabolomics relies on a comprehensive analysis of the set of metabolites or "metabolome" produced in response to particular environmental stimuli. Accordingly, the metabolome is the pool of metabolites, small molecules, within a cell, tissue, organ, biological fluid, or entire organism (Miller, 2007). Exposure of an organism to an external stressor will result in changes at the level of the metabolome (Ankley et al., 2006; van Ravenzwaay et al., 2007), and such changes may constitute a highly sensitive indicator of an external stress. Therefore, metabolomics has potential as a sensitive and rapid technique that can elucidate the relationships between metabolite levels and an external stressor, such as contaminant exposure, nutritional deficit or a disease.

The main advantage of metabolomics over traditional research is to overcome the bias associated to the assessment of predefined metabolites (Singh, 2006). Among the diverse applications of omic profiling methods to the environmental sciences, ecotoxigenomics addresses the response of organisms to pollutants based on the different sensitivity of species to toxicants (Spurgeon et al., 2008). Currently, the implementation of metabolomics in ecological risk assessment is still at an early stage, mostly applied as a screening tool to assess the potential toxic effects of pollutants or to determine the mode of action (MOA) of a toxicant. Otherwise, application of metabolomics for environmental monitoring allows the study of a large variety of species from relatively uncontrolled environments.

Bundy et al. (2009) highlight the challenge of identifying a large number of metabolites and the necessity of creating metabolite databases specifically dedicated to environmental issues. Multivariate statistical analysis has proved highly effective for metabolite identification. Thus, principal component analysis (PCA) is used to identify differences between metabolic profiles of organisms exposed to organic or inorganic pollutants (Jones et al., 2014; Kwon et al., 2012; Lankadurai et al., 2011). Besides, association between the metabolic profile and biological factors evaluated as markers for exposure to pollutants can be modelled by partial least squares (PLS) regression analysis (Ellis et al., 2012).

The implementation of metabolomics for the assessment of soil contamination is nevertheless at an early stage (Viant, 2009). A basic screening of published research in the web of science returns circa 100 items for the search "soil pollution-metabolomics", with a significant launch in 2011 (Figure 1), reduced to 21 records when the search is narrowed with the term "biomarkers", published in the period 2007-2013. However, emerging regulatory challenges demand the advance of toxicity testing. Toxicogenomics tools have been presented as an advanced from the current methodologies used for regulatory decision making in ecotoxicology, which entirely rely on whole animal exposures and adverse effects on survival, growth, and reproduction (Ankley et al., 2006). From the acquisition of reproducible metabolic profiles as response to the presence of specific pollutants in soil (Jones et al., 2008) to the application of metabolomics techniques to the study of the response of the entire community of a soil to factors such as pollution and climate change (Jones et al., 2014), the implementation of metabolomics in ecotoxicology is a sound answer to the current needs of society and the environment (van Ravenzwaay et al., 2012).

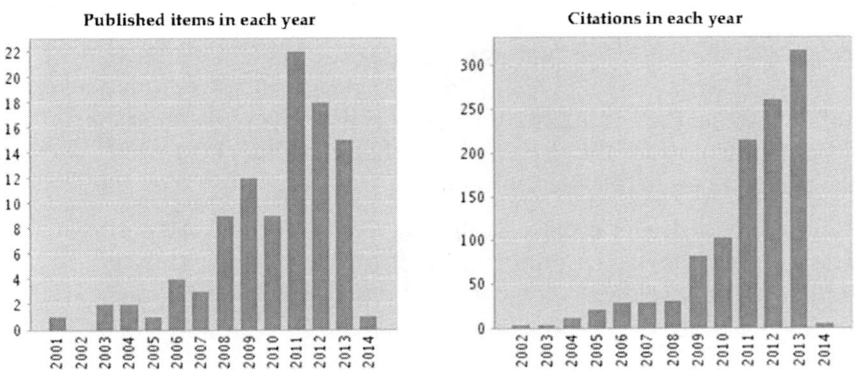

Figure 1. Citation report for the search "soil pollution-metabolomics" as obtained from Thomson Reuters (January 2014).

During the last decade a number of general revisions about the application of metabolomics in environmental health assessment have been published (Bundy et al., 2009; Miller, 2007; Snape et al., 2004; van Ravenzwaay et al., 2007; Viant et al., 2003). The present review specifically summarizes the most significant research concerning implementation of metabolomics in soil contamination assessment. The main objectives of this revision are i) to provide a systematized outline for the application of metabolomics in risk assessment of soil contamination, ii) to provide a rapid guide to the methodological approaches currently optimized and iii) to unify and simplify the knowledge currently available in the topic to provide an accessible tool for further advance in the implementation of metabolomics in risk assessment.

2. METHODOLOGICAL APPROACHES

2.1. Metabolites Isolation

Generally, metabolites are extracted from intact organisms (occasionally from selected relevant tissues) that have been exposed to the studied toxicant by moderate chemical extraction (Baylay et al., 2012; Yuk et al., 2010). The organisms commonly selected for toxicity testing are earthworms (Table 1), particularly the genus *Eisenia*, which are a classic model organism for toxicity assays (Sanchez-Hernandez, 2006; Van Gestel et al., 1992; van Gestel et al., 1989) and have been since long included in official guidelines (OECD, 1984, 2004). Earthworms ingest large amounts of soil and uptake a significant amount of contaminant through the skin. Therefore they are continuously exposed to contaminants. Extractions performed with methanol-chloroform (Baylay et al., 2012) or phosphate buffer solution (Yuk et al., 2010) on pulverized or lyophilised organisms are described to extract the maximum number of metabolites while allowing the performance of reliable analyses.

The isolated extracts usually might not require further sample treatment prior to analysis, which minimizes the introduction of artefacts but also facilitates the development of low cost, rapid methodologies.

2.2. Metabolites Determination: Chromatography, Spectroscopy And Spectrometry

The leading analytical techniques in metabolomics for soil contamination assessment are proton nuclear magnetic resonance spectroscopy (^{1}H NMR) and gas chromatography–mass spectrometry (GC-MS), as thoroughly reported in Table 1, both allowing the identification of compounds at molecular level in the analysed substances. Several authors have also implemented high pressure liquid chromatography (HPLC) and ultra high pressure liquid chromatography (UPLC) coupled with mass spectrometry detector (MS) for the assessment of biological responses to soil contamination with heavy metals (Hédiji et al., 2010; Hughes et al., 2009). Overall, these analytical techniques allow the determination and identification of the metabolites that foremost represent the metabolic alterations related to the toxic effects of organic or inorganic contaminants in soil.

2.3. Metabolomics Data Analysis

The general approach to data analysis in metabolomics can be summarized in three main stages: explorative, supervised and biological interpretation (Smilde et al., 2010). The explorative phase aims to find groups, clusters and outliers in metabolites and samples studied while the supervised discriminates two or more groups to make predictive models and to find biomarkers (Amiard-Triquet et al., 2012; Dallinger-Marianne, 2000; van Ravenzwaay et al., 2007). Multivariate methods are currently preferred, although univariate and semi-univariate methods have been commonly used for selecting biomarkers. For instance, the lysosomal system was identified as a particular target for the toxic effects of pollutants in soil organisms. However, it is nonspecific as a marker and only included in a

Table 1. Metabolic responses of test organism following exposure to selected environmental contaminants in contact tests.

Technique	Organic toxicant	Animal model tested	Biomarkers of contaminant exposure	Reference
^{1}H NMR GC/MS	2-fluro-4-methylaniline	*E. veneta*	[D][2-hexyl-5-ethyl-3-furansulfonate, maltose], [I]inosine monophosphate	(Bundy et al., 2002)
^{1}H NMR GC/MS	3-triflurομethyaniline	*E. veneta*	[I]lactate	(Lenz et al., 2005)
^{1}H NMR GC/MS	3-trifluromethylaniline	*E. veneta*	[I][Ala, Gly, Asn, glucose, citrate, succinate]	(Warne et al., 2000)
^{1}H NMR GC/MS	3-fluro-4-nitrophenol	*E. veneta*	[D][acetate, malonate], [I][succinate, trimethylamine-N-oxide]	(Bundy et al., 2001)
^{1}H NMR GC/MS	3,5-difluroaniline	*E. veneta*	[D]2-hexyl-5-ethyl-3-furansulfonate, [I]inosine monophosphate	(Bundy et al., 2002)
^{1}H NMR GC/MS	4-fluroaniline	*E. veneta*	[D][Maltose, hexyl-5-ethyl-3-furansulfonate]	(Bundy et al., 2002)
^{1}H NMR	Aroclor 1254	*E. fetida*	No significant changes	(Fitzpatrick et al., 1992)
^{1}H NMR	Caffeine	*E. fetida*	[I]fumarate	(McKelvie et al., 2011)
^{1}H NMR	Carbamazephine	*E. fetida*	[D][Fumarate, Glu, Val, [D]Leu]	McKelvie et al. (2011)
^{1}H NMR	Carbaryl	*E. fetida*	[D][Phe, Tyr, Lys, Ala, Val, Leu]	(Heimbach, 1988) (Edwards and Bater, 1992)
^{1}H NMR	Chlorpyrifos	*E. fetida*	No significant changes	(Yu et al., 2006)
^{1}H NMR	Chlorpyrifos	*L. rubellus*	[I]fumarate	(Baylay et al., 2012)

^{1}H NMR spectroscopy, GC–MS	Chlorpyrifos	*C. elegans*	^{I}[Ala, betaine, ornithine], ^{D}[choline, Glu, Gly, Ile, lactate, n-buterate, taurine]	(Baylay et al., 2012)
^{1}H NMR	Dimethyl phthalate	*E. fetida*	^{D}[Phenylalanine, Alanine, Leucine, Valine]	(Drewes and Vining, 1984)
^{1}H NMR GC/MS	DTT and Endosulfan	*E. fetida*	^{I}[maltose, Ala, Leu]	(McKelvie et al., 2009)
1-D & 2-D ^{1}H NMR spectroscopy	Endosulfan	*E. fetida*	Significant fluctuations in glutamine/GABAeglutamate cycle metabolites and spermidine	(Yuk et al., 2013)
1-D & 2-D ^{1}H NMR spectroscopy	Endosulfan Sulfate	*E. fetida*	Significant fluctuations in glutamine/GABAeglutamate cycle metabolites and spermidine	(Yuk et al., 2013)
^{1}H NMR	Estrone	*E. fetida*	^{D}[Adenine, Glu]	McKelvie et al. (2011)
^{1}H NMR GC-MS analysis	Imidacloprid/ Thiacloprid	*L. rubellus*	^{D}Toxicological endpoints (survival, weight loss, and reproduction)	Baylay et al. (2012)
^{1}H NMR	Naphthalene	*E. fetida*	^{D}[Ala, Leu, Val, Lys]	(Brown et al., 2009)
^{1}H NMR	Nonylphenol	*E. fetida*	^{D} [adenine, Glu]	McKelvie et al 2011
^{1}H NMR	Polybrominated diphenyl ethers (PBDE) 209	*E. fetida*	^{D}Maltose, ^{I}[Lys, Glu]	McKelvie et al. (2011)
^{1}H NMR	Perfluorooctanoic acid	*E. fetida*	^{I}[succinate, HEFS, Glu], ^{D}[leu, Val, LyS, Phe, Arg, maltose, ATP]	(Lankadurai et al., 2012)
^{1}H NMR	Perfluorooctane sulfonate	*E. fetida*	^{I}[succinate, HEFS, Glu], ^{D}[leu, Val, Lys, Phe, Arg, maltose, ATP]	(Lankadurai et al., 2012)
^{1}H NMR	PHA	*E. fetida*	Leu, Val, Ala, Lys and maltose changed in response to PAH exposure.	(Brown et al., 2009)
^{1}H NMR	Phenanthrene	*E. fetida*	^{I}[Ala, betaine, Scyllo- and myo-inositol, cholesterol, phosphatidylcholine], ^{D}Glu	(Lankadurai et al., 2011) Brown et al. (2010)

^{1}H NMR	Polychlorinated biphenyl	*E. fetida*	^{I}ATP	McKelvie et al. (2011) (Whitfield Åslund et al., 2011)
^{1}H NMR	Poly brominated diphenyl ethers	*E. fetida*	^{I}[lys, Glu], ^{D}maltose	McKelvie et al. (2011)
^{1}H NMR GC/MS	Pyrene	*L. rubellus*	^{D}[lactate, tetradecanoic acid, hexadecanoic acid, octadecanoic acid], ^{I}[Ala, Leu, Val, Ile, Lys, Tyr, methionine]	(Jones et al., 2008)
1-D & 2-D ^{1}H NMR spectroscopy	Rifluralin	*E. fetida*	^{I}[Ala, Gly, maltose, ATP]	Yuk et al. (2011)
^{1}H NMR	Thiacloprid	*L. rubellus*	No significant changes were reported	Baylay et al. (2012)
1-D & 2-D ^{1}H NMR spectroscopy	Trifluralin	*E. fetida*	^{I}[Ala, Gly, ATP], ^{D}maltose	Yuk et al. (2011)
^{1}H NMR spectroscopy and UPLC-MS	Cd	*C. elegans*	^{I}phytochelatins, ^{D}cystathionine	(Hughes et al., 2009)
^{1}H NMR, HPLC-PDA	Cd	*S. lycopersicum*	^{I}choline, ^{D}[glucose, citrate, malate, glutamine, asparagine, Phe, Tyr, Val, Ile, trigonelline]	(Hédiji et al., 2010)
^{1}H NMR, GC-MS analysis	Chlorpyrifos + Ni	*L. rubellus*	^{I}[Phe, glucose, malate, arachidonic acid, fumarate, Lys, Tyr, monosaccharides, monophosphorylated form of inositol, succinate, uracil, ethanolamine, Pro, putrescine], ^{D}myo-inositol,	(Baylay et al., 2012)
^{1}H NMR spectroscopy, GC–MS	Chlorpyrifos + Ni	*C. elegans*	^{I}[Ala, creatine, His, lactate, betaine, carnosine], ^{D}[choline, Gly, Ile, leu, lys, Val]	(Jones et al., 2012)
^{1}H NMR	Cu (II)	*L. rubellus*	^{I}His	(Gibb et al., 1997)

^{1}H NMR spectroscopy, GC–MS	Ni	*C. elegans*	^{I}[Asn, choline, Gln, lactate, succinate], ^{D}[Ala, Ile, Gly, Val]	Jones et al. (2012)
^{1}H NMR	Pb, Zn	*S. salsa*	i) Exposure to Pb - ^{D}Tyr. ii) Exposure to Zn - ^{I}[Val, Ile, Leu, Thr, Ala, Asn, Phe, fosfocoline], ^{D}[acetate, Gly, glucose, fumarate and ferulate] iii) Exposure to Pb & Zn - ^{I}[Ala, Asn, Tyr, Phe], ^{D}[acetate, succinate, Asp, malonate, fructose, glucose, fumarate and ferulate]	(Wu et al., 2013)
^{1}H NMR	Mixed metals (Cadmium, copper, lead and zinc)	*L. rubellus*	^{I} [maltose, His]	(Bundy et al., 2004)
^{1}H NMR	Tellurite	*P. pseudoalcaligenes*	^{I}[Thr, Leu, Tyr, betaine, Ser, Lys, Ile, Ala, Arg, Val, glutathione, adenosine], ^{D}[Lactate, Gly]	(Tremaroli et al., 2009)
^{1}H NMR	Titanium dioxide	*E. fetida*	^{I} [Phe, Tyr, Lys, Glu, Ala, lactate, Val, Leu], ^{D}maltose	(Whitfield Aslund et al., 2012)

D = decrease; I = increase

suite of biomarkers among diverse soil invertebrate species can provide the necessary specificity for risk assessment purposes (Kammenga et al., 2000). Finally, the biological interpretation seeks the links between metabolome data and underlying metabolic networks through metabolite set enrichment, pathway analysis and metabolic network inference (Trygg et al., 2006). Thus, finding metabolite relationships is essential to determine comprehensive and meaningful metabolic changes as biological response to environmental stimuli (Ellis et al., 2012; Morrison et al., 2007). Accordingly, such extensive evaluation of the impact of pollutants in the metabolism of target organisms is the approach that can add value to the assessment of soil health and viability of soil organisms undergoing stress from pollution.

3. METABOLOMICS BIOINFORMATICS

Information processing by bioinformatics tools and computational biology methods has become essential for solving complex biological problems in genomics, proteomics, and metabolomics. Understanding "omics" data requires both common statistical and computational based methods due to the multi-dimensional and complexity level of the data.

Data-analytical methods for the study of biological systems as developed in the field of computational biology provide a suit of indispensable tools to survey the outcome of metabolomics studies. First, computational biology allows a fast screening of the large biological and chemical data sets generated (Shulaev, 2006), and therefore the identification of the most relevant metabolites, i.e. compounds specifically representative of the metabolic changes in the model system following exposure to different concentrations of organic and inorganic toxicants. As a result of the large number of variables (metabolites) studied, metabolomics studies encompass a significant statistical power for the systematic detection of biological responses to environmental changes (van Ravenzwaay et al., 2012). Second, the mathematical models developed in computational biology allow the identification of relationships between the external stimuli and the metabolic response (Zhang et al., 2010). Third, the implementation of computational algorithms to structural biology makes possible to discover the structure-function of new macromolecular compounds, the functional enzymatic conversion and changes in their activity, as well as their molecular interaction and relationship with others compounds in the pathways where they are involved (Jimenez-Lopez et al., 2013). Moreover, it is possible to detect patterns in such biological responses and establish significant dose-response relationships. Besides, pattern recognition reduces the metabolomics data from hundreds of variables to two or three components that are orthogonal to each other. Overall, this advance of computational biology has been possible due to three significant technological breakthroughs: high-information-content data streams, novel bio-statistical methods, and the computational power to analyse these data.

Data processing and statistical analyses are commonly performed using multivariate (typically a principal component analysis (PCA) and (or) partial least squares (PLS) regression analysis) and univariate (t-test) analyses (Brown et al., 2010; Jones et al., 2014; McKelvie et al., 2011; Yuk et al., 2013). These analyses are performed in combination with the quantification and identification of the metabolites. Subsequently, biological interpretation of the data is necessary for understanding the link between the external stimulus and the metabolic response of the organisms.

Principal component analysis is the most widely used multivariate statistical approach in metabolomics, used to explain the overall variability in a data set via a a set of uncorrelated variables called principal components (PCs), which are linear combinations of the original variables (Trygg et al., 2006). The organization of samples in PCA scores plots is based on the similarities between their metabolic profiles. Thus, PCA allows for dimensional reduction of the data into a low dimensional plane, such as PC1 versus PC2. The scores plot (e.g., PC1 versus PC2) allows for a visual examination of the relationship between the samples based on their metabolic profiles. In a 1-D PCA loadings plot, the contribution (or weight) of each metabolite to the discrimination of the sample classes along one component is represented by the intensity of the metabolite peak. In the 2-D PCA loadings plot discrimination is performed by selecting the points that are scattered further away from the tight cluster of points found near the origin.

Other widely used multivariate statistical tools in metabolomics are PLS regression analysis and PLS discriminant analysis (PLS-DA). Both PLS-regression and PLS-DA are methods for samples classification, with pre-defined variables added to maximize the separation between the sample classes and to construct predictive models. The predefined variables for PLS- regression are measurable quantities such as the contaminant exposure concentration. Validation methods such as the leave-one-out cross validation are used to test the robustness of the models generated by PLS-regression, PLS-DA, OPLS, and OPLS-DA (Whitfield Åslund et al., 2011).

Although metabolomics studies mostly use multivariate statistics, univariate statistical analyses can contribute to the information gained from a study. Thus, t-tests can be used to assess the significance of the separation between the controls and stressed organisms in PCA and PLS-DA scores plots. Also, t tests can be used to determine which metabolites in the ^{1}H NMR spectra of the treatment class increased or decreased significantly relative to the controls.

4. BIOMARKERS

The somewhat secondary significance of biological responses for soil contamination assessment was customarily associated to the limitation of biomarkers as measurable responses to contaminants, which classically could only provide an indication of exposure to contaminants in soil (Sanchez-Hernandez, 2006). The development of metabolomics, considered an "emerging field" as late as mid-2010, has provided the tools for the determination of multiple biomarkers across different levels of biological organization, and therefore a better assessment of the ecological consequences of contamination. Since the creation of the first metabolomics web database, METLIN (Smith et al., 2005), 60,000 metabolites has been incorporated, a rapid development closely related to the evolution of mass spectrometry instrumentation and data analysis tools. Currently, the number of databases and metabolites registered is continuously increasing. Table 2 summarizes some of the most relevant databases operative and the corresponding website is also indicated. Further information on metabolomics databases can be obtained from the metabolomics society (http://www.metabolomicssociety.org). For instance, ChemSpider is an aggregated database of organic molecules containing more than 20 million compounds from many different providers. At present the database contains information from such diverse sources as a marine natural products database, ACD-Labs chemical databases, the EPA's DSSTox databases and from a series of chemical vendors. It has extensive search utilities and most compounds have a large number of calculated physico-chemical property values.

One of the goals in bioinformatics is to establish automated and efficient ways to integrate large, biological datasets from multiple sources. This objective is challenging because data sources are heterogeneous in terms of their functions, structures, data access methods and dissemination formats. In addition, the enormous quantity of information produced by "omics" is handled via computers that systematically analyze and store the accumulating sequence, structure and function data. Databases are essential in metabolomics because they provide a rapid and specific tool to identify the compounds isolated from an organism exposed to a particular environmental challenge. Thus, the KNApSAcK package provides tool for analysing datasets of mass spectra as well as for retrieving information on metabolites by entering the name of a metabolite, the name of an organism, molecular weight or molecular formula. A list of metabolites that are associated to a taxonomic class can be obtained by search with the taxonomic name, from which information of individual metabolites can be retrieved. The NIST Chemistry WebBook provides access to chemical and physical property data for chemical species. The data provided in the site are from collections maintained by the NIST Standard Reference Data Program and outside contributors. Data in the NIST Chemistry WebBook can be found by direct searches for chemical species or indirect searches based on related data. Specific databases are also being developed, such as LIPID MAPS, currently the largest database of lipid molecular structures. Otherwise, SetupX combines mass spectrometric and biological metadata, which is a step forward in the organization of information generated by metabolomics analysis.

Table 2 . Selected metabolomic databases.

METLIN	http://metlin.scripps.edu/index.php
LIPID MAPS	http://www.lipidmaps.org/
KEGG	http://www.genome.jp/kegg/pathway.html
ChemSpider	http://www.chemspider.com/
SetupX	http://fiehnlab.ucdavis.edu/projects/binbase_setupx
KNApSAcK	http://kanaya.naist.jp/KNApSAcK/
NIST	http://webbook.nist.gov/chemistry/
MassBank	http://www.massbank.jp/
HMP	http://www.hmdb.ca/
IIMDB	http://metabolomics.pharm.uconn.edu/iimdb/

Metabolomic databases are thus accompanied by accurate description of the biological study design and accompanying metadata reporting on the laboratory workflow from sample preparation to data processing.

Currently, standard analyses focus on the determination of amino acids, mono- and disaccharides, lipids/fatty acids, short chain fatty acids and small phenolics. Accordingly, it is possible to already launch the standardization of metabolomics analysis. For instance, the Northwest Metabolomics Research Center (University of Washington) has established a relevant list of target compounds to evaluate biological responses to changes in the environment. The list of compounds is summarized in Table 3.

The information of metabolites and metabolic pathways has been obtained from the website of Kyoto Encyclopedia of Genes and Genomes (Kegg, http://www.genome.jp/kegg/). According to the research results summarized in Table 1, the implementation of metabolomics in the assessment of soil contamination indicates that contaminants in soil affect several of the major metabolic pathways in living organisms (Table 3), including glycolysis, trycarboxylic acids cycle and amino acids metabolism. Moreover, data analysis indicates an overall reduction in the production of the associated metabolites. For instance, the interference in amino acids specialized pathways results in a decreased synthesis of purine and pyrimidine nucleotides (Brown et al., 2010; McKelvie et al., 2011). These nucleotides are essential for the production of the energy (ATP molecules) that drive most of the enzymatic reactions in living organisms, but also protein synthesis is consequently hampered, which explain the negative effect in processes such as antioxidant activity.

Table 3. Summary of metabolites and metabolic pathways representative of biological responses to environmental stimuli.

Metabolic Pathways	Number of Metabolites
Alanine, aspartate and glutamate metabolism	15
Arginine and proline metabolism	23
Butanoate metabolism	18
Citrate cycle (TCA cycle)	11
Cysteine and methionine metabolism	14
Fatty acid metabolism	3
Glutathione metabolism	14
Glycine, serine and threonine metabolism	21
Glycolysis / Gluconeogenesis	16
Histidine metabolism	13
Lysine biosynthesis	7
Lysine degradation	6
Nitrogen metabolism	9
Oxidative phosphorylation	6
Pentose phosphate pathway	10
Phenylalanine metabolism	10
Phenylalanine, tyrosine and tryptophan biosynthesis	8
Purine metabolism	30
Pyrimidine metabolism	30
Pyruvate metabolism	10
Synthesis and degradation of ketone bodies	4
Tryptophan metabolism	15
Tyrosine metabolism	18
Valine, leucine and isoleucine biosynthesis	11
Valine, leucine and isoleucine degradation	5

Another emerging group of biomarkers, as highlighted in several studies, are lipids (Rochfort et al., 2009; Sanchez-Hernandez, 2006). Rochfort et al., (2009) indicate that lipophilic extracts can be used in field based metabolomics experiments to investigate different treatment effects on earthworms. Lipid metabolism is highly sensitive to environmental contaminants (Vega-López et al., 2013), with increasing production of lipoprotein vesicle and lipid peroxidation rate during early stages of the biological response to the presence of a toxicant (Lankadurai et al., 2011). Relatedly, earthworm esterases has been proposed as biomarkers for pesticide contamination in soil (Sanchez-Hernandez, 2010). Esterases are directly involved in the natural tolerance of earthworms to pesticides, and can therefore be used as specific biomarkers, but furthermore, their characterization by metabolomics approach might help to select the appropriate earthworm species for regulatory toxicity testing. Overall, the increasing specificity of the research performed in ecotoxigenomics will allow a realistic and meaningful incorporation of biological responses in ecological risk assessment.

5. OXIDATIVE STRESS IN CONTAMINATED SOIL

The induction of the oxidative stress response by the presence of toxic compounds in the environment is a primary mechanisms of defence, although prolonged exposure to contaminants is likely to overwhelm this short-term defence (Regoli et al., 2002).

Metabolites such as proline possibly detoxify the ROS under stress in vivo (Smirnoff, 1993). Exposure of plants to both redox active, for example, Cu and Hg, and other metals, for example, Cd and Zn, induces the generation of free radicals that leads to oxidative stress. This represents one of the major causes of toxicity particularly due to redox metals. The cells are equipped with an elaborate network of antioxidative enzymes and low molecular weight metabolites which mitigate the oxidative stress. Proline scavenges different free radicals in certain in vitro generation and detection systems.

Proline quenches ROS and reactive nitrogen species (RNS), which relieves the oxidative burden from the glutathione system. Moreover, polyamines also have an antioxidative role by quenching the accumulation of $O_2^{\cdot-}$ probably through inhibition of NADPH oxidase (Paschalidis and Roubelakis-Angelakis, 2005). This may facilitate phytochelatin synthesis and enhance metal tolerance (Siripornadulsil et al., 2002).

Overall, oxidative defence response to toxicity or other environmental stress involves the generation of oxygenated metabolites from exposed organisms and activation/inhibition of the production of antioxidants enzymes and metabolites such as glutathione. The depletion of antioxidants for prolonged exposures might result in the decrease of the response effectiveness and eventual imbalance between generation and elimination of reactive oxygen species. Depletion of glutathione appears to be a major mechanism in short-term heavy metal toxicity (Schutzendubel and Polle, 2002). In accordance with this hypothesis, a good correlation between glutathione contents and tolerance index was observed with 10 pea genotypes differing in Cd sensitivity (Metwally et al., 2005). High GSH concentrations in hyperaccumulator *T. Goesingense* coincided with high constitutive activity of serine acetyl transferase (SAT); SAT catalyses the acetylation of L-Ser to OAS which in turn provides the carbon skeleton for Cys biosynthesis. Elevated GSH levels in *T. Goesingense* also coincided with the ability both to hyperaccumulate Ni and to resist its damaging oxidation effects.

The significance of glutathione and the metal-induced phytochelatins (PCs) in heavy metal tolerance has been studied intensely (Rauser, 1995). However, PCs are important for detoxification of only a limited set of metals such as Cd^{2+}, Cu^{2+} and AsO_2^{2-} while Zn^{2+} and Ni^{2+} are poor inducers of PCs and exhibit low binding affinity. Most other metals lack significant binding.

Evaluation of metabolites related to oxidative response constitutes a relevant group of target compounds for risk assessment. Although oxidative response to soil contamination has been classically addressed in plants, the study of this response in soil microorganisms is already being introduced in ecotoxicology as a fundamental part of the biological response of soil microorganisms to soil contamination (Boer et al., 2013; Tremaroli et al., 2009). Accordingly, Boer et al. (2013) describe the attenuation of the oxidative response for springtails in laboratory tests, which constitutes and early detection of soil pollution, and standardized test have been developed.

6. METABOLITES RELATED TO SOIL CONTAMINATION WITH ORGANIC COMPOUNDS

The importance of the identification of biomarkers and metabolic pathways specifically related to soils contamination with a particular pollutant or group of pollutants has been already highlighted through this chapter. From the information summarized in Table 1 and Table 3 it is possible to infer that soil contamination with organic compounds, namely pesticides o polycyclic aromatic hydrocarbons, abates essential metabolic pathways such as the trycarboxylic acid cycle and the oxidative stress response, while lipid metabolism appears to be enhanced. However, the advance in the application of bioinformatics is providing further progress in terms of identification of specific biomarkers for risk assessment of individual target compounds. Thus, toxicity of endosulfan has been directly related with alterations of the GABA-glutamine cycle (Yuk et al., 2013), while chlorpyrifos depresses the Cori cycle and reduces the production of phospholipids, as indicated by lower levels of choline (Jones et al., 2012). Baylay et al. (2012) specifically relates chlorpyrifos toxicity to increased levels of fumarate, an intermediate of the trycarboxylic acid cycle. Research conducted with the same earthworm (*E. fetida*) and other families of organic compounds revealed a different metabolic response (Brown et al., 2010; Lankadurai et al., 2012), confirming the capability of metabolomics to discriminate the metabolic pathways involved in the response to a particular toxic compound. Moreover, the results strongly suggest that sets of biomarkers might be soon sufficiently reliable as for their implantation in in toxicity standardized test.

The relevance of these and future studies on the development of risk assessment strategies is aggravated by the inherent risk of soil contamination for human health. Soil contaminants may be responsible for health effects costing millions of euros. Health problems range from cancer (arsenic, asbestos, dioxins), to neurological damage and lower IQ (lead, arsenic), kidney disease (lead, mercury, cadmium), and skeletal and bone diseases (lead, fluoride, cadmium).

Overall, few studies have been conducted on the toxicity of complex chemical mixtures in soils. The effects of the soil and organisms within it upon organic pollutants are unknown. The data currently available correspond mostly to short-term studies and high level exposure of these chemicals, which is less relevant to the potential low-level, long term health impacts on living organisms near to contaminated soil.

7. METABOLITES RELATED TO SOIL CONTAMINATION WITH HEAVY METALS

The uptake of excess metal ions is toxic to most organisms, and the biochemical impact of metal ions on the cells varies with the chemistry of the element as their chemical nature. In plants, phytotoxicity of heavy metals in most parts can be attributed to symplastic accumulation of heavy metals, such as the cytosol and chloroplast stroma. Metal-induced changes in development are the result of either a direct and immediate impairment of metabolism or signaling processes that initiate adaptive or toxicity responses that need to be considered as active processes of the organism. Transport processes have been recognized as a central mechanism of metal detoxification and tolerance (Hall, 2002; Hall and Williams, 2003).

Some metals, for example, Zn and Cu, are essential for normal plant growth and development as they serve as structural and functional components of specific proteins. Other metals, for example, Cd and Pb, have no known function in plants although a Cd requirement for carbonic anhydrase from marine diatoms has been reported (Lane and Morel, 2000).

Upon exposure to metals, organisms often synthesize a set of diverse metabolites that accumulate to concentrations in the millimolar range, particularly specific amino acids, such as proline and histidine, peptides such as glutathione and phytochelatins (PC), and the amines spermine, spermidine, putrescine, nicotianamine, and mugineic acids that can be detected as response to these metals exposure. The advance of toxicogenomics in relation to organic contaminants is significantly ahead of the equivalent research in metal contaminated soil (Table 1). Nevertheless, research conducted up to date has yielded a number of biomarkers representative of the biological response of soil microorganisms to metals toxicity. Thus, soil contamination with Pb has been related with an enhancement of lipid metabolism (Sanchez-Hernandez, 2006) and more directly with reduction of tyrosine levels (Wu et al., 2013). Otherwise, Cd toxicity promotes the secretion of phytochelatins in *C. elegans*, likely at the expenses of the sulphur metabolism, as suggested by the reduction in cystathionine (Hughes et al., 2009), while the response of tomato plants to Cd involves several biochemical pathways (Hédiji et al., 2010). These examples illustrate the genuine specificity of biological reactions to different metals but also the variation in representative biomarkers among different organisms. Accordingly, exposure of *C. elegans* to Ni (Jones et al., 2012) yields a different metabolome than Cd since different biochemical pathways are affected.

In plants, data currently available demonstrate the significance of nitrogen-containing metabolites beyond phytochelatins and glutathione in plant response and acclimation to heavy metals. The various metal ions have specific chemical properties and induce distinct responses of adaptation and damage development. Thus, accumulating N-metabolites display a variety of functions, i.e. metal ion chelation, antioxidant defence, protection of macromolecules, and possibly signalling.

Proline is an extensively studied molecule in the context of plant responses to abiotic stresses. Up-regulation of proline is often encountered in plants under heavy metal stress, comparable to what occur under other abiotic stresses. When compared at equal toxic strength, proline accumulation decreased in the order Cd > Zn > Cu (Schat et al., 1997). In addition, it has been suggested different functions of proline under metal-stress, being involved in osmoregulation, metal chelation, antioxidant, and regulator of specific functions in plant morphogenesis.

Furthermore, Ni-hyperaccumulation has been specifically linked to histidine production (Krämer, 2005), particularly for *Saccharomyces cerevisiae* (Pearce and Sherman, 1999). The beneficial role of high histidine levels has been shown in transgenic *Arabidopsis thaliana* which accumulated about 2-fold higher histidine levels than wild- type plants and showed more than 10-fold increased biomass production in the presence of toxic Ni in the growth medium (Wycisk et al., 2004).. Moreover, cell surface-engineered yeast displaying a histidine oligopeptide (hexa-His) has been shown to adsorb 3–8 times more copper ions than the parent strain, being more resistant to Cu than the parent (Kuroda et al., 2002).

Otherwise, polyamine contents are altered in response to the exposure to heavy metals. Weinstein et al. (1986) showed an increment in putrescine content in Cd-treated oat seedlings and detached oat leaves with a marginal rise in spermidine and spermine content. They influence a variety of growth and development processes in plants and have been suggested to be a class of plant growth regulators and to act as second messengers (Kakkar and Sawhney, 2002). It has been suggested that they could stabilize and protect the membrane systems against the toxic effects of metal ions, particularly the redox active metals.

Overall, the number of studies remains rather scarce, and the preliminary results available in the literature merely constitute a launching platform for this promising research field.

8. FUTURE PERSPECTIVES

The main objective of metabolomics implementation in soil risk assessment is to meet the continuously increasing demand of safety data from human and ecological risk assessments. Accordingly, regulatory programs worldwide are currently incorporating tests with end-points that involve the effects of chemicals and the impact in specific metabolic pathways (Ankley et al., 2006). Toxicological end-points can be general biological responses such as survival or weight loss (Baylay et al., 2012), but specific biomarkers provide the accuracy that was classically elusive for test with living organisms

Several issues immediately arise from the summary here presented, such as the need to perform field toxicological test, with natural soils rather than use artificial soils, as was the case with some of the studies listed in Table 1. Ecotoxigenomics can also benefit from the incorporation of further analytical techniques. Techniques based on mass spectrometry are certainly required to understand the mechanisms involved in the alteration of metabolic pathways as response to toxicants. However, for screenings which merely require the detection of differences between metabolic phenotypes, optical methods such as FT-IR would be suitable, particularly if extremely high sample throughput is required (Bundy et al., 2009). Although no data was available in the existing literature, Figure 2 illustrates the change in the fingerprint of organic compounds in a soil amended with different sources of carbon collected 10 after the application. While some of the groups of compounds might be merely related to the sources of carbon added, the variations in the signal associated to polysaccharides (600-1000 cm^{-1}) can be associated to changes in the metabolic fingerprint of the soil system and therefore linked to microbiological activity in soil. Overall, the introduction of these results seeks to encourage further characterization of families of compounds in intact soil (or functional pools such as aggregates) in relation with soil processes, an approach that can find immediate application in the assessment of biological responses to toxic compounds in soil.

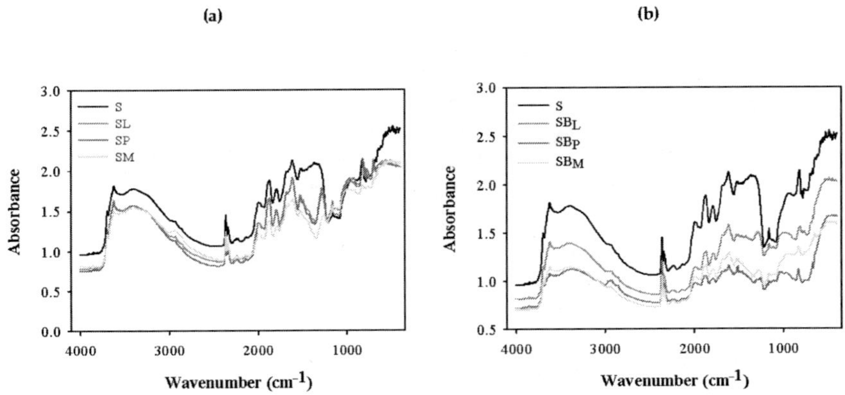

Figure 2. Absorption spectra obtained by Fourier transform infrared spectroscopy (FTIR) for an agricultural soil (S), soil amended with fresh residues (a): dry leaf litter (SL), peanut shell (SP), maize residue (SM), and soil amended with biochar (b) derived from those feedstocks (BL, BP or BM). Spectra presented (after 10 d incubation) are the average of 5 spectra obtained for different samples of each treatment. Hernandez-Soriano et al., unpublished data.

The variability of biological responses has been one of the main obstacles for their implementation in standardized risk assessment. However, the examination of changes in biological processes by accurate analytical techniques and powerful statistical tools has launched a new era in our understanding of the soil processes. The possibility of identifying the most sensitive metabolites for a certain toxicant and develop a tailored standardized test is the ultimate goal pursued.

9. ACKNOWLEDGEMENTS

MCH-S thanks The University of Queensland for a postdoctoral research fellowship. JCJ-L thanks the European research program Marie Curie (FP7-PEOPLE-2011-IOF) for his PIOF-GA-2011-301550 grant.

The funders had no role in study design, data collection and analysis, decision to publish, or preparation of the manuscript.

REFERENCES

1. Amiard-Triquet, C., Amiard, J.C., Rainbow, P.S., 2012. Ecological Biomarkers: Indicators of Ecotoxicological Effects. Taylor & Francis.
2. Ankley, G.T., Daston, G.P., Degitz, S.J., Denslow, N.D., Hoke, R.A., Kennedy, S.W., Miracle, A.L., Perkins, E.J., Snape, J., Tillitt, D.E., Tyler, C.R., Versteeg, D., 2006. Toxicogenomics in Regulatory Ecotoxicology. Environmental Science & Technology 40, 4055-4065.
3. Baylay, A.J., Spurgeon, D.J., Svendsen, C., Griffin, J.L., Swain, S.C., Sturzenbaum, S.R., Jones, O.A.H., 2012. A metabolomics based test of independent action and concentration addition using the earthworm Lumbricus rubellus. Ecotoxicology 21, 1436-1447.
4. Boer, M., Ellers, J., Gestel, C.M., Dunnen, J., Straalen, N., Roelofs, D., 2013. Transcriptional responses indicate attenuated oxidative stress in the springtail Folsomia candida exposed to mixtures of cadmium and phenanthrene. Ecotoxicology 22, 619-631.
5. Brown, S.A.E., McKelvie, J.R., Simpson, A.J., Simpson, M.J., 2010. 1H NMR metabolomics of earthworm exposure to sub-lethal concentrations of phenanthrene in soil. Environmental Pollution 158, 2117-2123.
6. Brown, S.A.E., Simpson, A.J., Simpson, M.J., 2009. 1H NMR metabolomics of earthworm responses to sub-lethal PAH exposure. Environmental Chemistry 6, 432-440.
7. Bundy, J., Davey, M., Viant, M., 2009. Environmental metabolomics: a critical review and future perspectives. Metabolomics 5, 3-21.
8. Bundy, J., Spurgeon, D., Svendsen, C., Hankard, P., Weeks, J., Osborn, D., Lindon, J., Nicholson, J., 2004. Environmental Metabonomics: Applying Combination Biomarker Analysis in Earthworms at a Metal Contaminated Site. Ecotoxicology 13, 797-806.
9. Bundy, J.G., Lenz, E.M., Bailey, N.J., Gavaghan, C.L., Svendsen, C., Spurgeon, D., Hankard, P.K., Osborn, D., Weeks, J.M., Trauger, S.A., Speir, P., Sanders, I., Lindon, J.C., Nicholson, J.K., Tang, H., 2002. Metabonomic assessment of toxicity of 4-fluoroaniline, 3,5-difluoroaniline and 2-fluoro-4-methylaniline to the earthworm Eisenia veneta (rosa): Identification of new endogenous biomarkers. Environmental Toxicology and Chemistry 21, 1966-1972.
10. Bundy, J.G., Osborn, D., Weeks, J.M., Lindon, J.C., Nicholson, J.K., 2001. An NMR-based metabonomic approach to the investigation of coelomic fluid biochemistry in earthworms under toxic stress. FEBS Letters 500, 31-35.
11. Dallinger-Marianne, J., 2000. Biomarkers in terrestrial invertebrates for ecotoxicological soil risk assessment. Reviews of Environmental Contamination and Toxicology 164, 93.
12. Drewes, C.D., Vining, E.P., 1984. In vivo neurotoxic effects of dieldrin on giant nerve fibers and escape reflex function in the earthworm, Eisenia foetida. Pesticide Biochemistry and Physiology 22, 93-103.
13. Edwards, C.A., Bater, J.E., 1992. The use of earthworms in environmental management. Soil Biology and Biochemistry 24, 1683-1689.
14. Ellis, J.K., Athersuch, T.J., Thomas, L.D.K., Teichert, F., Perez-Trujillo, M., Svendsen, C., Spurgeon, D.J., Singh, R., Jaerup, L., Bundy, J.G., Keun, H.C., 2012. Metabolic profiling detects early effects of environmental and lifestyle exposure to cadmium in a human population. Bmc Medicine 10.
15. Fitzpatrick, L.C., Sassani, R., Venables, B.J., Goven, A.J., 1992. Comparative toxicity of polychlorinated biphenyls to earthworms Eisenia foetida and Lumbricus terrestris. Environ Pollut 77, 65-69.
16. Gibb, J.O.T., Holmes, E., Nicholson, J.K., Weeks, J.M., 1997. Proton NMR spectroscopic studies on tissue extracts of invertebrate species with pollution indicator potential.

Comparative Biochemistry and Physiology Part B: Biochemistry and Molecular Biology 118, 587-598.
17. Hall, J.L., 2002. Cellular mechanisms for heavy metal detoxification and tolerance. J Exp Bot 53, 1-11.
18. Hall, J.L., Williams, L.E., 2003. Transition metal transporters in plants. J Exp Bot 54, 2601-2613.
19. Hédiji, H., Djebali, W., Cabasson, C., Maucourt, M., Baldet, P., Bertrand, A., Boulila Zoghlami, L., Deborde, C., Moing, A., Brouquisse, R., Chaïbi, W., Gallusci, P., 2010. Effects of long-term cadmium exposure on growth and metabolomic profile of tomato plants. Ecotoxicology and Environmental Safety 73, 1965-1974.
20. Heimbach, F., 1988. A comparison of laboratory methods for toxicity testing with earthworms, In: Edwards, C.A., Neuhauser, E.F. (Eds.), Earthworms in waste and environmental management. SPB Academic Publishing, The Hague, The Netherlands, pp. 329-335.
21. Hughes, S.L., Bundy, J.G., Want, E.J., Kille, P., Stürzenbaum, S.R., 2009. The Metabolomic Responses of Caenorhabditis elegans to Cadmium Are Largely Independent of Metallothionein Status, but Dominated by Changes in Cystathionine and Phytochelatins. Journal of Proteome Research 8, 3512-3519.
22. Jimenez-Lopez, J., Kotchoni, S., Hernandez-Soriano, M., Gachomo, E., Alché, J., 2013. Structural functionality, catalytic mechanism modeling and molecular allergenicity of phenylcoumaran benzylic ether reductase, an olive pollen (Ole e 12) allergen. Journal of Computer-Aided Molecular Design 27, 873-895.
23. Jones, O.A.H., Sdepanian, S., Lofts, S., Svendsen, C., Spurgeon, D.J., Maguire, M.L., Griffin, J.L., 2014. Metabolomic analysis of soil communities can be used for pollution assessment. Environmental Toxicology and Chemistry 33, 61-64.
24. Jones, O.A.H., Spurgeon, D.J., Svendsen, C., Griffin, J.L., 2008. A metabolomics based approach to assessing the toxicity of the polyaromatic hydrocarbon pyrene to the earthworm Lumbricus rubellus. Chemosphere 71, 601-609.
25. Jones, O.A.H., Swain, S.C., Svendsen, C., Griffin, J.L., Sturzenbaum, S.R., Spurgeon, D.J., 2012. Potential New Method of Mixture Effects Testing Using Metabolomics and Caenorhabditis elegans. Journal of Proteome Research 11, 1446-1453.
26. Kakkar, R.K., Sawhney, V.K., 2002. Polyamine research in plants – a changing perspective. Physiologia Plantarum 116, 281-292.
27. Kammenga, J.E., Dallinger, R., Donker, M.H., Kohler, H.R., Simonsen, V., Triebskorn, R., Weeks, J.M., 2000. Biomarkers in terrestrial invertebrates for ecotoxicological soil risk assessment. Rev Environ Contam Toxicol 164, 93-147.
28. Krämer, U., 2005. Phytoremediation: novel approaches to cleaning up polluted soils. Current Opinion in Biotechnology 16, 133-141.
29. Kuroda, K., Ueda, M., Shibasaki, S., Tanaka, A., 2002. Cell surface-engineered yeast with ability to bind, and self-aggregate in response to, copper ion. Applied Microbiology and Biotechnology 59, 259-264.
30. Kwon, Y.-K., Jung, Y.-S., Park, J.-C., Seo, J., Choi, M.-S., Hwang, G.-S., 2012. Characterizing the effect of heavy metal contamination on marine mussels using metabolomics. Marine Pollution Bulletin 64, 1874-1879.
31. Lane, T.W., Morel, F.M.M., 2000. A biological function for cadmium in marine diatoms. Proceedings of the National Academy of Sciences 97, 4627-4631.
32. Lankadurai, B.P., Simpson, A.J., Simpson, M.J., 2012. 1H NMR metabolomics of Eisenia fetida responses after sub-lethal exposure to perfluorooctanoic acid and perfluorooctane sulfonate. Environmental Chemistry 9, 502-511.
33. Lankadurai, B.P., Wolfe, D.M., Simpson, A.J., Simpson, M.J., 2011. 1H NMR-based metabolomics of time-dependent responses of Eisenia fetida to sub-lethal phenanthrene exposure. Environmental Pollution 159, 2845-2851.
34. Lenz, E.M., Weeks, J.M., Lindon, J.C., Osborn, D., Nicholson, J.K., 2005. Qualitative high field 1H-NMR spectroscopy for the characterization of endogenous metabolites in earthworms with biochemical biomarker potential. Metabolomics 1, 123-136.

35. McKelvie, J.R., Wolfe, D.M., Celejewski, M.A., Alaee, M., Simpson, A.J., Simpson, M.J., 2011. Metabolic responses of Eisenia fetida after sub-lethal exposure to organic contaminants with different toxic modes of action. Environmental Pollution 159, 3620-3626.
36. McKelvie, J.R., Yuk, J., Xu, Y., Simpson, A.J., Simpson, M.J., 2009. H-1 NMR and GC/MS metabolomics of earthworm responses to sub-lethal DDT and endosulfan exposure. Metabolomics 5, 84-94.
37. Metwally, A., Safronova, V.I., Belimov, A.A., Dietz, K.J., 2005. Genotypic variation of the response to cadmium toxicity in Pisum sativum L. J Exp Bot 56, 167-178.
38. Miller, M.G., 2007. Environmental Metabolomics: A SWOT Analysis (Strengths, Weaknesses, Opportunities, and Threats). Journal of Proteome Research 6, 540-545.
39. Morrison, N., Bearden, D., Bundy, J., Collette, T., Currie, F., Davey, M., Haigh, N., Hancock, D., Jones, O.H., Rochfort, S., Sansone, S.-A., Štys, D., Teng, Q., Field, D., Viant, M., 2007. Standard reporting requirements for biological samples in metabolomics experiments: environmental context. Metabolomics 3, 203-210.
40. OECD, 1984. Earthworm, acute toxicity tests. Guideline for testing chemicals. No. 207. OECD, Paris, France.
41. OECD, 2004. Earthworm reproduction test. Guideline for testing chemicals. No. 222. OECD, Paris, France.
42. Paschalidis, K.A., Roubelakis-Angelakis, K.A., 2005. Sites and regulation of polyamine catabolism in the tobacco plant. Correlations with cell division/expansion, cell cycle progression, and vascular development. Plant Physiol 138, 2174-2184.
43. Pearce, D.A., Sherman, F., 1999. Toxicity of copper, cobalt, and nickel salts is dependent on histidine metabolism in the yeast Saccharomyces cerevisiae. J Bacteriol 181, 4774-4779.
44. Rauser, W., 1995. Phytochelatins and related peptides. Structure, biosynthesis, and function. Plant Physiology 109, 1141-1149.
45. Regoli, F., Gorbi, S., Frenzilli, G., Nigro, M., Corsi, I., Focardi, S., Winston, G.W., 2002. Oxidative stress in ecotoxicology: from the analysis of individual antioxidants to a more integrated approach. Marine Environmental Research 54, 419-423.
46. Rochfort, S.J., Ezernieks, V., Yen, A.L., 2009. NMR-based metabolomics using earthworms as potential indicators for soil health. Metabolomics 5, 95-107.
47. Sanchez-Hernandez, J.C., 2006. Earthworm Biomarkers in Ecological Risk Assessment, In: Ware, G., Whitacre, D., Albert, L., Voogt, P., Gerba, C., Hutzinger, O., Knaak, J., Mayer, F., Morgan, D.P., Park, D., Tjeerdema, R., Yang, R.H., Gunther, F. (Eds.), Reviews of Environmental Contamination and Toxicology. Springer New York, pp. 85-126.
48. Sanchez-Hernandez, J.C., 2010. Environmental applications of earthworm esterases in the agroecosystem. Journal of Pesticide Science 35, 290-301.
49. Schat, H., Sharma, S., Vooijs, R., 1997. Heavy metal-induced accumulation of free proline in metal-tolerant and a nontolerant ecotype of Silene vulgaris. Physiologia Plantarum 101, 477-482.
50. Schutzendubel, A., Polle, A., 2002. Plant responses to abiotic stresses: heavy metal-induced oxidative stress and protection by mycorrhization. J Exp Bot 53, 1351-1365.
51. Shulaev, V., 2006. Metabolomics technology and bioinformatics. Briefings in Bioinformatics 7, 128-139.
52. Singh, O.V., 2006. Proteomics and metabolomics: The molecular make-up of toxic aromatic pollutant bioremediation. PROTEOMICS 6, 5481-5492.
53. Siripornadulsil, S., Traina, S., Verma, D.P., Sayre, R.T., 2002. Molecular mechanisms of proline-mediated tolerance to toxic heavy metals in transgenic microalgae. Plant Cell 14, 2837-2847.
54. Smilde, A.K., Westerhuis, J.A., Hoefsloot, H.C.J., Bijlsma, S., Rubingh, C.M., Vis, D.J., Jellema, R.H., Pijl, H., Roelfsema, F., Greef, J., 2010. Dynamic metabolomic data analysis: a tutorial review. Metabolomics 6, 3-17.
55. Smirnoff, N., 1993. The role of active oxygen in the response of plants to water deficit and desiccation. New Phytologist 125, 27-58.

56. Smith, C.A., O'Maille, G., Want, E.J., Qin, C., Trauger, S.A., Brandon, T.R., Custodio, D.E., Abagyan, R., Siuzdak, G., 2005. METLIN: a metabolite mass spectral database. Ther Drug Monit 27, 747-751.
57. Snape, J.R., Maund, S.J., Pickford, D.B., Hutchinson, T.H., 2004. Ecotoxicogenomics: the challenge of integrating genomics into aquatic and terrestrial ecotoxicology. Aquatic Toxicology 67, 143-154.
58. Spurgeon, D.J., Morgan, A.J., Kille, P., 2008. Current research in soil invertebrate ecotoxicogenomics, In: Hogstrand, C., Kille, P. (Eds.), Comparative Toxicogenomics, pp. 133-163.
59. Tremaroli, V., Workentine, M.L., Weljie, A.M., Vogel, H.J., Ceri, H., Viti, C., Tatti, E., Zhang, P., Hynes, A.P., Turner, R.J., Zannoni, D., 2009. Metabolomic investigation of the bacterial response to a metal challenge. Appl Environ Microbiol 75, 719-728.
60. Trygg, J., Holmes, E., Lundstedt, T., 2006. Chemometrics in Metabonomics. Journal of Proteome Research 6, 469-479.
61. Van Gestel, C.A.M., Dirven-Van Breemen, E.M., Baerselman, R., Emans, H.J.B., Janssen, J.A.M., Postuma, R., Van Vliet, P.J.M., 1992. Comparison of sublethal and lethal criteria for nine different chemicals in standardized toxicity tests using the earthworm Eisenia andrei. Ecotoxicology and Environmental Safety 23, 206-220.
62. van Gestel, C.A.M., van Dis, W.A., van Breemen, E.M., Sparenburg, P.M., 1989. Development of a standardized reproduction toxicity test with the earthworm species Eisenia fetida andrei using copper, pentachlorophenol, and 2,4-dichloroaniline. Ecotoxicology and Environmental Safety 18, 305-312.
63. van Ravenzwaay, B., Cunha, G.C.-P., Leibold, E., Looser, R., Mellert, W., Prokoudine, A., Walk, T., Wiemer, J., 2007. The use of metabolomics for the discovery of new biomarkers of effect. Toxicology Letters 172, 21-28.
64. van Ravenzwaay, B., Herold, M., Kamp, H., Kapp, M.D., Fabian, E., Looser, R., Krennrich, G., Mellert, W., Prokoudine, A., Strauss, V., Walk, T., Wiemer, J., 2012. Metabolomics: A tool for early detection of toxicological effects and an opportunity for biology based grouping of chemicals—From QSAR to QBAR. Mutation Research/Genetic Toxicology and Environmental Mutagenesis 746, 144-150.
65. Vega-López, A., Ayala-López, G., Posadas-Espadas, B.P., Olivares-Rubio, H.F., Dzul-Caamal, R., 2013. Relations of oxidative stress in freshwater phytoplankton with heavy metals and polycyclic aromatic hydrocarbons. Comparative Biochemistry and Physiology Part A: Molecular & Integrative Physiology 165, 498-507.
66. Viant, M.R., 2009. Applications of metabolomics to the environmental sciences. Metabolomics 5, 1-2.
67. Viant, M.R., Rosenblum, E.S., Tjeerdema, R.S., 2003. NMR-Based Metabolomics: A Powerful Approach for Characterizing the Effects of Environmental Stressors on Organism Health. Environmental Science & Technology 37, 4982-4989.
68. Warne, M., Lenz, E.M., Osborn, D., Weeks, J., Nicholson, J., 2000. An NMR-based metabonomic investigation of the toxic effects of 3-trifluoromethyl-aniline on the earthworm Eisenia veneta. Biomarkers 5, 56-72.
69. Weinstein, L.H., Kaur-Sawhney, R., Rajam, M.V., Wettlaufer, S.H., Galston, A.W., 1986. Cadmium-induced accumulation of putrescine in oat and bean leaves. Plant Physiol 82, 641-645.
70. Whitfield Åslund, M., Simpson, A., Simpson, M., 2011. 1H NMR metabolomics of earthworm responses to polychlorinated biphenyl (PCB) exposure in soil. Ecotoxicology 20, 836-846.
71. Whitfield Aslund, M.L., McShane, H., Simpson, M.J., Simpson, A.J., Whalen, J.K., Hendershot, W.H., Sunahara, G.I., 2012. Earthworm sublethal responses to titanium dioxide nanomaterial in soil detected by (1)H NMR metabolomics. Environ Sci Technol 46, 1111-1118.
72. Wu, H., Liu, X., Zhao, J., Yu, J., 2013. Regulation of Metabolites, Gene Expression, and Antioxidant Enzymes to Environmentally Relevant Lead and Zinc in the Halophyte Suaeda salsa. Journal of Plant Growth Regulation 32, 353-361.

73. Wycisk, K., Kim, E.J., Schroeder, J.I., Krämer, U., 2004. Enhancing the first enzymatic step in the histidine biosynthesis pathway increases the free histidine pool and nickel tolerance in Arabidopsis thaliana. FEBS Letters 578, 128-134.
74. Yu, Y.L., Wu, X.M., Li, S.N., Fang, H., Zhan, H.Y., Yu, J.Q., 2006. An exploration of the relationship between adsorption and bioavailability of pesticides in soil to earthworm. Environmental Pollution 141, 428-433.
75. Yuk, J., McKelvie, J.R., Simpson, M.J., Spraul, M., Simpson, A.J., 2010. Comparison of 1-D and 2-D NMR techniques for screening earthworm responses to sub-lethal endosulfan exposure. Environmental Chemistry 7, 524-536.
76. Yuk, J., Simpson, M.J., Simpson, A.J., 2013. 1-D and 2-D NMR-based metabolomics of earthworms exposed to endosulfan and endosulfan sulfate in soil. Environmental Pollution 175, 35-44.
77. Zhang, Q., Bhattacharya, S., Andersen, M.E., Conolly, R.B., 2010. Computational Systems Biology and Dose-Response Modeling in Relation to New Directions in Toxicity Testing. Journal of Toxicology and Environmental Health, Part B 13, 253-276.

CHAPTER 17

Effect of Crop Rotation and Nitrogen Fertilization on the Quality and Quantity of Soil Organic Matter

Agnieszka Rutkowska[1] and Dorota Pikuła[1]

[1] *Department of Plant Nutrition and Fertilization, Institute of Soil Science and Plant Cultivation-State Research Institute, Poland*

1. INTRODUCTION

Soil organic matter (SOM) is one of the most important features of the soil. Its characteristic depends on a variety of biotic and abiotic variables of the ecosystem, such as climate, soil texture, mineral composition, quantity of organic residues and other factors. Currently, in an era of rapidly changing civilization, leading to changes in climate and soil conditions, SOM content becomes increasingly important, not only for the proper functioning of ecosystems, but also for socio-economic development of many regions of the world [1, 2]. In the first half of the past century, there were hardly contradictions between the agricultural cultivation and the environment. The substance circulations were closed, animal production comparatively small and mainly regularly allocated. The mineral fertilization was only used to a slight extent. A fundamental change has taken place during the last decades. With the increasing use of mineral fertilizers, yields have increased by more than 100% thus the quantity of roots and harvest residues on the field has increased strongly as a source of organic matter. However, in agricultural practice they are commonly removed from the field after harvest resulting in SOMdecrease.

During the last ten years in EU countries, the progressive degradation of SOM is observed. Thus, this issue was reflected in the EU soil strategy (COM (2002) 179), on which the actual reduction of soil organic matter content was listed as one of the most important problems. In Poland, the reduction of SOM content in soils became a problem particularly significant.

Nearly the whole territory of Poland (99,7%), covering about 313 thousands square kilometres is situated in the Baltic Sea basin. This territory is drainaged by two big rivers Vistula and Odra and seven small rivers flowing directly to the sea. Natural farming condition in Poland are poor, due to prevalence of light, sand-derived soils (60% very light and light soils) and unfavourable climate. Due to soil texture and acid or very acid reaction more than 60% of the soils in Poland can be classified as soils witha relatively low content of organic matter. According to the newest survey 7,6% of arable soils show low content of SOM (below 1%), 47,1% of soils presentmedium content (1,1-2,0%), 29,3%, high content (2,1-3%) and only 3% of soils show very high content of organic matter (above 3%) [3]. The percentage of acid and very acid soils is very high and exceeds 50%, and soil acidity seems to be one of the most important factors leading to degradation of the quantity and quality of organic matter [4]. The summaryof the last four-years period of agrochemical soils monitoring program reveals that 20,2% of soils are vey acid (pH_{KCl} below 4,5) and 29,4% acid (pH_{KCl} 4,5-5,5) [4].

Besides, the intensification of soil use combined with the simplified crop rotation and predominance of cereals together with an expansion of farming systems based on crop production with a reduced number of livestock or without animals intensifies the process of organic matter degradation [5- 7].

The turn over of organic substance are time dependent and become most apparent after decades. Therefore, long-term field experiments are a necessary tool for tracking changes in organic carbon content in the soil. In Europe, the most famous long-term experiences with testing of different mineral and natural fertilisers and/or cultivation of various plant species were held at Rothamsted (England), Halle and Bad Laustädt (Germany), Prague-Ruzyne (Czech Republic) and Skierniewice (Poland) [8].

The Institute of Soil Scienceand Plant Cultivation National Research Institute (IUNG-PIB) in Puławy is also involved in several kinds of long term field studies, including SOM content monitoring, since 1979. A special trait of these experiments is that crop rotations included plants enriching and exhausting soil from humus, have been apermanent factor for the last 33 years.

The objectives of presented paper was to access the impact of mineral nitrogen fertilization, manure application and crop rotation on the quantity and quantity soil organic matter in long term field experiment.

2. COMPOSITION OF SOIL ORGANIC MATTER AND METHODS OF HUMUS SUBSTANCES FRACTIONATION

Humic substances (HS) is the major organic constituents of soil. Humic substances are a mixture of particles that differ in their structure, mass, size, elemental composition and properties. The first record of humic acids extration by Achard dates back to 1786, and the first major study was done by Sprengel in 1820 [in 9]. Despite such a long research history on humic substances, there has been constant dispute over the structure of humus. There is an increasing interest in HS, because of their capacity for complexing metal ions, involvement in the organic geochemical cycle, affecting bioavailability and ecological effects of nutrient in water and stabilizing soil fertility.

The content and quality of humus are directly and indirectly determined by physical, chemical, biological and environmental properties. SOM quality is determined mainly by the percentage share of humic substances in soil. Humic substances can be subdivided into three major fractions: humic acids, fulvic acids and humins. Electron microscope observations reveal the humic acids of different soils to have a polymeric structure, appearing in form of rings, chains, and clusters. The size of their macromolecules ranges from 60 to 500 A and is mainly decided by the occurring humification process, which also exerts an influence on their spatial structure. Some of the main features of humic substances are shown in Figure 1 [10].

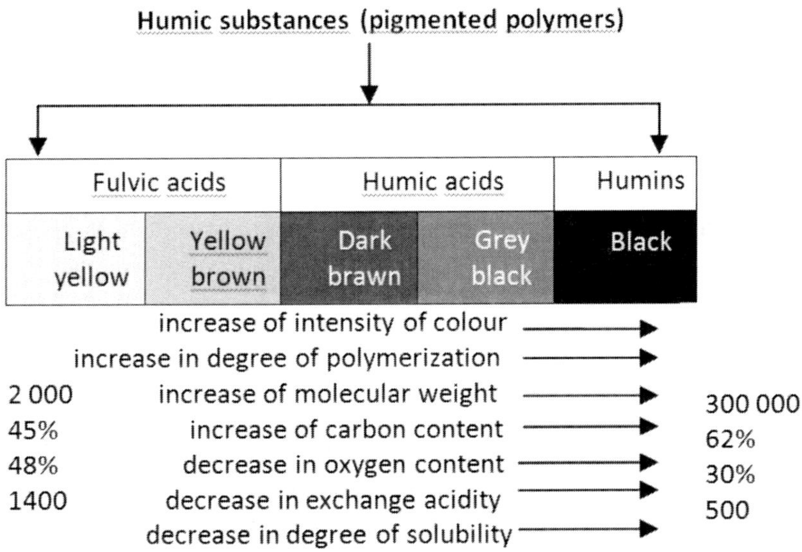

Figure 1. Chemical properties of humic substances.

Humic acids (HA) comprise a mixture of weak aliphatic and aromatic organic acids, containing carboxyl and phenolate groups. HA are not soluble in water under acidic conditions but soluble in water under alkaline conditions and precipitated from aqueous solutions when the pH decreases below 2. HA tend to be more aromatic and more prone to precipitation under the acid conditions common in soils, making them less mobile.

Fulvic acids (FA) are a humic substances soluble in water at any pH. They areless aromatic than humic acids. This part of humus remains in solution after removal of humic acids by acidification. FA have an oxygen content twice that of humic acids thus revealing a more acidic character than that of HA. The exchange capacity of FA is more than double that of HA.

Humins are the fraction of humic substances, not extracted from soil with either a strong base or a strong acid. Humins present within the soil are the most resistant to decomposition (slow to break down) of all the humic substance.This fraction plays a key role in soil fertility by improving structure and soil water capacity or providing a reservoir for the plant nutrients.

There are different ways to separate humic substances into each particular fraction. Every technique has its own advantages and limitations. The reagents used for extraction of humic acids from the soil are listed in table 1.

The most popular method is to use NaOH to extract humic acids from the soil. It is evident that extraction from soil with NaOH solution leads to the recovery of approximately two-thirds of the soil organic matter [6]. The amount of organic matter extracted from soil with the caustic alkali increase with time of extraction. Humic acids extracted by alkali solutions are characterized by high purity what is what is necessary for further physicochemical analyzes (UV-VIS, NMR and IR) [11].

Another reagent used for extraction of soil humusis sodium pyrophosphate ($Na_4P_2O_7$). Its advantage is that the auto oxidation process that occurs during the extraction of humic material is less intensive than with NaOH. However, this reagent allows to isolate only humus compounds loosely linked with the mineral fraction of the soil.

Table 1. Reagents commonly used for extration of organic constituents from soil [6]

Type of material	Extractant	Organic matter extracted
Humic substances	NaOH Mild extractants	to 80%
	$Na_4P_2O_7$ and other Organic chelates:acetyloacetone, cupferron, hydroxyquinonline Formic acid (HCOOH)	to 30% to 30% to 55%

The most common methods of humus fractionation are Schnitzers method and he standard, prior to extraction from the soil is determined by total organic carbon, by one of the recommended methods.

Schnitzer and Turin methods based on NaOH reagent are suitable for soils without carbonate thus the carbonate determination in soil is needed before [12-13]. In both methods, diluted sodium hydroxide is used after decalcification in mineral acid solution, which burnts the conjunctions of humic acids with calcium. Afterwards, humic and fulvic acids pass to the solution during alkaline extraction. On the base of the difference between the quantity of humus fractions before and without decalcification, the calculation of humic and fulvic acids associated with calcium is possible.

The separated fractions of humic substances have been recognized as the main indicators of soil fertility. The quality of SOM can be evaluated by determining the ratio of humic acids to fulvic acids (HA:FA) or carbon of humic acids to carbon of fulvic acids (CHA:CFA). It is widely described that fertile soils are characterized by higher humus content and CHA:CFA ratio>1.On the agricultural lands, soil humus properties are mostly determined by post- harvest residue left after the harvest of crops [8, 14]. Legumes increase organic matter content in soil. This is a consequence of chemical composition of organic material. Cereal straw contains more lignin, and legumes one more cellulose and nitrogen. Thus, the mineralization process of legumes residues occurs faster and the C:N ratio is narrower as compared to other crops [15].

The percentage of humus which occurs in the various humic fractions varies considerably from one soil type to another (Table 2) [16].

Table 2. Humic acid/fulvic acid ratios of some surface soils

Soil	Humic acid/ Fulvic acid ratio	Soil	Humic acid/ Fulvic acid ratio
Chernozemordinary	2.0 – 2.5	Gray forest	1.0
Chernozem deep	1.7	Sod podzolic	0.8
Chestnut dark	1.5 – 1.7	Tundra	0.3

A further attempt is made to investigate the properties of humic acids focusing on their spectroscopic characterization by UV–VIS and fluorescence spectra. This method is an important tool for determining the differences in humic substances structure, maturity and condensation degree [16, 17]. Measuring of optical properties of humic acids (absorbance and absorbance ratio) of alkaline soil extract allows to determine the degree of humification process and humic substances quality. Humic substances from various types of soil differ in the absorbance ratio. The measurement of absorbance is made in the wavelength range 280-665. The absorbance at wavelength 280 nm indicates the high content of lignin, 465 nm - young humic substances components associated with the first phase of humification process. Absorbance at 665 nm is related to well humified components [18].

The intensity of absorbance at wave length 280, 465 and 665 nm is used to calculate the ratiosA 2/4 (the ratio of absorbance in the wave length 280 nm to 465 nm), A 2/6 (280 nm to 665 nm) and A4/6 (465 nm to 665 nm). The absorbance ratio allows to recognize the structure of humus components. Larger values of 465 nm to 665 nm (A4/A6 ratio) are associated with the presence of smaller size organic molecules or more aliphatic structures and usually with higher content of functional groups [19].

2. MATERIALS AND METHODS

The study was conducted on the basis of a three factorial long-term field experiment carried on since 1979 at the Experimental Station Grabów of the Institute of Soil Science and Plant Cultivation in Puławy, on typical soil in Poland, classified as light loamy and sand texture according to USDA soil classification. The experiment was conducted with two crop rotations (I factor): A – recognized assoil exhausting from humus (potatoes, winter wheat, spring barley and corn for silage) and B –considered to enrich soil with humus (potatoes, winter wheat and mustard* as aftercrop for ploughing, spring barley with undersown** and clover with grass mixture). The experiment was performed in the split – block layout in two cycles moved by one year (Table 3).

Table 3.. The scheme of the experiment; ▭ Soil sampling term for SOM quality evaluation

Rotation	Year	Crop rotation A		Crop rotation B	
		Cycle 1	Cycle 2	Cycle 1	Cycle 2
I	1980	potatoes		potatoes	
	1981	winter wheat	potatoes	winter wheat*	potatoes
	1982	spring barley	winter wheat	spring barley**	winter wheat
	1983	maize for silage	spring barley	clover-grasses mixture	spring barley
	1984		maize for silage		clover-grasses mixture
II	1984	potatoes		potatoes	
	1985	winter wheat	potatoes	winter wheat*	potatoes
	1986	spring barley	winter wheat	spring barley**	winter wheat
	1987	maize for silage	spring barley	clover-grasses mixture	spring barley
	1988		maize for silage		clover-grasses mixture
III	1988	potatoes		potatoes	
	1989	winter wheat	potatoes	winter wheat*	potatoes
	1990	spring barley	winter wheat	spring barley**	winter wheat
	1991	maize for silage	spring barley	clover-grasses mixture	spring barley
	1992		maize for silage		clover-grasses mixture
IV	1992	potatoes		potatoes	
	1993	winter wheat	potatoes	winter wheat*	potatoes
	1994	spring barley	winter wheat	spring barley**	winter wheat
	1995	maize for silage	spring barley	clover-grasses mixture	spring barley

	1996		maize for silage		clover-grasses mixture
	1996	potatoes		potatoes	
	1997	winter wheat	potatoes	winter wheat*	potatoes
	1998	spring barley	winter wheat	spring barley**	winter wheat
V	1999	maize for silage	spring barley	clover-grasses mixture	spring barley
	2000		maize for silage		clover-grasses mixture
	2000	potatoes		potatoes	
	2001	winter wheat	potatoes	winter wheat*	potatoes
VI	2002	spring barley	winter wheat	spring barley**	winter wheat
	2003	maize for silage	spring barley	clover-grasses mixture	spring barley
	2004		maize for silage		clover-grasses mixture
	2004	potatoes		potatoes	
	2005	winter wheat	potatoes	winter wheat*	potatoes
	2006	spring barley	winter wheat	spring barley**	winter wheat
VII	2007	maize for silage	spring barley	clover-grasses mixture	spring barley
	2008		maize for silage		clover-grasses mixture

In each crop rotation, five rates of manure (II factor) were applied under potatoes: 0, 20, 40, 60 and 80 t· ha^{-1} every four years.

These factors were permanent in the history of the experiment.

Since 1984 (the third rotation) in the both crop rotations mineral fertilization has been set (III factor). Each crop was supplied with four rates of nitrogen fertilizers in accordance with table 4.

The paper presents the results of permanent experiment in which the quantity of soil organic matter was determined through 33 years. The quality of soil organic matter was evaluated on the basis of the soil samples (0-30 cm) collected after seventh rotation in the both cycles 1 and 2(Table 3). The average values for the cycles of described parameters were considered. SOM was expressed as the percentage content of organic carbon in soil.

Table 4. Mineral fertilization for crop rotation A and B

Crop rotation	Plant	Mineral fertilization [kg ha^{-1}]					
		N0	N1	N2	N3	P_2O_5	K_2O
A	Potatoes	0	45	90	135	54	160
	Winter wheat	0	40	80	120	54	100
	Spring barley	0	40	80	120	54	85
	Maize for silage	0	45	90	135	54	120
B	Potatoes	0	45	90	135	54	160
	Winter wheat	0	40	80	120	54	100
	Spring barley	0	30	60	90	54	85
	Clover with grass	0	40	80	120	54	115

In the experiment numerous parameters were evaluated, but for the paper purposes the follows ones were determined: organic carbon content by direct method using Analyzer C-MAT 5500 and fractional composition of organic matter by Schnitzer method. The content of organic carbon of separated fraction was calculated as follow:

- C_d –carbon in solution after decalcification
- CHA+FA – sum of humic and fulvic acids in extracts obtained with 0,5 m NaOH
- CF – carbon of fulvic acids in solutions, following humic acids precipitation
- CHA – carbon of humic acids calculated from the difference:
- CHA = CHA+FA – CFA

The fractional composition was expressed as the percentage share of respective fraction in the total organic carbon pool (TOC).

Optical parameters of humic acids (HU) were measured in the UV-VIS, and afterwards, A_4/A_6 ratios were calculated.

The absorbance at 280, 465 and 665 nm of a solution (pH 8,3), containing at least 1 mg of in $NaHCO_3$was measured in a UV –VIS spectrometer Perkin Elmer Lambda 20. The ratio of $A_4/A_6$was used do characterize SOM according to Kononova (1966) [16].

Statistical processing of the results was performed using Statgraphics 5 Plus package.

The data were processed by ANOVA, for each crop rotation, manure and mineral fertilization. There were proofed significant effect of crop rotation and manure application on both organic carbon content in soil and SOM quality. However, these parameters did not been affected by mineral fertilization.

The average values for treatments with different rates of manure and mineral N fertilizers describe the effect of experimental factors on soil organic carbon quantity. SOM quality was evaluated by the average values for crop rotations (A, B), and for the extreme treatments – without mineral nitrogen (N0), and the highest N rate (N3) as well as for manure, rates (1 and 5). Furthermore, the treatment with the highest mineral and manure doses was included (see explanations under table 5).

3. RESULTS

3.1. The Quantity of Som

The analysis of variance demonstrated the importance of main effects and random effect (years of study) as well as the synergies of all the experimental factors on Corg (P-Value 0,0000 for all tests). The content of organic carbon in soil through 33 years of the experiment is presented by Figure 2. Figure 3 illustrates the impact of manure application on Corg. content in soil under cultivation of plants exhausting and enriching soil in organic matter. The results are the average for treatments with different rates of manure and mineral nitrogen fertilizers.

To understand the figure 1, it is needed to consider that for the first two four-years rotations, up to 1984, the experimental scheme included only manure application. At the beginning of the experiment in 1979, the initial organic carbon content amounted to 0, 74% (Figure 2). After eight years, in crop rotations with clover grass mixture (B), organic carbon content oscillated around the initial value and amounted to 0,78%. Meanwhile, in crop rotation without legumes (A) Corg. value dropped to 0,72%.

In the following years (since the third rotation), in fields under plants exhausting soil with SOM (A), organic carbon quantity decreased regularly through the experiment and after 33 years dropped to 0, 61%. On the contrary, in crop rotation with clover – grass mixture, the tendency to stabilization organic carbon quantity in soil was observed with the highest value 0,79% in 1988 and the lowest one 0,72% in 2004. The disturbances in Corg. content detected through the experiment in both crop rotations could be caused by both climatic conditions and spatial field variability.

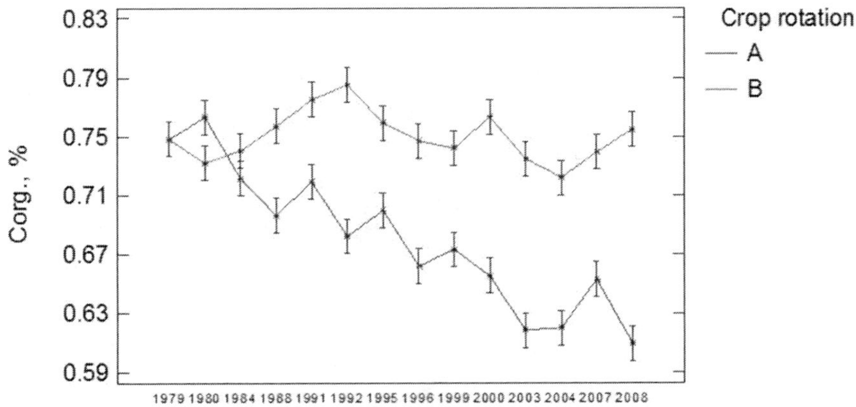

Figure 2. Effect of crop rotation on organic carbon content in soil through 1979-2008

The results illustrated by figure 3 indicated, that the effect of manure on soil organic carbon content was strongly linked to the crop rotation. In the both crop rotations, manure applicationin creased soil organic carbon. However, in crop rotations with plants exhausting soil from SOM even the highest manure rates 60 and 80 t ha^{-1} was not sufficient to secure Corg. content at the initial level (Figure 3) over the years, what confirms the results presented by Figure 2. Meanwhile, in crop rotation with legumes, only 20 t ha^{-1} manure per hectare was required to obtain such content.

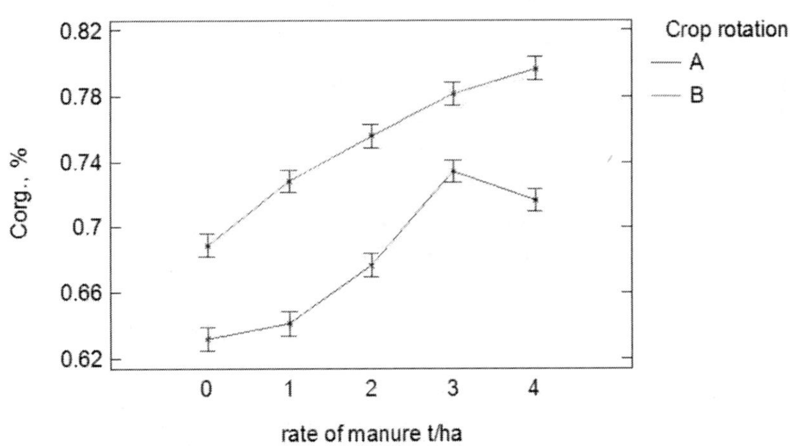

Figure 3. Effect of interaction manure application x crop rotation on soil organic carbon content

3.2. The Quality of Som

The properties of soil organic matter were evaluated on the base of its fractional composition. Table 5 shows the group composition of humus after 29 years of the experiment done by Schnitzer's method.

Table 5. Explanations: A – crop rotation with plants exhausting soil from humus B – crop rotation with plants enriching soil with humus1 – treatment without manure 5 – treatment with 80 t manure ha^{-1} N0-treatment without mineral fertilizers N3 – 120 kg N ha^{-1} Fractional composition of SOM

Combination	C_{org} %	C_d %	CHA+CFA	C_{FA}	C_{HA}	C_H	$C_{HA}:C_{FA}$
A1N0	0,55	0,033	47,3	20,0	27,3	52,7	1,37
A1N3	0,61	0,034	41,0	16,4	24,6	59,0	1,50
A5N0	0,74	0,039	43,3	14,9	38,4	56,7	1,91
A5N3	0,74	0,039	52,7	18,9	33,8	47,3	1,79
B1B0	0,75	0,035	48,0	20,7	27,3	52,0	1,32
B1N3	0,80	0,035	43,8	18,8	25,0	56,2	1,33
B5N0	0,78	0,042	48,7	21,8	26,9	51,3	1,23
B5N3	0,82	0,033	50,6	22,0	28,6	49,4	1,30

As it has been described in Material and Methods, the results for selected treatments were listed. Constantly, the values of organic carbon content in soil samples were presented.

Humus content and its quality in the soil depended on manure application and the choice of plant species for crop rotation. There was found that crop rotation B and manure fertilization generally increased organic C content in the soil. Crop rotation influenced the strong to quality of SOM (Figure 4).

Both manure application and selection of plants with legumes in crop rotation strongly modified the fractional composition of SOM and led to an increase of the sum of humic acids and fulvic acids in soil as well as the decrease in the amount of humins. Manure fertilization increased the ratio of humic acids in soil where as crop rotation with legumes affected a larger proportion of fulvic acids in comparison with humic acids. As a consequence, in treatments with regular, one to four years manure application, the ratio CHA:CFA was higher as compared with only mineral N fertilization. Similarly, soil in fields under clover with grasses cultivation are characterized by the lower ratio of humic acids to fulvic acids.

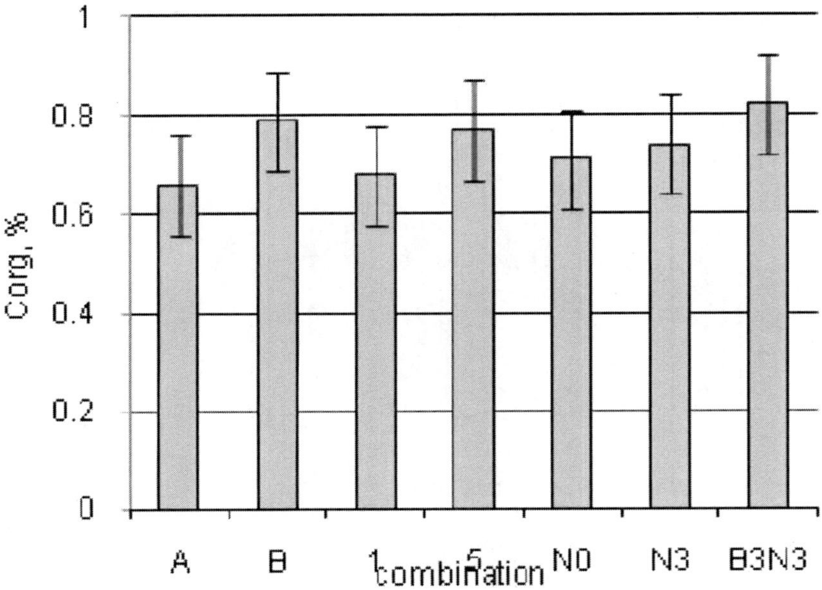

Figure 4. Effect of manure application, mineral fertilization, and crop rotation on Corg. content(explanations as under table 3). Error bars indicate standard deviation

In the described experiment, the selection of plant species for crop rotation was recognized as crucial for the parameter. Humus under plants which enhance soil with organic matter revealed much lower values of CHA:CFA ratio as compared with the exhausting ones (Figure 6). These differences will be discussed in the next part of the paper.

The isolated humic acids were treated by 1 mg of in $NaHCO_3$ and the liquid samples were subjected to UV - VIS radiation by the spectrometer for measurement within the range of visible and UV light. Then, the ratio of UV absorbance in different wave-lengths were analyzed. The coefficients of absorbance of humic acids were listed in table 6.

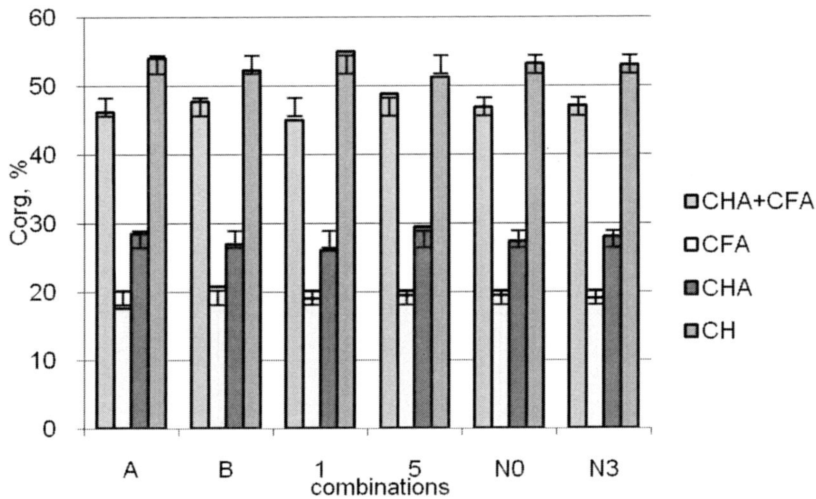

Figure 5. Fractional composition of humic substances (explanations as under table 3).

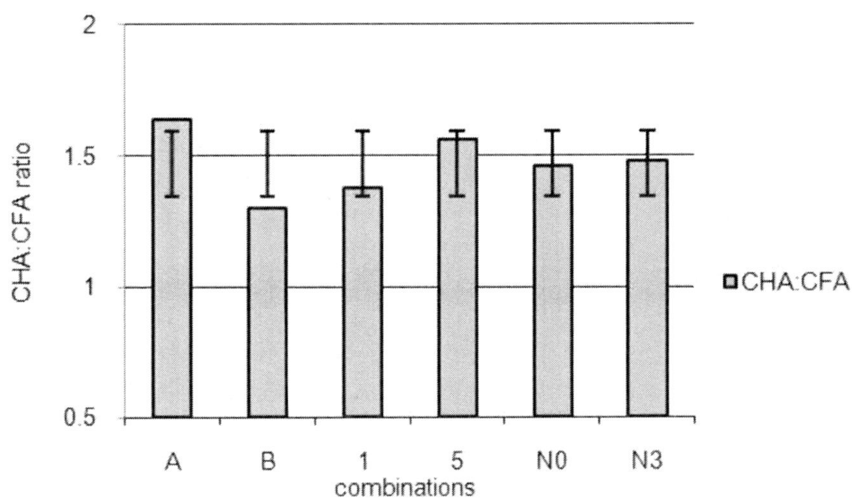

Figure 6. The ratio of humic acids to fulvic acids (CHA:CFA).Error bars indicate standard deviation

Table 6. Explanations: A – crop rotation with plants exhausting soil from humus B – crop rotation with plants enriching soil with humus 1 – treatment without manure 5 – treatment with 80 t manure ha^{-1} N0- treatment without mineral fertilizers N3 – 120 kg N ha^{-1} 1)– wave-lengths (nm) 2) – absorption ratio

Combination	A280[1)]	A465[1)]	A665[1)]	A280/465[2)]	A280/665[2)]	A465/665[2)]
A1N0	2,16	0,401	0,0871	5,39	24,8	4,60
A1N3	2,09	0,384	0,0864	5,44	24,2	4,44
A5N0	1,85	0,325	0,0695	5,69	26,6	4,68
A5N3	2,03	0,318	0,0657	6,38	39,0	4,84
B1B0	2,08	0,346	0,0736	6,01	28,3	4,70
B1N3	1,96	0,318	0,0650	6,61	32,0	4,89
B5N0	2,04	0,306	0,30585	6,66	34,9	5,23
B5N3	2,09	0,339	0,0657	6,20	31,8	5,16

The coefficients of absorbance of humic acids

The soil with manure application and mineral fertilization revealed the spectrum 280 nm in both crop rotations, which indicates the early stage of humification process characterized by high lignin content. However, there was no clear evidence in relation of fertilization and crop rotation on the intensity of the spectrum which amounted from 2,00 to 2,07 (Figure 5).

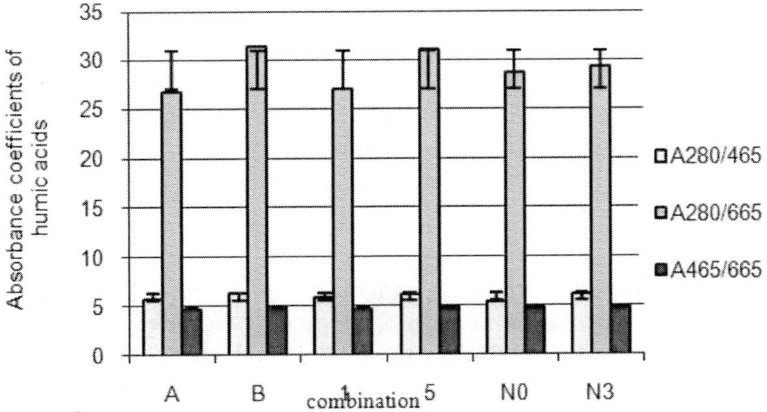

Figure 7. Absorbance coefficients of humic acids. Error bars indicate standard deviation

4. DISCUSSION

Carbon plays a significant role both for agriculture and environment protection. The organic matter is a major component for soil formation. It considerably determines soil fertility and soil properties that are relevant to the yield. In agriculture land, both soil organic matter quantity and quality are determined by natural, agroclimatic condition as well as soil cultivation technology. The relevance of soil organic matter has been long neglected, and for decades the crop production was focused on mineral fertilization. During the last ten years in EU countries, aprogressive lost of soil organic matter has been observed. In Poland, due to simplification in crop production with prevalence of cereal monocultures as well as crop residue removal and reduced manure production intensify the process of natural SOM degradation. An of an adequate farm management relies in maintaining a sustainable balance of soil organic matter. A balance can be achieved by selection of species of cultivated plants, their share in the crop structure, and the quality of organic fertilizers. For different crop species the amount of residues left in filed varies.It can be estimated that the weight of cereals crop residues is about 3-fold greater than the roots, and legumes with grasses by up to 6-fold. Moreover, a different duration and degree of shading the soil surface and the number of tillage performed might affects the mineralization of organic compounds in humus.

The impact of all these factors on soil organic matter transformations could be accounted only in long-term perspective because of gradual character of the processes. The results of described 33 years experiment confirm the positive effect of organic fertilization and crop rotation included legumes – grass mixtures on SOM reproduction. For the research presented, the effect of crop rotation had a larger impact on organic carbon accumulation in soil that the effect of manure applications. Those results could be explained by the fact that clover mixture with grasses results ina greater amount of humus in soil compared with manure application. Moreover, carbon from manure is more prone to undergo mineralization process [20].

The reduction of Corg in soil in the following years relative to the initial value, is not significantly related with manure rates and can be explained by a rather slow depleting of the total pool of humus in agricultural soil as the effect of tillage intensity and regular removal of straw from the field [21, 22]. The results from the long term field experiment indicate that an external source of organic matter such as manure applied regularly might not assure a positive balance of SOM. The results confirmed that including legumes into crop rotation might provide an effective method to preserve the optimal level of humus.

The role of mineral nitrogen fertilization remains unclear. Some authors have described that applicaton of high amounts of mineral fertilizers might accelerate mineralization process and therefore diminish organic carbon. Other authors [23, 24] propose that prolonged application of mineral N fertilisers on lessives and brown soils may cause a decrease of carbon content in soil of about 21% compared with soils not supplied with mineral fertilizers. Numerous papers indicate that the inhance combined application of manure and mineral nitrogen fertilizers enhances the mineralization process of humus [25-27]. However, there exists a strong consensus on the literature that mineral nitrogen has beneficial effect for humus stabilization [8].

In the own research, the significant effect of mineral nitrogen fertilization did not be proven.

A distinctive/significant difference was observed in the evolution of organic carbon content in soil over time for crop rotation that included plants recognized to deplete humic substances or plants that can enrich the humic content in soil.

Soil cultivation and management practice, particular manure application and mineral fertilization, tillage systems and crop rotation induced an important discussion about humus quality. According to stability and decomposability, soil organic carbon is divided into the stable and labile forms. The stable forms are represented by the total carbon content, humic substances sum, humic acids sum and fulvic acids sum [10, 28]. The labile carbon is soluble in water, more active and undergoes short time changes in the soil. Körschnes [29] claims that this part of SOM is a useful tool to characterize the soil management practice. The objectives of own experiment was both quantitative changes in Corg content in soil and quality of humus according to Schnitzer method. That approach enables to point out the most proper soil management regarding to organic and mineral fertilization, and selection of plant species for crop rotations to obtain the positive balance of SOM and soil fertility measured by humus quality [29-31]. SOM quality is determined not by the absolute quantity of humus acids but its percentage share in humus. Humic and fulvic acids content underlies the calculation HA:FA ratio. The ratio describes humus quality and its stability as well. It is commonly accepted that fertile soils are characterized by the ratio HA:FA >1. Soils fertilized with composts, slurry and manure display higher HA:FA ratio than soils supplied with mineral fertilizers [32]. The research presented provides similar results. Furthermore, the strong impact of crop rotations on the ratio was found. Humus of the soil under clover with grasses mixture (crop rotation B) revealed lowest value of the ratio than with maize for silage (crop rotation A) which amounted to 1.30 and 1.64 respectively. The literature reports that the lowest ratio of humic acids to fulvic acids in soil under legumes cultivation, comes off from the great carbon amount left by these plants, which does not impact on the ratio [8]. Besides, legumes affect the rise of fulvic acids fraction and the lower HA:FA ratio as the result.

Fractional composition of humus was modified both by crop rotation and manure application. Legumes left a great amount of residues abundant with carbon and nitrogen - the source energy for soil microorganisms, which promote the process of humus transformation with the prevalence of mineralization one [32-35]. Hence, soils under legumes cultivation are characterized by the lower HA:FA ratio. Nevertheless, the further detailed analyzes of UV-VIS proved the good quality of soil under legumes. It demonstrates that fluorescence, although poorly utilized for humic acids research, is a powerful tool to contribute to the knowledge and the action of the organic matter in soil.

UV-VIS absorption technique is applied for a long time for the characterization of humic substances. It is commonly accepted that absorbance at 280 nm is an indicative of lignin content where as absorbance at 465 nmcan be related to organic matter in early stages of decomposition, and absorbance at 665 nm with advanced humification process of SOM. The data presented in the literature indicate [14, 17, 35] that for a UV- VIS spectrum of "young" humic acids the maximum signal at 280 nm will disappear with advanced humification process. Such spectrum is observed in soils fertilized with slurry and mineral fertilizers. For the research presented, the soil treated with manure revealed a maximum at 280 nm for both crop rotations. However, results on signal intensity at 280 were no conclusive for fertilization and crop rotation.

Higher values of the A465/665 absorbance ratio were recorded in the soil with crop rotation B and in the soil fertilized with manure, especially the highest rates. It is a clear evidence that there were "young" humic acids of a lower condensation level of aromatic structures that predominate in the structure of humic acids as compared with HA in the advanced humification process [14, 18, 36].

Manure applications as compared to mineral nitrogen fertilization increases the content of lignins in the particles of humic acids (higher values of the A280/465 absorbance ratio) (Figure 5). Manure promoted a formulation of humus with greater share of aliphatic structure (lower absorption values) and higher (higher values of A465/665).

Maintaining or increasing soil organic matter (SOM) is justified both from an agronomic and a climatic perspective because it affects the capacity of the soil to sustain crop growth, is an important factor in decreasing soil compaction and erosion, and is also a source and possible sink of atmospheric CO_2-C [37]. Understanding the processes that control SOM dynamics is the key to SOM management. Land use and agricultural management practices such as crop rotation, soil tillage and organic amendments can affect SOM by influencing both the quantity and quality of crop residues that are returned to the soil; they also influence the rate of decomposition of added residues and native SOM [38-39].

5. CONCLUSIONS

1. The results of 32 – years of the field experiment show that the most important factor which stabilizes organic carbon content in agricultural soils is crop rotation with legumes. This effect has not been obtained even by systematic application of the very high rates of manure.
2. Because of intensified soil microbial activity, the ratio of humic acids to fulvic acids was lower in soil under legumes cultivations as compared with manure application. The further UV – VIS spectral analysis of humic acids indicated a high quality of humus of soils under clover with grasses mixture.
3. Mineral fertilization has not modified both soil organic carbon content and humus quality.

REFERENCES

1. E. R. Jones Rusco, G. Bidoglio, 2001 Organic matter in the soils in Europe: Present status and future trends. European Soil Bureau, JRS Ispra, Italy
2. M. Komatsuzaki, H. Ohta, 2007 Soil management practices for sustainable agro-ecosystems. Technical Raport. Sustainability Science 2 103 120
3. H. Terelak, S. Krasowicz, T. Stuczyński, 2000 Srodowisko glebowe Polski i racjonalne użytkowanie rolniczej przestrzeni produkcyjnej. Pamiętnik Puławski 120 2
4. P. Ochal, 2012 Wykorzystanie syntetycznego wskaźnika do oceny stanu agrochemicznego gleb w Polsce. PhD thesis. IUNG-PIB Puławy
5. S. Mercik, M. Stepień, W. Stępień, T. Sosulski, 2005 Dynamic of organic carbon content insoil depending on long-term fertilization and crop rotation. Roczniki Gleboznawcze; t LVI 3/4 53
6. A. Kusińska, 1999 Zasoby i skład humusu glebowego pod niektórymi gatunkami roślin w dwóch systemach uprawy. Zeszyty Problemowe Postępów Nauk Roliczych 465 319
7. J. Pranagal, 2004 Wpływ systemu uprawy na zawartość węgla organicznego w glebie. Annales UMCS Sec. E, 59 1 1 10
8. S.S. Gonet, 1989 Właściwości kwasów humusowych w warunkach zróżnicowanego nawożenia. ATR Bydgoszcz, Rozprawy 33 4 55
9. M. Susic, 2003 Structure and origin of humic acids and their relationsfip to kerogen, bitumen, petroleum ans coal.http://humicacid.wordpress.com/structure-and-

origin-of-humic-acids-and-their-relationship-to-kerogen-bitumen-petroleum-and-coal/
10. F. J. Stevenson, 1982 Humus chemistry genesis, composition, reactions. Willey Interscience, New York
11. H. Dziadowiec, S. Gonet, 1999 Przewodnik metodyczny do badań materii organicznej gleb. Prace Komisji Naukowej Polskiego Towarzystwa Gleboznawczego 42 43
12. S. M. Griffith, M. Schnitzher, 1975 Analytical characteristics of humic and fulvic acids extracted from tropical volcanic soils. Soil Science Society of America Proceedings 89 861
13. Y. Chen, N. Sanesi, M. Schnitzer, 1978 Chemical and physical characteristics of humic and fulvic acids extracted from soils of Mediterranean region. Geoderma 20 87
14. S. S. Gonet, B. Dębska, 1999 Properties of humic acids producted during decomposition of plant residues in soil. RostlinnaVyroba 45 10 455 460
15. K. Zawiślak, J. Adamiak, J. Tyburski, 1988 Dynamika substancji organicznej i składników mineralnych w warstwie uprawnej gleby pod wieloletnimi monokulturami. Cz. I. Gatunki o większych wymaganiach glebowych. Zeszyty Problemowe Postępów Nauk Rolniczych 331 227
16. M. Kononova, 1966 Soilorganic matter. 2a ed. Pergamon Press Oxford.
17. D. S. Orlov, 1985 Humus Acids of Soils A.A. Balkema, Rotterdam
18. K. Kumada, 1987 Chemistry of soil organic matter Developments in Soil Science 17.Japan Sc. Soc. Press Tokyo, Elsevier Amsterdam.
19. F. J. Stevenson, 1994 Humus Chemistry: genesis, composition, reactions. 2nd ed. Wiley, New York
20. R. Antil, 2011 Predicting Nitrogen and carbon mineralization of composted manure and sevage sludge in soil. Compost Science & Utilization 19 1 33
21. M. Fotyma, K. Filipiak, 2006 The influence of long term application on FYM and nitrogen fertilizers on the yield and uptake of nitrogen by crops grown in two rotations. Nawozy i Nawożenie- Fertilizers and Fertilization 1 26 71 89
22. D. Pikuła, 2012 The yield of winter wheat depending on weather and nitrogen supply. Nawozy i Nawożenie- Fertilizers and Fertilization 41 187 197
23. J. Janowiak, 1995 Wpływ nawożenia obornikiem z dodatkiem słomy i zróżnicowanych dawek azotu na właściwości materii organicznej. Zeszyty Problemowe Postępów Nauk Rolniczych 421 145 150
24. H. Dziadowiec, J. Jończak, A. Czarnecki, M. Kejna, 2003 Wieloletnia dynamika zawartości węgla organicznego w poziomie ornopróchnicznym gleb intensywnie rolniczo użytkowanych. [W].Zintegrowany monitoring środowiska przyrodniczego, W. Bochenek, E.Gil (red.). IOŚ, Instytut Geografii i Przestrzennego zagospodarowania PAN, Stacja Naukowo-Badawcza 166 172
25. M. Adamus, J. Drozd, E. Stanisławska, 1989 Wpływ zróżnicowanego nawożenia organicznego i mineralnego na niektóre elementy żyzności gleby. Roczniki Gleboznawcze 40 1 101 110
26. W. Łoginov, J. Andrzejewski, J. Janowiak, 1991 Rola nawożenia organicznego w utrzymaniu zasobów materii organicznej w glebie. Roczniki Gleboznawcze 42 3-4 19 25
27. H. Panak, G. Nowak, 1989 Wpływ intensywnego nawożenia mineralnego na rozkład materii organicznej w glebie. Roczniki Gleboznawcze 40 1 39
28. A. Piccolo, J. S. C. C. Mbagwu, 1990 Effects of different organic waste amendments on soil microagregates stability and molecular sitzes of humic substances. Plant Soil 123 27
29. M. Körschens, 1996 Long-term data sets from Germany and Eastern Europe. In: POWISON D.S., Smith P., Smith J.N., (eds): Evaluation of soil organic matter models. 1 ,Springer-Verlang, Berlin;
30. M. Körschens, 1999 Soil organic matter balance, optiumum content in soils. In: Proc. Oral Presentation Organic Matter Balance and Optimum Organic Matter Stock in Soil. VURV, Praha 1 24

31. M. Körschens, 2000 Carbon and nitrogen dynamics as well as nitrogen utilization in dependence on soil texture. K.Skogs-o. Lantbr-akad. Tidskr 139 8 43
32. B. Dębska, 2004 Właściwości substancji humusowych gleby nawożonej gnojowicą. PhD thesis ATR Bydgoszcz
33. A. Dziamski, J. Żarski, Z. Stypczyńska, 2000 Effect of irrigation and nitrogen fertilization on the mass of roots of spring barley and their distribution in very light soil. Zeszyty Naukowe Nr 226, Agriculture 45 25
34. K. Kondratowicz-Maciejewska, 2007 Susceptibility of organic matter to oxidation and soil microbiological activity under conditions of varied crop rotation systems and fertilization. Polish Journal of soil science 15/1 89 98
35. D. J. Mc Kenney, S. W. Wang, C. F. Drury, W. I. Findlay, 1993 Denitrification and mineralization in soil amended with legume,grass and corn residues. Soil Science Society of America 57 1013
36. D. Kalembasa, S. Kalembasa, A. Amberger, 1999 Spectroscopic characterization of organic compounds extracted from slurries by 0,1 M NaOH. Humic Substances in Ecosysystems 3 55 57
37. K. Paustian, J. Six, E. T. Elliot, H. W. Hunt, 1999 Management options for reducing CO2 emissions from agricultural soils Biogeochemistry 48 147
38. E. G. Gregorich, M. R. Carter, D. A. Angers, C. M. Monreal, B. H. Ellert, 1994 Towards a minimum data set to assess organic matter quality in agricultural soils. Canadian Journal of Soil Science 74 367
39. R. J. Haynes, M. H. Meare, 1996 Aggregation of organic matter storage in meso-ternmal, humid soils. In: Structure and Organic Matter Storage in Agricultural Soils (eds M.R. Carter & B.A. Stewart). Advances in Soil Science. CRC Levis, Boca Raton, Florida.

Index